Green Energy and Technology

For further volumes:
http://www.springer.com/series/8059

Daphne Mah · Peter Hills
Victor O. K. Li · Richard Balme
Editors

Smart Grid Applications and Developments

Springer

Editors
Daphne Mah
Department of Geography
Hong Kong Baptist University
Hong Kong
People's Republic of China

Victor O. K. Li
Department of Electrical and Electronic Engineering
The University of Hong Kong
Hong Kong
People's Republic of China

Peter Hills
The Kadoorie Institute
The University of Hong Kong
Hong Kong
People's Republic of China

Richard Balme
Paris School of International Affairs and Centre d'Etudes Europeennes
Sciences Po
Paris
France

ISSN 1865-3529 ISSN 1865-3537 (electronic)
ISBN 978-1-4471-7072-3 ISBN 978-1-4471-6281-0 (eBook)
DOI 10.1007/978-1-4471-6281-0
Springer London Heidelberg New York Dordrecht

Softcover reprint of the hardcover 1st edition 2014

Printed on acid-free paper

Springer is part of Springer Science+Business Media (www.springer.com)

Acknowledgments

This book is very much a collaborative effort. We wish to extend special thanks to all our contributing authors for contributing complementary knowledge from their areas of expertise, for being so responsive to our requests, and for submitting their work in a timely manner. This has certainly made our task as Editors that much easier.

We also wish to thank the staff at Springer for supporting our initial concept for the book, for guiding us through the process of compiling the manuscripts, and providing assistance in the production of this book.

We owe special thanks to Ms. Kaboo Leung, who was a research assistant in the Kadoorie Institute of the University of Hong Kong and latterly in the Department of Geography at Hong Kong Baptist University, for her excellent support throughout the preparation of the book. Without her continuous assistance and her attention to detail it would have been difficult to complete the project.

We would also like to record a special note of appreciation for the project funding received from the Hong Kong Baptist University (through Faculty Research Grants from the Faculty of Social Sciences) and the University of Hong Kong (through the Initiative on Clean Energy and Environment). These grants helped to fund the research on which several of the chapters are based.

Finally, we gratefully acknowledge the general support that we have received from the University of Hong Kong, Hong Kong Baptist University, and Sciences Po which has allowed us to develop and extend our work on smart grids.

July 2014

Daphne Mah
Peter Hills
Victor O. K. Li
Richard Balme

Contents

About the Editors

Richard Balme is Professor and Scientific Director for the Master in International Public Management at the Paris School of International Affairs (PSIA), and Senior Fellow at the Centre for European Studies of Sciences Po. He is also Visiting Professor at the School of Public Policy and Management, Tsinghua University in Beijing. He teaches public policy analysis, comparative politics, and international relations. Among his publications are European Governance and Democracy: Power and Protest in the European Union, (with D. Chabanet, Rowman and Littlefield 2008) and Europe–Asia Relations: Building Multilateralisms, (with B. Bridges, Palgrave, 2008). His recent research is focused on comparative environmental governance.

Peter Hills is currently Chair Professor in the Kadoorie Institute at the University of Hong Kong. He joined the University in 1982 from the University of Nottingham, where he was a Lecturer for 6 years. He was formerly Director of the Kadoorie Institute (2008–2012), and prior to that Director of the Centre of Urban Planning and Environmental Management (1989–2008). He has been actively involved in energy-related research since the late 1970s and has published widely on such topics as the coal industry in the United Kingdom, offshore oil and gas in China, national energy planning in Asia, the environmental impacts of energy systems, and global climate change. He has served as a consultant on energy issues to various international agencies including the Asian Development Bank, the UN Economic and Social Commission for Asia and the Pacific, the International Labour Office, and the Asia and Pacific Development Centre. He has also served as a member of government energy advisory bodies in Hong Kong.

Victor O. K. Li received SB, SM, EE, and ScD degrees in Electrical Engineering and Computer Science from MIT in 1977, 1979, 1980, and 1981, respectively. He is Chair Professor of Information Engineering and Head of the Department of Electrical and Electronic Engineering at the University of Hong Kong (HKU). He has also served as Assoc. Dean of Engineering and Managing Director of Versitech Ltd., the technology transfer and commercial arm of HKU. He served on the board of China.com Ltd., and now serves on the board of Sunevision Holdings Ltd. and Anxin-China Holdings Ltd., listed on the Hong Kong Stock Exchange. Previously, he was Professor of Electrical Engineering at the University of Southern California (USC), Los Angeles, California, USA, and Director of the

USC Communication Sciences Institute. Sought by government, industry, and academic organizations, he has lectured and consulted extensively around the world. He has received numerous awards, including the PRC Ministry of Education Changjiang Chair Professorship at Tsinghua University, the UK Royal Academy of Engineering Senior Visiting Fellowship in Communications, the Croucher Foundation Senior Research Fellowship, and the Order of the Bronze Bauhinia Star, Government of the Hong Kong Special Administrative Region, China. He is a Registered Professional Engineer and a Fellow of the IEEE, the IAE, and the HKIE.

Dr. Daphne Mah is currently Assistant Professor at Department of Geography of Hong Kong Baptist University (HKBU) and an Honorary Assistant Professor at the Kadoorie Institute at the University of Hong Kong (HKU). Dr. Mah obtained her M.Sc. degree in Environmental Management from the University of Nottingham in the UK, and completed her Ph.D. study in 2010 at HKU, investigating the role of local states, policy capacity, and the development of wind energy policies in Xinjiang, Shanghai and Guangdong. Before joining HKBU in 2012, Dr. Mah was a journalist in Ming Pao Daily News, the head of the campaigns team of Friends of the Earth (Hong Kong), a senior research associate and a part-time lecturer at City University of Hong Kong, a post-doctoral research fellow and a lecturer at HKU.

About the Contributors

Benjamin Blunier studied Electrical Engineering at the University of Technology of Belfort-Montbéliard (UTBM), France where he received the M.Sc. and Ph.D. degrees in 2004 and 2007, respectively. He studied fuel cell systems modeling and especially their air management and control for hybrid and electric vehicles. Since 2008, he has been an associate professor at UTBM. His current research interests include fuel cells systems, electric, hybrid and plug-in hybrid vehicles, and intelligent energy management in smart grids and micro-grids.

John W. M. Cheng received his B.Eng. (Sask.), M.Eng. (McGill), and Ph.D. (McGill) in Canada. He has over 20 years of industrial and research experiences including low-carbon technologies, climate change and impacts to electric systems, energy and environmental policies, smart grids, energy storage, equipment health monitoring, renewables resource assessment and integration, power system reliability, military systems, and dispatcher training simulator. Dr. Cheng is a registered Professional Engineer (P.Eng.) in Ontario, a member of HKIE, IEEE, and CIGRE. He has also been serving as the secretary of WEC-Hong Kong since 2010.

Yi Ding is an Associate Professor in the Department of Electrical Engineering, Technical University of Denmark (DTU), Denmark. Before he joined DTU, he had held academic and research positions in University of Alberta, Canada and Nanyang Technological University. His research interests include power systems reliability/performance analysis incorporating renewable energy resources, smart grid performance analysis, and engineering systems reliability modeling and optimization.

Ahmad Faruqui is a Principal with The Brattle Group which specializes in analyses and strategy relating to the customer. He has helped design, monitor, and evaluate energy efficiency investments for a wide range of electric and gas utilities and testified before a dozen state and provincial commissions and legislative bodies. He has also worked for the Alberta Utilities Commission, Edison Foundation, the Edison Electric Institute, the Electric Power Research Institute, the Federal Energy Regulatory Commission, the Ontario Energy Board and the World Bank. His work has been cited in publications such as The Economist, The New York Times, and USA Today, and he has appeared on Fox News and National

Public Radio. The author, co-author, or editor of four books and more than 150 articles, he holds a Ph.D. in Economics from The University of California at Davis and BA and MA degrees in Economics from The University of Karachi.

Jan Ove Gjerde was born in Hareid in 1958. He received an M.Sc. in Electrical Engineering from The Norwegian Institute of Technology in 1983. He has long experience in R&D from the energy sector, both as a senior research scientist at Sintef Energy Research and as a department head at ABB Corporate Research, Power System and Components Department.

Currently he is Vice President of R&D at Statnett responsible for developing R&D strategies for using on new concepts for Risk Management, WAMS, WACS and WAPS, and the overall R&D Programs at Statnett.

Jasper Ip is a Senior Research Assistant of the Department of Geography at Hong Kong Baptist University. He has been researching on sustainable energy policies such as smart grids, renewable energy since 2008.

Mr. Ip received his Bachelor of Science Degree in Applied Science (B.Sc. (Hons)) in Hong Kong Baptist University in 1996. Before joining the Department of Geography at HKBU in 2012, Mr. Ip has been working on energy policies and renewable energy applications for 10 years. He was a senior research assistant in the Kadoorie Institute in the University of Hong Kong on energy policies research, a senior research assistant at the Centre on Renewable Energy at the University of Hong Kong on solar energy applications, a project manager of an engineering company to develop wind and solar energy projects and a project officer of Friends of the Earth (Hong Kong) on wind resource assessment, solar energy applications, and energy education.

John Juliano has been a consultant in the Utilities industry for over 20 years. He has worked in business strategy, operations strategy, financial analysis, and technology assessment across the energy value chain. John has written or cowritten over two dozen papers on utility issues. He is currently working with the IBM Institute for Business Value on research related to utility business models and electricity consumer behavior.

Elias Kyriakides received the B.Sc. degree from the Illinois Institute of Technology in Chicago, Illinois in 2000, and the M.Sc. and Ph.D. degrees from Arizona State University in Tempe, Arizona in 2001 and 2003 respectively, all in Electrical Engineering. He is currently an Assistant Professor in the Department of Electrical and Computer Engineering at the University of Cyprus, and a founding member of the KIOS Research Center for Intelligent Systems and Networks. He is the Action Chair of the ESF-COST Action IC0806 “Intelligent Monitoring, Control, and Security of Critical Infrastructure Systems” (IntelliCIS) (2009–2013). His research interests include synchronized measurements in power systems, security and reliability of the power system network, optimization of power system operation techniques, modelling of electric machines, and renewable energy sources.

Kaboo Po Yi Leung received a B.Sc. (Hons) in Economics (2010) from the University of Washington, Seattle, focusing on environmental economics. With her professional career in energy consulting, Kaboo specializes in statistical analysis for energy projects, ranging from efficiency to Combined Heat and Power. She also involves in various research projects to analyze energy issues from an economic perspective and analyze energy policies in areas such as nuclear power and energy market reforms across different countries.

William Lilley is a division head in Saudi Aramco's Kingdom Economic and Energy Department. William leads a multidisciplinary team that examines the wide-scale impact of energy and economic evolution on the company's business and proposes actions for the company to help drive the economic prosperity of the Kingdom.

While working in Newcastle for Australia's national science agency the CSIRO, William developed its multi-million dollar 'Intelligent Grid' research program, which investigated the value of distributed energy for Australia.

On secondment to the Department of Climate Change and Energy Efficiency, William was a Principal Advisor on the 'Smart Grid Smart City' initiative before returning to Newcastle as Program Leader of the Future Grid project, which focused on nationally significant savings from optimal energy network design.

William holds a Bachelor of Science (Honours) and Doctor of Philosophy (Environmental Science) from the University of Newcastle.

Amy Poh Ai Ling received a BBA (2006) and MS (2008) from National University of Malaysia, and a Ph.D. (2012) in Advanced Mathematical Sciences from Meiji University. She was awarded Role Model Student Award (2006) and Excellent Service Award (2010) from National University of Malaysia, and Excellent Student Award (2012) from Meiji University. She worked with Sony EMCS and Erapoly Sdn. Bhd. She is currently a JSPS Research Fellow affiliate with Department of Technology Management for Innovations, Graduate School of Engineering, University of Tokyo as a postdoctoral. She is involved in the smart grid demand respond team, working on the forecasting trend with statistical data analysis and in the infrastructure design development with appropriate modeling tools. She has an enthusiasm for statistical calculation, smart grid, and safety studies. She is currently Chair for Malaysia Japan Academic Scholar Conference 2013.

Takashi Matsuyama received B.Eng., M.Eng., and D.Eng. degrees in Electrical Engineering from Kyoto University, Japan, in 1974, 1976, and 1980, respectively. He is currently a professor in the Department of Intelligence Science and Technology, Graduate School of Informatics, Kyoto University. He served as the director of the Academic Center for Computing and Media Studies (2002–2006), the director general of the Institute for Information Management and Communications (2005–2010), and a vice president of Kyoto University (2008–2010).

He has been studying cooperative distribute sensing-control-reasoning systems for over 30 years. Their application fields include knowledge-based image understanding, visual surveillance, 3D video, human-computer interaction, and smart energy management. He wrote more than 100 journal papers and more than 20 books and won more than 10 best paper awards from Japanese and international academic societies including the Marr Prize at ICCV'95 and the commendation for Science and Technology by the Minister of Education, Culture, Sports, Science and Technology in 2009.

Kerry D. McBee is pursuing his Ph.D. degree in the Department of Engineering at Colorado School of Mines, Golden, Colorado, which is where he received his B.Sc. degree in 1999. He received his M.Sc. degree in Electric Power Engineering at Rensselaer Polytechnic Institute, Troy, New York, in 2000. During his career he has focused on power quality, reliability, forensic engineering, and distribution design for companies such as NEI Power Engineers, Peak Power Engineering, Knott Laboratory, and Xcel Energy. His field of interest is Smart Grid implementation affects upon distribution engineering and utility operations.

Peter Meibom is head of department at the Danish Energy Association and Adjunct Professor at the Technical University of Denmark. He received the M.Sc. degree in Mathematics and Physics from the University of Roskilde, Roskilde, Denmark, and the Ph.D. degree from the Technical University of Denmark, Denmark. He has worked with the modeling of energy systems characterized by a large share of renewable energy sources in the system.

Abdellatif Miraoui is a President of Cadi Ayyad University (Marrakech—Morocco). He is a Full Professor of Electrical Engineering (Electrical Machines and Energy) at Belfort-Montbéliard University of Technology (UTBM) since 2000. His special interest includes Fuel Cell energy, energy management (Ultra capacitor, batteries, etc.) in transportation, design and optimisation of electrical propulsions/tractions. Professor A. Miraoui was a Vice President—Research Affairs of UTBM from 2008 to 2011. He was Director of the Electrical Engineering department (in UTBM) from 2001 to 2009, the head of "Energy Conversion and Command" Research Team (38 researchers on 2007) and the editor of the International Journal on Electrical Engineering Transportation (IJEET). He is the author of over 80 journals and 150 international conference papers.

A. Miraoui is the author of four textbooks (three in French and one in English) about Fuel Cells. He is a Member of several International Journals and Conferences Committees. He is a Member of IEEE Power Electronics Society (PELS), Industrial Electronics Society (IES), and IEEE Vehicular Technology Society (VTS). Professor A. Miraoui is Doctor Honoris Causa of Cluj-Napoca Technical University, Romania. In 2007, He received a high distinction from French Higher Education Ministry "Chevalier dans l'Ordre des Palmes Académiques". He was also distinguished as Honorary Professor from University of Brasov-Romania.

Phuong H. Nguyen was born in Hanoi, Vietnam in 1980. He received the BE degree in Electrical Engineering at Hanoi University of Technology, Vietnam, in 2002 and the M.Eng. degree in Electrical Engineering at the Asian Institute of

Technology, Thailand, in 2004. From 2006 he started the Ph.D. program at Eindhoven University of Technology, the Netherlands with "Electrical Infrastructure of the Future" project. Since the end of 2010 he received his Ph.D. and was employed at the same group as a postdoctoral researcher. His research interests include distributed state estimation, control and operation of the power system, multi-agent system, and their applications in smart grids.

Jacob Østergaard is a Professor and Head of Centre for Electric Power and Energy, Department of Electrical Engineering, Technical University of Denmark. His research interests cover intelligent energy systems with focus on integration of renewable energy, control architectures for power systems, and flexible demand. Professor Østergaard serves in several professional organizations, boards, and steering committees. He is head of the experimental facilities, PowerLabDK. In 2009 he received the IBM Faculty Award.

Luke Reedman is the Stream Leader for Energy Modelling in CSIRO's Energy Flagship. An energy economist by training, Luke leads a multidisciplinary research program to quantify plausible energy pathways for the Australian energy sector, encompassing research into the Future Grid.

Luke led the economic program within the successful Intelligent Grid project. A key to the success of this groundbreaking report was estimating the economic value of wide-scale uptake of distributed energy in Australia at around $130 billion by 2050.

In July 2011, Luke was acknowledged as having "made invaluable contributions to the economic modelling" that underpinned the Australian Government's Clean Energy Future plan. Other significant work includes modelling for the 2012 Energy White Paper and the 2011 Garnaut Review on Climate Change.

Luke holds a Bachelor of Economics (Honours, Class 1) in Economics from the University of Newcastle, a Graduate Diploma of Applied Finance and Investment from the Securities Institute of Australia and a Doctor of Philosophy (Economics) from the University of Newcastle.

Paulo F. Ribeiro received the Ph.D. from the University of Manchester, Manchester, UK. He is an Associate Professor of Electrical Engineering at the Technical University of Eindhoven, Eindhoven, The Netherlands. His research interests include power electronics, power quality, system modeling and simulation, energy storage, integration of renewables, smart grids, engineering education, and philosophy of technology.

Robin Roche received the Master's degree in Electrical Engineering from the University of Technology of Belfort-Montbéliard (UTBM), France, in 2009. Since then, he has been working toward the Ph.D. degree, also in Electrical Engineering, at UTBM's Research Institute on Transportation, Energy and Society (IRTES). His research interests include intelligent energy management in power systems, smart grids, applications of computational intelligence, energy storage, and hybrid vehicles, as well as the societal and economic impacts of these technologies.

Allan Schurr is a 25-year veteran of the energy industry, spanning generation, delivery, marketing, technology, strategy, and regulatory affairs. Allan is responsible for IBM strategy development in the energy and environment arena. As part of this role, he is responsible for partner ecosystem management, policy advocacy, emerging solutions to accelerate the development of an intelligent utility network and the integration of renewable energy sources and distributed energy assets.

Sanem Sergici is a Senior Associate in The Brattle Group's Cambridge, MA office with expertise in electricity markets, applied econometrics, and industrial organization. At Brattle, the focus of Sergici's work has been on assisting electric utilities, regulators, and wholesale market operators in their strategic questions related to energy efficiency, demand response, and customer behavior in the context of Smart Grid. Sergici has significant expertise in the design and evaluation of a variety of demand response and behavior-based energy efficiency programs; development of load forecasting models; ratemaking for electric utilities; energy litigation and resource planning. She has spoken at several industry conferences and published in several industry journals.

Sergici received her Ph.D. in Applied Economics from Northeastern University in the fields of applied econometrics and industrial organization. She received her MA in Economics from Northeastern University, and BS in Economics from Middle East Technical University (METU), Ankara, Turkey.

Marcelo G. Simões received a BS and MS from University of São Paulo, Brazil, and a Ph.D. from The University of Tennessee, USA, in 1985, 1990, and 1995, respectively, and his D.Sc. degree (Livre-Docência) from the University of São Paulo in 1998. He is Associate Professor with Colorado School of Mines, where he has been establishing research and education activities in the development of intelligent control for high-power-electronics applications in renewable- and distributed energy systems. He is currently Past-Chair for the IEEE IAS/IACC and Co-Chair for the IEEE IES Smartgrid Committee. He has been involved in activities related to the control and management of smartgrid applications since 2002 with his NSF CAREER award "*Intelligent Based Performance Enhancement Control of Micropower Energy Systems.*"

Yuqian Song received the B.Eng. in Electronic and Electrical Engineering in Harbin Institute of Technology, Harbin, China, and M.Sc. in the University of York, UK. He is currently a Ph.D. student in the Department of Electrical and Electronic Engineering at the University of Hong Kong. His research interests include power system modeling, power system operation and optimization, electricity market, and carbon reduction policies.

Poul Ejnar Sørensen is a Professor and Team Leader in Wind Power Integration and Control at Department of Wind Energy, Technical University of Denmark. His research interests include wind power variability, dynamic simulation models of wind power, power system stability, wind power control, and power quality. He is convener of the IEC working group preparing a new standard series IEC 61400-21 for Electrical simulation models of wind power.

Sid Suryanarayanan received the Ph.D. in Electrical Engineering in 2004 from Arizona State University. His research interests are in the design, operation, and economics of finite-inertia systems and integration of renewable energy systems to electric grids. He is an assistant professor in the Department of Electrical and Computer Engineering at Colorado State University where he serves as the site director of the Center for Research and Education in Wind (CREW), and as a 2011–2012 Resident Faculty Fellow of the School of Global Environmental Sustainability (SoGES).

Anthony Szatow has a broad professional and academic background spanning roles in consulting, research, and advocacy, and the disciplines of economics, social science, and engineering. His focus is on making Energy for the People an energy service company known for its creative, but pragmatic approach to getting clean energy solutions done. Energy for the People works with property developers, communities, and building owners to ensure they get the right energy technologies, at the best price, delivered commercially to maximize value. He is dedicated to ensuring the rapid uptake of clean energy regardless of policy or regulatory settings. Szatow is a father of two, part-time (backyard) farmer and carpenter, living the quarter acre dream in Castlemaine with his better half, Juliette Anich. From time to time, he gets roped into stirring up food revolutions.

Michael Valocchi, Global Energy and Utility Industry Leader for IBM Global Business Services, is responsible for the development and execution of the strategy to deliver consulting services and directing industry thought leadership. Michael has over 24 years of experience leading projects related to smart grid, mergers/acquisitions, technology, and regulatory strategies. He has also written or co-written numerous papers and articles on the utility industry. In addition, Michael has served on the editorial board of Montgomery Research Institute's Energy and Utilities Project for the past six editions.

Joost van der Vleuten was born in the Netherlands with a life-long interest in nature, sustainable development and working in different cultural environments. He studied Tropical Forestry and holds a Master's degree in Natural Resources Management.

He worked for the Dutch Ministry of Foreign Affairs as part of a program that supports governments and non-governmental organizations working on the interface of environmental management, economic development, and political empowerment. He lived and worked in Botswana, the Philippines and Ghana.

Gender duties brought him to Hong Kong where he worked for a variety of nongovernmental organizations including the Kadoorie Institute (University of Hong Kong) and Hong Kong Baptist University.

Luigi Vanfretti is Associate Professor (Tenured) and Docent at the Electric Power Systems Department of KTH Royal Institute of Technology. He was conferred the Swedish Title of Docent in 2012 and was an Assistant Professor in the same department from 2010 to 2013.

Since 2011, he has served as advisor to the R&D Division of Statnett SF, where he is now a Special Advisor in Strategy and Public Affairs. His duties include architectural analysis for synchrophasor data transfer, communications, and application systems to be utilized in Smart Transmission Grid applications, as well as providing inputs into R&D strategy development.

He received the Electrical Engineering Degree from Universidad de San Carlos de Guatemala in June 2005, and received the Francisco Vela award. During Fall 2005, he was a visiting researcher at the EEE Department of The University of Glasgow, Scotland. He received his MS in 2007 and Ph.D. in 2009, both in Electric Power Engineering, from the ECSE Department at Rensselaer Polytechnic Institute (RPI). During Spring 2010 he was a Post-Doctoral Research Associate at RPI working on synchrophasor data applications. He was awarded the Charles M. Close Doctoral Prize from RPI in May, 2010 for his research and teaching work at ECSE.

He is a member of the IEEE Power and Energy Society where he serves as Vice-Chair of the IEEE Working Group on Power Systems Dynamics Measurements and the Task Force on Open Source Software. He has actively contributed to the North American Synchrophasor Initiative (NASPI) Working Group both as a presenter and as a meeting organizer for the Vancouver 2010 Meeting.

His research interests are modeling, dynamics, stability, and control of power systems; applications of PMU data for Smart Transmission Grids, and open source software for power system engineering.

Dirk Van Hertem graduated as an M.Eng. in 2001 from the KHK, Geel, Belgium and as an M.Sc. in Electrical Engineering from the KU Leuven, Belgium in 2003. In 2009, he has obtained his Ph.D. degree, also from the KU Leuven. In 2010, Dirk Van Hertem was a member of EPS group at KTH Royal Institute of Technology, in Stockholm, Sweden where he was the program manager for controllable power systems for the EKC competence center at KTH.

Since spring 2011 he is back at the University of Leuven where he is an assistant professor in the ELECTA group. His special fields of interest are power system operation and control in systems with FACTS and HVDC and building the transmission system of the future, including offshore grids and the supergrid concept. He is an active member of both IEEE (PES and IAS) and Cigré.

Qiuwei Wu obtained the Ph.D. degree from Nanyang Technological University, Singapore, in 2009, in Power System Engineering. He was a senior R&D engineer with Vestas Technology R&D Singapore Pte. Ltd. from March 2008 to October 2009. He was a PostDoc with Centre for Electric Technology (CET), Department of Electrical Engineering, Technical University of Denmark (DTU from

November 2009 to October 2010, and was an Assistant Professor with the same centre from November 2010 to August 2013. He has been an Associate Professor with the same center since September 2013.

Jiahai Yuan received his Ph.D. degree on Management Science in 2006 and has been an Associate Professor with School of Economics and Management, North China Electric Power University since 2009. From March of 2011 to 2012, he held a position of visiting scholar with Erb Institute for Global Sustainable Enterprise, University of Michigan, Ann Arbor. His research interest is on energy economics, planning and policy, with a special interest on energy issue in China. He has authored (or co-authored) more than 40 papers in journals and conference proceedings, among which 15 papers are published in peer-reviewed journals with international reputation as Energy Economics, Energy Policy, Renewable and Sustainable Energy Reviews, etc. His research covers a wide range of interrelated topics, including empirical study on the relation among economic growth, energy consumption and GHG emissions in China, China's national energy policy, integrated resource strategic planning (IRSP) theory and application, sustainable transition of power system.

Jin Zhong received the B.Sc. from Tsinghua University, China, and Ph.D. degree from Chalmers University of Technology, Sweden, all in Electrical Engineering. She is currently an Associate Professor in the Department of Electrical and Electronic Engineering at the University of Hong Kong (HKU), and the Director of Smart Power Grid Lab of HKU. Her research areas include smart grid, renewable energy integration, power system operations, and electricity market.

Part I
Introduction

Introduction and Overview

Daphne Mah, Peter Hills, Victor O. K. Li and Richard Balme

Abstract The policy challenges associated with global warming, the prospect of increasingly expensive fossil fuels, and the recent re-emergence of serious concerns about the safety of nuclear power after the Fukushima accident in Japan are encouraging many western and Asian economies to develop smart grids (SGs) as a component of their energy policy portfolios. This introductory chapter first discusses the evolving definitions of SGs, and the five major applications of these technologies. We will then provide an energy sector outlook, highlighting the major energy developments in the near future that may affect SG deployment. We will conclude with an overview of the objectives and structure of the book.

1 Introduction

The policy challenges associated with global warming, the prospect of increasingly expensive fossil fuels, and the recent re-emergence of serious concerns about the safety of nuclear power after the Fukushima accident in Japan are encouraging

---

D. Mah (✉)
Department of Geography, Hong Kong Baptist University, Hong Kong,
People's Republic of China
e-mail: daphnemah@hkbu.edu.hk

P. Hills
The Kadoorie Institute, The University of Hong Kong, Hong Kong,
People's Republic of China

V. O. K. Li
Department of Electrical and Electronic Engineering, The University of Hong Kong,
Hong Kong, People's Republic of China

R. Balme
Paris School of International Affairs and Centre d'Etudes Europeennes,
Sciences Po, Paris, France

D. Mah et al. (eds.), *Smart Grid Applications and Developments*,
Green Energy and Technology, DOI: 10.1007/978-1-4471-6281-0_1,

many western and Asian economies to develop smart grids (SGs) as a component of their energy policy portfolios. SGs are often compared to the Internet and smart phones as one of the major transformational technologies that can potentially reshape economies and societies [1]. Although SGs are defined in various ways in different countries and across institutions [1, 2, 3, 4], they are typically taken to mean a modern grid concept that uses advanced information technology to update and modernize existing power grids to improve reliability, security, and efficiency of electricity supply systems [2].

SGs are widely regarded as one of the key building blocks of a sustainable energy future [3, 5]. These future grids will look very different from those of today. Traditional grids are typically centralized and fossil fuel-based. These smarter grids are essential to enable the wider use of renewable energy and plug-in electric cars as well as to accelerate energy-efficiency efforts—which are the core elements for more sustainable energy systems [2]. They can also reduce transmission and distribution losses, and enhance grid reliability. They are therefore expected to bring economic, environmental, and social benefits in many important ways: revitalizing the economy and providing green jobs, reducing capital expenditure on energy infrastructure [2], reducing power disturbance costs to economies [2] as well as facilitating a low-carbon energy future [2, 6].

SG is a subject that is highly dynamic and rapidly evolving. As an emerging energy-related technology in its nascent phase of deployment [7], SGs involve the integration of a broad range of state-of-the-art technologies that include wind and other renewable energy, electric vehicles (EV), car batteries, microgrids, computer networking, and communication systems [4]. This is an area where rapid innovations in technology are taking place. SGs are highly dynamic because they are also a key to both demand-side (e.g., energy saving and energy efficiency) and supply-side (e.g., renewable energy) management of energy systems, making such technologies a fertile arena in which many possible solutions for a more sustainable future may be located. SGs are an area where many substantive developments are rapidly unfolding. Many countries and cities are piloting SGs—at different scales from small pilots to large demonstration projects, and with varying scope to test technologies, business models, and consumer acceptance [2]. What also adds to the dynamic nature of SGs is their stakeholder landscape. The development of these SGs is a long-term process and a cross-sector effort that requires visionary and strategic planning, and collaboration among policymakers, utilities, the business sector, consumers, and other stakeholders [5, 8].

There has been strong international interest in SGs. Drivers of SG policy initiatives are many: rising energy costs with growing demand for energy, increasing awareness of global climate issues, growing need for energy efficiency, and rapid innovations in technology are just some of them [2]. Many countries and cities have made considerable progress in recent years to develop SGs at various scales and with different objectives in mind. The US national policy framework for SGs announced in June 2011, the European Technology Platform established in Europe in 2005, the four large-scale demonstration projects launched in Japan in 2010, the Korean smart grid vision announced in 2008, and China's smart grid initiatives formulated in 2009 are just some examples of these endeavors (For more details about these international developments, please refer to Part 4 International Case Studies of this book).

2 What Are SGs?

2.1 In What Ways Are SGs "Smarter"?

SGs are regarded as essential for a sustainable energy transformation. In what ways are they "smarter" than conventional systems? How can SGs overcome problems that cannot be solved by traditional grids?

Today's electric grids are using technologies that were state-of-the-art more than a century ago. These traditional grids typically consist of centralized power plants, transmission, and distribution lines. They are highly dependent on fossil fuels, and one-directional [9]. However, these aging grids face a number of challenges, including coping with the continued growth in demand, the need to integrate various renewable energy sources, which may involve intermittent availability, and EV, and the need to improve security of supply [3].

SGs are "smarter" in two ways. First, they have the ability to manage the two-way flow of electricity and information to optimize supply and demand. Traditional electric grids have one-way communication between utilities and customers: that is there is a one-way flow of information from customers to the grid (through meters) and a one-way flow of energy from the grid to customers [5, 9]. In contrast, SGs enable two-way flow of information (through a variety of interfaces) and energy (through distributed generation and storage) [9]. This is achieved through smart metering technologies and sensors that are installed throughout transmission and distribution grids, and which are linked to integrated communication networks to collect and consolidate data [1, 5]. This ability of two-way communication is fundamental to SG operations. Customers, for example, can proactively monitor and manage their electricity use, and can even sell back to the grid surplus renewable electricity that is produced at home [9].

SGs are "smarter" also in the sense that they are capable of integrating a wide variety of energy sources and energy customer services—which are now separately managed in traditional power systems—in highly interconnected electricity systems [2, 5, 7]. SGs also integrate a variety of interfaces, including home energy management systems (HEMS), building energy management systems (BEMS), and advanced metering infrastructure (AMI). SGs can coordinate the needs and capabilities of different generators, grid operators, end users, and electricity market stakeholders to operate all parts of the system efficiently [3]. All these components require the integration of SGs to achieve scale benefits and cost effectiveness [3].

2.2 Five Major Applications of SGs

SGs are complex systems that may provide five major applications:

- Smart systems: SGs can improve resilience to disruptions, attacks, and natural disasters [2]. This can be achieved through advanced sensors and computer-based remote controls. These sophisticated communication technologies and automation can help prevent disruptions rather than simply react to them, and therefore limit outages and network losses [5]. SGs can also identify and fix problems faster [1].
- Smart renewables: Today's grids are mostly designed for centralized supply sources and are therefore less accommodating to renewable resources that are intermittent and widely distributed [2]. SGs can accommodate a variety of generation, including renewable energy resources such as wind and photovoltaic solar, and other forms of distributed generation such as small-scale combined heat and power, and energy storage [1, 3]. SGs are regarded as essential to mainstreaming renewables because through state-of-the-art modeling and decision support tools, wind forecasting, and contingency analysis, for example, can be improved and these can enhance the integration of these intermittent sources into the power system [2, 3].
- Smart consumers: In SG systems, consumers are no longer passive purchasers [1]. SGs can inform and empower consumers to proactively manage their consumption [1]. Consumers can be provided with devices and information to manage their energy usage, and to reduce demand in response to peak load [1]. This can be achieved through smart meters and smart appliances that are connected with sensors to collect electricity consumption data, and which is essential to enable dynamic pricing and consumer participation in demand-side management [7]. Power companies can introduce a variety of demand response programmes. Direct load control programmes, for example, are mainly offered to residential and small commercial customers—in which consumers can choose to allow a programme operator to remotely turn off their appliances or equipment at short notice [9]. Consumers in this way can help reduce peak consumption and reduce the costs of generating expensive power to meet peak demand [2]. Real-time pricing programmes and other dynamic pricing programmes can be introduced to provide incentives for reducing peak load consumption [9]. Interruptible supply contracts, on the other hand, are offered mostly to large industrial or commercial customers. With these contracts, customers curtail their consumption in case of predefined grid contingencies [13].
- Smart transport: EV and plug-in hybrid EV can have a major role to play in reducing emissions. SGs can better manage vehicle charging so that rather than increasing peak loads, the charging can be carried out more strategically, when for example electricity demand is low or when the production of renewable electricity is high. In the long run, SGs can use EV as batteries to store renewable and other sources of electricity for later use [3].

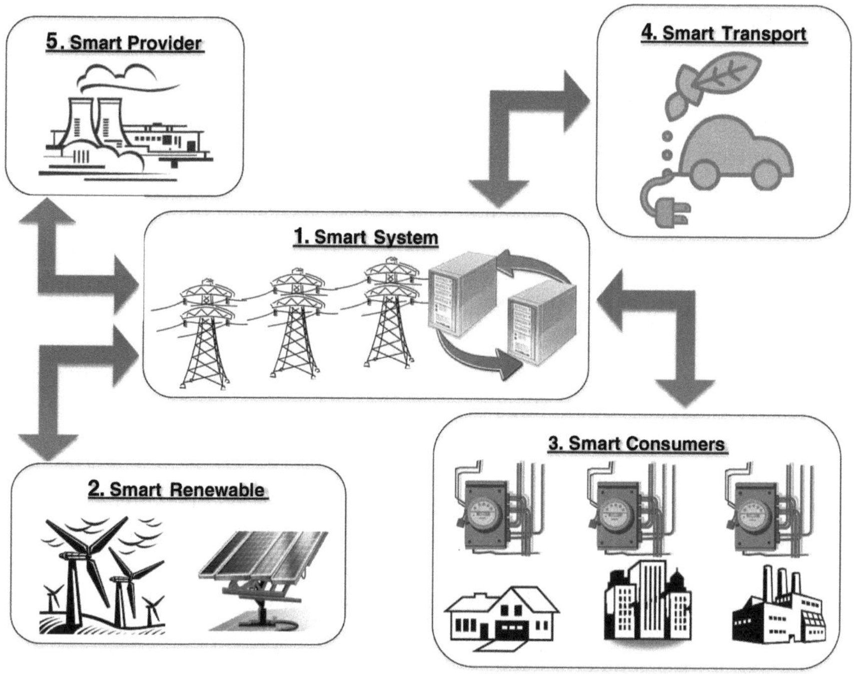

Fig. 1 Five major applications of smart grids. *Source* authors, Computer Networks and Security Lab [31], and HD PLCMAG [32]

- Smart electricity service providers: Utility companies will not be the only significant players in SGs. SGs create new markets as these technologies are conducive to new products and energy services, and also new market players. Energy efficiency and intelligent appliances, smart meters, new sensing and communications capabilities, and passenger vehicles are some examples of these new products [2]. SGs, therefore, tend to bring major changes in the market place—they rely on numerous third parties, including energy-service providers and brokers to provide core and additional services [1]. Energy-service providers, such as home energy monitoring service and energy-service companies (ESCOs) can analyze customer energy usage and provide customized energy services to meet customer needs. They can also perform direct load control or provide financial incentives for customer-responsive demand in homes and businesses [14] (Fig. 1).

3 Energy Sector Outlook

A recurring theme in this book is that the overall context for energy production and use is progressively moving in a direction that will support SG initiatives. This context reflects not just the global concern with climate change and the need to use energy more wisely and efficiently but also the kinds of energy that will be produced and the nature of the power grids that will deliver electricity to end users. A brief commentary on some important global energy and energy-related trends is, therefore, needed to contextualise SG initiatives. We emphasize, however, that this is not a book about energy policy. Rather, our focus is on the way in which thinking about SGs is developing, how the concept is being translated into practice and the kinds of technical, policy, and governance issues that surround such grid transformations.

Clearly, international thinking on energy and energy policy over the past 20 years has been profoundly influenced by global climate change concerns. The latest assessment by the intergovernmental panel on climate change (IPCC) published in September 2013 concludes that human influence on the climate system is clear, and it is extremely likely that human influence has been the dominant cause of the observed warming since the mid-twentieth century [14]. While CO_2 emissions in some countries have either slowed or declined, most notably the USA as a result of the switch away from coal to natural gas in electricity production, global energy-related CO_2 emissions increased by 3.2 % over the 2010 figure to reach 31.2 Gt in 2011. Much of this increase was the result of the increasing use of coal in China and India. Effective and widespread decoupling of emissions and economic growth has yet to be achieved [15].

Global electricity demand is expected to continue to increase—by 70 % in the period to 2035—and approximately half of this increase in demand is expected to arise in the coal-dominated energy economies of China (38 %) and India (13 %) [15].

In terms of sectoral demand for electricity, industry will remain the largest electricity consumer, accounting for 2.3 % increase in demand per year and over 40 % of total demand in 2035 [15]. The other major demand sectors will remain the residential and service sectors [15]. The transport sector is expected to experience a significant and substantial increase in demand for electricity as a result of the deployment of EV [15]. Hybrid-electric (HEV) and EV sales passed 1,000,000 and 100,000 in 2012 at growth rate of 43 % and more than 200 %, respectively [16]. Japan and the United States of America continue to lead the electric vehicle market [16]. Increasingly, governments around the world are providing various types of support for EV deployment including vehicle tax breaks and programmes to encourage the installation of charging stations. In India, for example, the government plans to get 6 million EVs and HEVs on the road by 2020, with half of the investment cost (US$4.2 billion) supported by government funding [16].

In terms of energy production, output from various forms of renewables continues to grow at a rapid pace. In 2012, solar PV electricity output grew by 42 % and wind power by 19 % [16]. The share of renewable in world electricity generation is projected to grow from around 20 % in 2010 to 28 % by 2020 and almost 60 % by 2050 [16]. The diffusion of renewables has already spread well beyond the developed OECD economies such as the USA and those in Europe and is well established in developing economies such as Brazil, China, and India. In fact, non-OECD countries account for 53 % of world renewable electricity production [16]. Furthermore, China has itself become a major player in the renewables technology area and is the world's largest solar panel producer [17].

Nuclear power continues to represent a key element of the generation mix in a number of countries and although the Fukushima accident resulted in a widespread revisiting of nuclear expansion plans around the world new nuclear projects increased from four in 2011 (the year of the accident) to seven in 2012 [16]. Nuclear seems to be slowly regaining public support since 2011, a trend that is the most pronounced in the USA, China, and France, although local support for nuclear continues to fall in Japan, Poland, and Spain [16]. Nonetheless, as a non-carbon-emitting option (at least insofar as the process of generating electricity is concerned) nuclear will doubtless continue to figure on the energy policy agenda in coming decades.

While much attention is naturally given to the fuel and technology mix options used to generate electricity, the means by which electricity is distributed and delivered to end users is just as important an issue. As the focus of this book itself indicates, power grids themselves have moved center stage and are now seen as a critical element in progress toward a low-carbon economy. The scale of investment in power grids in coming decades will be enormous, possibly up to $17 trillion of investment is expected worldwide for grid infrastructure upgrades to meet growing energy demand by 2035 [15].

In a world still hungry for more energy and with a range of difficult choices facing policy-makers, local communities, and power utilities regarding the fuel/technology mix to be deployed as well as how best to manage the demand for electricity and to ensure its efficient use, the context for energy policy-making will continue to pose difficult and serious challenges. Our concern here is with just one part—albeit a very significant component—of this bigger system but one that will nonetheless play a critical role in shaping electricity system management and consumer usage patterns in future decades.

4 The Development of SGs at the Global Level

The prospects for future SG development are very favorable. Key drivers such as the continuing growth in the deployment of renewable energy technologies, especially wind and solar power, growth in electric transport, and overall increases in the demand for electricity look set to create a positive environment for SGs over the period to 2020 and beyond. By 2020, renewable capacity is expected to increase by around 300 %, electric transport, including cars, by 45 %, and electricity demand itself by 27 % [16].

Between 2008 and 2012 the number of smart meter installations around the world grew more than fivefold from 46 to 285 million. The number is expected to reach 1 billion by 2018 [16]. A number of European countries have been making rapid progress in smart meter deployment: more than 36 million have been installed in Italy, two million in Denmark, and over two million in France. Installations in France are expected to reach 35 million by 2035 [18].

But it is not only in the hardware area that progress has been made. Although by no means as widely deployed as smart metering, dynamic pricing of electricity is beginning to appear but despite a series of pilot programmes since the late 1990s, recent policy initiatives concerning this pricing mechanism have been limited in scale and primarily confined to the United States. US-based programmes tend to take on a mandatory form such as the hourly pricing system in New York [19] and the mandatory time-of-use and optional critical-peak pricing in California [21].

Consumer engagement in matters related to power access provision has also proceeded although again at a far from uniform pace. Consumers in European countries are typically more engaged in electricity markets and cost reductions due to the widespread liberalization of electricity markets between 1995 and 2007 [22]. In the United States of America, progress with market reforms various markedly between the different states. For example, it has been relatively slow in California and but more progressive in Texas. Overall, however, in a growing number of Western countries liberalization has allowed consumers to make proactive decisions regarding their choice of electricity provider [22]. This is not, however, the case in Asia-Pacific countries. Here, customer choice and participation remain low because most electricity markets remain highly regulated [22]. Nonetheless, there are a few examples of community-based SG projects that have taken place with active consumer involvement. These include the four smart city demonstration projects in Japan, which involved more than 5,000 households, and the SG test-bed project on Jeju Island in South Korea.

Some examples of international collaboration in the SG area have also emerged in recent years. These include STRONGrid which involves five European countries, NASPI between Canada and the USA, and other demonstration projects in Italy, South Korea, and India [16].

SGs are not only attracting widespread international attention but are also the focus of very substantial capital investment as countries update and modernize their power grids. In 2008, SG-related investment was estimated at around US$3.4 billion. By 2012, this figure had reached nearly US$14 billion and is expected to grow to US$25 billion by 2018 [16]. Over half of the SG investment in 2012 took place in the USA (36 %) and China (26 %), while the global cumulative costs and benefits of SGs may reach US$738 billion and US$3,179 billion, respectively, by 2020 [16].

5 Policy Support Mechanisms and SG Initiatives

The more widespread diffusion of SG technologies in coming decades will depend on a wide variety of factors. There will be a number of crucial policy drivers for SGs that will provide a positive environment for such developments and which will facilitate grid enhancements. Some of the most important are linked to government initiatives to better manage energy demand and to enhance energy efficiency. This will be particularly important in non-OECD countries which will account for a very high proportion of primary energy consumption growth over the next two decades. Between 2011 and 2030, energy consumption in non-OECD countries is projected to grow by 61 % and is expected to account for 65 % of total global consumption [20].

A key element in creating a favorable context for SG initiatives will be governments' determination to pursue energy-efficiency initiatives. This is already underway in various developed economies with the USA, for example, committed to improving building and appliance efficiency through extensive energy-efficiency standards [23]. In the European Union, an Energy-Efficiency Directive has been introduced to support its 20 % efficiency improvement target by 2020 and beyond [23].

Elsewhere, developing economies are grappling with the problem of managing energy demand more effectively. China, for example, incorporated energy consumption caps, energy intensity reduction targets, and CO_2 per capita reduction targets in its 12th Five-year plan and has set out energy saving programmes in its energy efficiency plans [23].

Over the next two decades, it is widely expected that renewables will account for an increasing proportion of energy consumption. BP's Energy Outlook 2030 sees renewables growing at 7.6 % per annum over the period to 2030 [20]. More than 100 countries had national renewable electricity policies in place by 2010 [16]. In some countries, such as Germany, Italy, and Spain, financial incentives for renewables have been reduced as they become more competitive, while in other countries, such as Japan, China, and South Korea, efforts are still being made to improve incentive schemes through feed-in tariffs, renewable portfolio standards, renewable energy certificates, tax incentives, and other mechanisms [16]. In some countries, these policies are being further reinforced by policies to phase out fossil fuel use over coming decades. For example, in 2011, Denmark announced its long-term goal of fossil-fuel independence by 2050 [24].

Another policy area of great significance to future SG developments will be the trajectory of the nuclear industry. While the Fukushima accident in 2011 encouraged some countries to reconsider their nuclear policies (e.g., Japan has shut down most of its nuclear plants, and France is considering reducing nuclear electricity from 79 to 50 % by 2025), other countries have continued to proceed with planned nuclear expansion programmes (e.g., U.S. began to build four reactors in 2012, China has resumed nuclear plant construction after a brief halt in the approval process for new plants in the immediate wake of Fukushima, while Russia and India continue to expand nuclear capacity by 15 to 20 GW each by 2025) [16].

Another critical element in relation to the further development of SGs will be development and innovation in business models. In light of the substantial capital investment required as well as the sizable potential global market for new energy products and services, the development of new business strategies to achieve economies of scale and risk sharing will require effective collaboration between the government and the business sector [25].

Moving now to the actual process of planning for SGs, there are clear indications that this has been gathering momentum since approximately 2009 although it is still far from being a global phenomenon. Early initiatives to establish long-term strategic plans/policies to facilitate SG deployment include roadmaps and other policy initiatives developed by countries such as Italy and the United States of America [16]. In the EU, SGs are seen as one of the keys to achieving the 20/20/20 European targets (i.e., 20 % increase in energy efficiency, 20 % reduction in CO_2, and 20 % renewables) set for 2020. The European Commission released mandates for smart meters in 2009, for EV in 2010, and for SGs in 2011 to provide reference architecture, standardization tools, and consistent standards for European Standards Organizations [26].

Elsewhere, in 2011, South Korea enacted the Smart Grid Stimulus Law to provide legislative support for SG planning and to tackle potential barriers [16], while in Japan, after the Fukushima accident in 2011, the Japanese government reviewed its Basic Energy Plan and consulted the public with energy mix scenarios—pursuing 0, 15, or 20–25 % nuclear [27, 28]. However, recent government changes have created uncertainty on Japan's position regarding nuclear power. In China, SGs are being addressed as one of the important elements of the current national energy plan [29].

This brief summary of some of the recent trends involving SGs suggests that while international interest in such grids has grown markedly over the past 5 years progress in the implementation of facilitating policies and the introduction of comprehensive grid enhancement programmes has been rather uneven. This is perhaps to be expected. SGs have been passing through what might be termed "the conceptual stage". The overall policy environment for SGs is, however, becoming increasingly favorable and supportive. A growing number of countries are moving to the stage of developing more extensive programmes for SG implementation. The key hardware and software elements are already available and we expect to see the diffusion of operational SGs rapidly gathering momentum over the period to 2020. But a cautious note on these trends is that the existence of competing priorities among stakeholders represents one of the major challenges to advancing SGs [30]. It is still uncertain whether SG developments will be able to introduce radical changes into existing power systems, or whether such developments may in fact result in only incremental changes, and re-inforce the dominant role of traditional, centralized power systems by making them smarter [30].

6 About This Book

With new technological and policy advances in SGs being announced almost from month to month, there is a need to contextualize and analyze these developments and to bring their impacts and significance to a wider audience. Hence, the basic rationale for this book. We caution, however, that the development of SGs is such a dynamic area that any book or paper can at best provide a snapshot of the situation at a given point in time. Given the rapid diffusion of SG technologies and developments in related areas such as ICT the picture will almost certainly look very different in 5 years time. Nonetheless, it is important to document and review the progress that has been made and to offer some observations on the possible trajectory of SG initiatives over the short to medium term.

6.1 Our Perspective

Meeting today's energy and climate challenges requires us to expand the scope and diversity of energy technologies as well as policy options. This requires new interdisciplinary, forward looking, and more participatory approaches to problem solving. More traditional approaches are unlikely to provide the innovative and imaginative approaches required to address such major challenges.

We have designed the structure and content of this book with four principal considerations in mind:

An interdisciplinary approach: SG technologies and their application are inherently interdisciplinary as they involve expertise not only from power system professionals but also from specialists in areas such as information systems, transport, and buildings. But the effective implementation of SGs requires more than technological advances. Policy supports, often involving new policy perspectives and approaches, are needed to make these technologies practical, useful, cost-effective, and socially relevant. This book focuses not only on the technological issues involved but also the perspectives of various stakeholders. It also provides a number of national case studies that demonstrate progress with SG implementation and provide insights into the opportunities, challenges, and barriers that currently exist. Coverage of these complementary dimensions allows this book to adopt an interdisciplinary approach to SGs from the perspectives of science and technology, economics, and governance and policies.

Global and Asian perspectives: Many SG developments are taking place in developed economies (including the USA, Europe, and Australia), industrialized countries in Asia (such as Japan and Korea) and in emerging countries such as China. Current studies on the topic of SGs have tended to focus primarily on Western developed economies. In an attempt to address this imbalance in coverage, the case studies presented this book include both Western developed economies (i.e., the US and Europe) and a number of Asian economies (i.e., China, Korea and Japan). This broader geographical perspective allows us to inject an Asian view to the global discussion in this important area.

Learning from international case studies: SGs are a new, highly dynamic and significant component of energy planning and management around the world. R&D breakthroughs and market applications, policy developments and piloting projects, successes and failures in our case study countries may reveal important similarities as well as interesting differences. This book aims to provide insights derived from these empirical experiences. Good practice demonstrated by policy initiatives in various countries is vitally important to policy-makers for the purposes of policy learning and diffusion.

Theoretically informed and empirically based: A number of chapters in this book focus on multi-stakeholder perspectives and international experiences. The analysis in these chapters is underpinned by key social science theories concerning governance, policy, and politics. This theoretical underpinning is intended to provide coherence in the analysis throughout the book.

6.2 Objectives of the Book

Our aim in compiling this book and drawing together the valuable work of our contributing authors is to provide the reader with an interdisciplinary, informative, and comprehensible introduction to SG technologies in the context of increasing interactions between technological innovation and sustainability, between government, business and society, and between economic, political and social developments. Clearly, it is not feasible to provide coverage of each and every SG-related issue but our objective here is to provide the reader with a comprehensive understanding of the key issues associated with SG development and to illustrate how such grids are rapidly taking shape in many parts of the world.

This book, therefore, showcases state-of-the-art R&D developments and policy experiences. All aspects of this book are designed to contribute to a better understanding of governance, institution, and policy challenges and help formulate policy recommendations for successful SG deployment.

We hope that this book will be of interest to and meet the needs of not only academics, but also government policy-makers, those in energy and environmental consultancies and utility planning, and in international institutions as well as NGOs. It is also intended to meet the needs of undergraduate and postgraduate students wishing to gain a more comprehensive understanding of the technical and policy backgrounds to SG applications.

6.3 Structure and Content: A Brief Overview

We have selected the chapters that appear in this book on the basis of the four considerations or guiding principles outlined earlier. Our contributing authors come from a variety of academic and professional backgrounds and their work draws on a variety of national contexts and experiences. All are either active researchers or managers/practitioners in the area of SGs. We hope that this direct involvement in SG work on the part of our authors will help to convey to readers not only the key technical and policy issues involved but also the significance and the dynamism of the field.

The book itself is organized into five sections: Part 1 provides an overview of SGs. Part 2 deals with the technological aspects of SGs. It introduces important technical characteristics of these grids and showcases state-of-the-art R&D developments for different SG technologies. It also highlights the interdisciplinary and wide-ranging nature of SG technologies. Part 3 integrates a multi-stakeholder perspective into our discussion of SGs. The aim here is to provide a better understanding of the interests involved, the ways in which different stakeholders use their power, and how conflicts among power utilities, policy-makers, market regulators, consumers, environmental NGOs and other stakeholders can emerge and be resolved, and the implications of these issues for policy-making. Part 4 offers a collection of international case studies of SG development to showcase policy experiences in both Western and Asian countries. Part 5 consists of a postscript which aims at sharing insights, identifying the lessons learnt, highlighting the policy implications of the discussions in the various chapters of the book.

Part 1, comprising two chapters, is an overview of SGs. In this chapter, the editors provide an historical overview of SGs. It discusses the basic (often varied) definitions of SGs, the evolution of SGs as a concept and as a practice. In chapter "A Holistic View on Developing Smart Grids for a Low-Carbon Future", John Cheng presents a holistic overview of the development of smart grids as an essential element for a low-carbon future. It examines how four key aspects, technology, economies, regulation/policy, and social acceptance, are complementary and therefore need to be considered in a holistic manner for the deployment of smart grids.

Part 2 includes four chapters related to SG technology development. In chapter "Statusand Prospects of European Renewable-based Energy Systems Facilitated by Smart Grid Technologies", the plans and status of renewable energy resource development and energy policy in Europe are introduced. The development of SG technologies in the European Union is also discussed. Chapter "Arcturus: An International Repository of Evidence on Dynamic Pricing" is a comprehensive study on dynamic pricing, reviewing results from 163 pricing treatments offered in seven countries located in four continents. It was found that customers respond as the peak to off-peak price ratio increases, by lowering their peak demand, and as the price continues to increase, they continue to lower their demand, but at a decreasing rate. In addition, the use of enabling technologies boosts the amount of demand response. Chapter "Microgrids and Distributed Energy Future" reviews the concepts and models of microgrids, describes its evolution from the traditional power system, and studies issues related to microgrid operations, including grid-connected and islanding operations, operation and control strategies, and economic issues. Existing microgrid projects from around the world are described. Chapter "Communication and Network Security Requirements for Smart Grid" focuses on the communication technology of SG. It is noted that deployment of SG requires technologies on sensing and measurement, advanced control methods, advanced components, improved interfaces and decision support, and integrated communications. To achieve the full functionalities of such SG enabling technologies, communication technology plays a fundamental role. This chapter describes a communication-oriented SG framework, identifies the communication requirements, and the security and privacy requirements.

Part 3 of the book, comprising three chapters, focuses on stakeholder perspectives in SG development. In chapter "Smart Grids: The Regulatory Challenges", Mah, Leung and Hills examine some of the regulatory challenges associated with SG systems. Working from the position that a successful transition to SGs requires not only technological advances but also the need to overcome various regulatory barriers, the authors review the changing regulatory context for grid development. They argue that SGs present new and different challenges for regulators. These include disincentives to utilities, pricing inefficiencies, and cyber security and privacy concerns. Drawing upon case studies from North America and Europe the authors demonstrate that a variety of regulatory initiatives have emerged to facilitate SG development from which they conclude that a mix of regulatory approaches will most likely be needed to achieve a successful transition to SGs. Furthermore, with the more complex and dynamic stakeholder landscape that has emerged in relation to SGs, the authors also suggest that future regulatory frameworks will need to be more open and participatory if they are to respond effectively to the challenge of SGs.

Chapter "*i*-Energy: Smart Demand-Side Energy Management", by Matsuyama, provides valuable insights into novel research that explores the potential for smart demand-side energy management. This chapter views SGs from a different perspective, arguing that these grids aim at achieving more effective energy management from the supplier's viewpoint but that what is also required is a consumer perspective. Matsuyama, therefore, proposes the concept of i-Energy which emphasizes the importance of energy management from the consumer's side. The chapter discusses four steps in the realization of the i-Energy concept: the Smart Tap Network for monitoring electricity consumption patterns, the Energy on Demand Protocol to achieve a priority-based best-effort supply mechanism, Power Flow Coloring to allow versatile power flow controls, and, finally, the Smart Community for bi-directional energy trading among households, offices, and factories.

Chapter "Switching Perspectives: Creating New Business Models for a Changing World of Energy", the final chapter in this part of the book, examines the need to develop new business models for the rapidly evolving energy scene. Valocchi, Juliano and Schurr argue that new technologies, policy changes, and more demanding consumers are driving the need for a revamp of traditional electric utility business models and the traditional "grow and build" mindset. These are rapidly becoming outdated and inappropriate. The foundations on which these traditional business models were based, namely one-way flows of power and information, declining costs with increased usage, passive consumers, easy access to cheap carbon fuels and regulatory protection from threats to core business interests, are being eroded. A new kind of "grow and build" era is upon us, one in which there must be a switch to new business models that will facilitate and support information exchange, consumer participation, and new services. The future is not so much about energy itself but information and services.

Part 4 provides a collection of international case studies of SG developments to showcase policy experiences across the Western and Asian countries. Our case studies of Europe (chapter "Smart Transmission Grids Vision for Europe: Towards a Realistic Research Agenda"), a comparative study of the USA and Europe (chapter "Comparison of Smart Grid Technologiesand Progress in the USA and Europe"), Japan (chapter "Towards Sustainable Energy Systems Through Deploying Smart Grids: the Japanese Case"), Korea (chapter "Governingthe Transition of Socio-Technical Systems: a Case Study of the Development of Smart Grids inKorea"), China (chapter "Developing Super Smart Grids in China: Perspective of Socio-Technical Systems Transition"), and Australia (chapter "Exploring the Value of Distributed Energy for Australia") provide a better understanding how countries differ in their deployment of SGs with different aspirations, approaches, and focus. These case studies, together, highlight the key issues, opportunities, and barriers to the large-scale deployment of SGs, which involve not only the technological aspects but also the financial as well as regulatory ones. Furthermore, the adoption of the same theoretical perspectives of the socio-technical systems in the case studies of Korea and China (chapters "Governingthe Transition of Socio-Technical Systems: a Case Study of the Development of Smart Grids in Korea" and "Developing Super Smart Grids in China: Perspective of Socio-Technical Systems Transition") has guided us to examine the complexity of the energy system transition, which can be enabled by the deployment of SGs, from the macro, meso, and micro levels.

Starting with Europe, in chapter "Smart Transmission Grids Vision for Europe: Towards a Realistic Research Agenda", Vanfretti et al. advocate for a realistic turn in SG research if the overwhelming challenges posed by their deployment are to be met. If SGs at the transmission level are to become a reality, there needs to be an alignment in the current research practices integrating complex interactions between policies, the regulatory background, technology maturity, and socially responsible and farsighted investment. The different time frames in which the transmission system is managed are pivotal in the appreciation of SGs. "Smarter" grids require adjustments in the operations and the operational planning phase as well. On the operational side, an increased use of AMI, for instance from phasor measurement units, can be expected to bring considerable advantages to the power system. Nevertheless, challenges such as standardization, big data management, and ICT requirements remain important. If these challenges can be met, the measurement data can be applied to improve system operations, for instance through the provision of improved control and protection functions. The uses of new ICT technologies, and especially the possibility of visualization, allow the operators to become better aware of the system state to take the appropriate actions. However, the main challenge may lie in development of new reliability concepts which can be implemented in realistic power systems to provide a maximum social welfare to the users, and for which adequate tools are still under development.

Chapter "Comparison of Smart Grid Technologies and Progress in the USA and Europe", by M. Godoy Simões and his colleagues, provides a comprehensive account of the major historical and technical events of SG developments in the USA and Europe. By contrasting these technical, legal as well as policy developments in these two major economies, the authors provide a full understanding of how and why SGs are developed in various pathways across places.

Chapter "Towards Sustainable Energy Systems Through Deploying Smart Grids: the Japanese Case", by Amy Poh Ai Ling, is a case study of Japan. Poh discusses the aspirations, concepts, and elements of SG deployment in Japan. Poh highlights that SGs are a part of the government's "go green" effort to elevate Japan to a leading country in green growth. Poh also provides an overview of Japan's major initiatives on SGs, which include smart communities, large-scale city-based pilot projects as well as overseas collaboration.

In chapter "Governing the Transition of Socio-Technical Systems: a Case Study of the Development of Smart Grids in Korea", Mah, van der Vleuten, Chi-man Ip and Hills show how driven by its "Green Growth Vision", Korea embarked on its SG initiatives in 2009. Within just 3 years, the country has made some important progress in the development of SGs. Grounded in a perspective of governance and innovation systems as a general analytical framework, the chapter first reveals the complexity of the socio-technical system by highlighting the breadth and depth of factors influencing SG development in Korea. The convergence of policies, business incentives, and consumer motivations is critical to drive changes in socio-technical systems. Various factors spanning from macroeconomic policies and global views, to partial electricity market reform and public distrust, and to experimentation in a demonstration project are found to be crucial in the socio-technical system for SGs in Korea. The presence of partial electricity market reform and public distrust has created barriers to develop some of the favorable conditions for change, such as policy consistency, second-order learning, and the development of financially viable business models, still lacking in Korea.

Chapter "Developing Super Smart Grids in China: Perspective of Socio-Technical Systems Transition" by Yuan Jia-Hai in turns considers the prospects for SG developments in China. It shows that technological development in renewable energies industries will not be sufficient to ensure a transition toward a cleaner and more efficient energy system for China. Indeed such a transition will require the separation of generation, transmission, distribution, and retail prices, which in turn, calls for a new perspective on the regulation of transmission and distribution network. Given that there is deep-rooted origin of the current tariff policy, pricing reform can be regarded as a cornerstone for the Chinese Government in its energy policy. The integration of medium-and-small scale renewable generation as well as pricing mechanism reform also calls for the gradual redefinition of the power grids and restructuring of the utility companies. This certainly poses the most difficult institutional puzzle for developing SG in China. Powerful grid companies would have no incentive to develop SG if it was destined to be dismantled. Therefore, the lock-in effect of energy systems is particularly active in China and a subtle design to synchronize the SG roadmap with power sector deregulation is necessary. Potential strategies to overcome institutional loopholes include empowering the National Energy Administration to be responsible for national strategy formulation, the creation of a think-tank to support policy innovation and the establishment of an industry-government consortium to provide integrated translation research and strong front end support. At the same time, restructuring of the power sector should be synchronized with the development of SG. Dismantling the Grid Companies into regional transmission companies and provincial distribution companies should provide public goods to private users. Leveling the playing field in both technology and investment and opening the power sector to private investors seems of critical importance.

The final chapter in this section examines the important role of distributed energy in smart grid diffusion in Australia. Based on the findings of a major government-funded study, William Lilley and his colleagues examine the opportunities of and key barriers for large-scale uptake of such distributed systems.

Acknowledgments We gratefully acknowledge the funding of our research by the Hong Kong Baptist University (through the Faculty Research Grants from the Faculty of Social Sciences, FRG2/12-13/057 and FRG1/13-14/051) and the University of Hong Kong (through the Initiative on Clean Energy and Environment).

References

1. DOE (2008) What the smart grid means to Amercians. The United States department of energy. http://energy.gov/sites/prod/files/oeprod/DocumentsandMedia/ConsumerAdvocates.pdf, Washington, DC
2. EAC (2008) Smart grid: enabler of the new energy economy. The electricity advisory committee, U.S. department of energy. http://energy.gov/sites/prod/files/oeprod/DocumentsandMedia/final-smart-grid-report.pdf, Washington, D.C
3. IEA (2011) Technology roadmap: smart grids. International Energy Agency. http://www.iea.org/papers/2011/smartgrids_roadmap.pdf, Paris
4. PwC (2010) Smart solutions for smart grids: capability statement. Price waterhouse Coopers, London
5. World Energy Council (2012) Smart grids: best practice fundamentals for a modern energy system. World Energy Council. http://www.worldenergy.org/documents/20121006_smart_grids_best_practice_fundamentals_for_a_modern_energy_system.pdf, London
6. DOE (2008) What a smart grid means to our nation's future. The United States department of energy. http://energy.gov/oe/downloads/what-smart-grid-means-our-nation-s-future, Washington, DC
7. DOE (2009) Smart grid system report. U.S. department of energy. http://www.oe.energy.gov/DocumentsandMedia/SGSRMain_090707_lowres.pdf, Washington, DC
8. Mah DN-Y, van der Vleuten JM, Ip JC-M, Hills PR (2012) Governing the transition of socio-technical systems: a case study of the development of smart grids in Korea. Energy Policy 45:133–141
9. Lott M, Seaman T, Upshaw C, (2011) The smart grid in texas: a primer. UT Austin. https://www.sgiclearinghouse.org/BusinessCases, Austin
10. DOE (2009) Smart grid system report 2009. The United States department of energy. http://energy.gov/sites/prod/files/2009%20Smart%20Grid%20System%20Report.pdf, Washington, DC
11. Mark D, Ostrowski K, Tai H, (2010) Introduction. MsKinsey & Company. http://www.mckinsey.com/client_service/electric_power_and_natural_gas/latest_thinking/mckinsey_on_smart_grid, Atlanta
12. Office Executive (2011) A policy framework for the 21st century grid: enabling our secure energy future. Executive Office of the President of the United State, Washington, DC

13. Cooke D (2011) Empowering customer choice in electricity markets—information paper. http://www.iea.org/publications/freepublications/publication/Empower.pdf. International Energy Agency, France
14. IPCC (2013) 27 September, 2013. Climate change 2013: the physical science basis—summary for policymakers. Working group I contribution to the IPCC fifth assessment report. Retrieved 3 Oct 2013, from http://www.climatechange2013.org/images/uploads/WGIAR5-SPM_Approved27Sep2013.pdf
15. IEA (2012) World energy outlook 2012. OECD Publishing, France. Retrieved 27 June 2013, from http://dx.doi.org/10.1787/weo-2012-en
16. IEA (2013) Tracking clean energy progress 2013: IEA input to the clean energy ministerial. International Energy Agency, France. Retrieved 26 June 2013, from http://www.iea.org/publications/freepublications/publication/TCEP_web.pdf
17. Worldwatch Institute (2010) Renewable energy and energy efficiency in China: current status and prospects for 2020. Worldwatch Institute. http://www.worldwatch.org/system/files/18220China%20Energy.pdf, Washington, DC
18. AEA (2012) European smart metering landscape report 2012. Österreichische Energieagentur—Austrian Energy Agency, Vienna. Retrieved 2 May 2013, from http://www.energyagency.at/fileadmin/dam/pdf/projekte/klimapolitik/SmartRegionsLandscapeReport2012.pdf
19. PSC (2006) Proceeding on motion of the commission regarding expedited implementation of mandatory hourly pricing for commodity service, order denying petitions for rehearing and clarification in part and adopting mandatory hourly pricing requirements, Case 03-E-0641 C.F.R
20. BP (2013) Statistical review of world energy 2013. British Petroleum. London
21. California Public Utilities Commission (2012) Decision adopting a dynamic pricing structure for residential and small commercial customers and denying the motion for approval of a settlement agreement (December 20)
22. IEA (2011) Empowering customer choice in electricity markets. International Energy Agency, Paris. from http://www.iea.org/publications/freepublications/publication/Empower.pdf
23. IEA (2013) Redrawing the energy-climate map. International Energy Agency, France. Retrieved 26 June 2013, from http://www.worldenergyoutlook.org/media/weowebsite/2013/energyclimatemap/RedrawingEnergyClimateMap.pdf
24. Danish Government (2011) Energy strategy 2050—from coal, oil and gas to green energy. Denmark: the Danish ministry of climate and energy. Retrieved 27 June 2013, from http://www.ens.dk/sites/ens.dk/files/dokumenter/publikationer/downloads/energy_strategy_2050.pdf
25. Giordano V, Fulli G (2012) A business case for smart grid technologies: a systemic perspective. Energy Policy 40:252–259. doi: http://dx.doi.org/10.1016/j.enpol.2011.09.066
26. Sporer R, (2013) First set of standards for the grid: where are we today? Paper presented at the smart grid standardization achievements (January 28), Brussel. Retrieved 28 June 2013, from http://ec.europa.eu/energy/gas_electricity/smartgrids/smartgrids_en.htm
27. CDD (2012) Deliberative poll on energy and environmental policy options. The Center for Deliberative Democracy, Stanford University, Stanford. Retrieved 25 April 2013, from http://cdd.stanford.edu/docs/2012/jp-energy-policy.pdf
28. DeWit A, Tetsunari I, Masaru K (2012) Fukushima and the political economy of power policy in Japan. In: Kingston J (ed) Natural disaster and nuclear crisis in Japan: response and recovery after Japan's 3/11. Routledge, New York, pp 156–172
29. NEA (2011) 12th five-year plan for energy technology (2011–2015). China National Energy Administration, Beijing. Retrieved 17 June 2013, from http://www.nea.gov.cn/131398352_11n.pdf
30. Stephens J, Wilson E, Peterson T, Meadowcroft J (2013) Getting smart? Climate change and the electric grid. Challenges 4:201–216. doi:10.3390/challe4020201
31. Computer Networks and Security Lab (2013) August 8 2013. Research Retrieved 30 Oct 2013, from http://cnslab.snu.ac.kr/twiki/bin/view/Main/Research
32. HD PLCMAG (2011) HD-PLC Alliance. Retrieved 30 Oct 2013, from http://hd-plc.org

A Holistic View on Developing Smart Grids for a Low-Carbon Future

John W. M. Cheng

Abstract This chapter provides a holistic overview of the development of smart grids as an essential element for a low-carbon future. It examines how four key aspects, namely technology, economics, regulation/policy, and social acceptance, are complementary and therefore need to be considered concurrently for the deployment of smart grids. This chapter emphasizes that there is no one-size-fit-all solution, and the choice of solutions differs across countries and localities. A good understanding of local features such as historical, cultural, resources, market, and regulatory factors that coexist with strong and consistent commitment, careful road mapping, learning through trials as well as supporting innovation and research are required for the effective transition toward different smart grid systems in a low-carbon world.

1 Introduction

Power grids are designed to interconnect generation and load entities for the purpose of delivering electricity in a safe, economical, reliable, and socially acceptable manner. Engineering marvels have made today's power grids one of the most complex and valuable systems humans have ever built [18]. Our industrialization, technological advancements, social well-being, and high living standards are based on the electricity delivered through the power grids, without which we would not have developed so rapidly in the past century.

Disclaimer: This chapter and other materials mentioned herein reflect my own views and do not necessarily reflect the views of CLP Research Institute or any other organization with which I am affiliated.

J. W. M. Cheng (✉)
CLP Research Institute, Kowloon, Hong Kong SAR, China
e-mail: john.cheng@clp.com.hk

D. Mah et al. (eds.), *Smart Grid Applications and Developments*,
Green Energy and Technology, DOI: 10.1007/978-1-4471-6281-0_2,

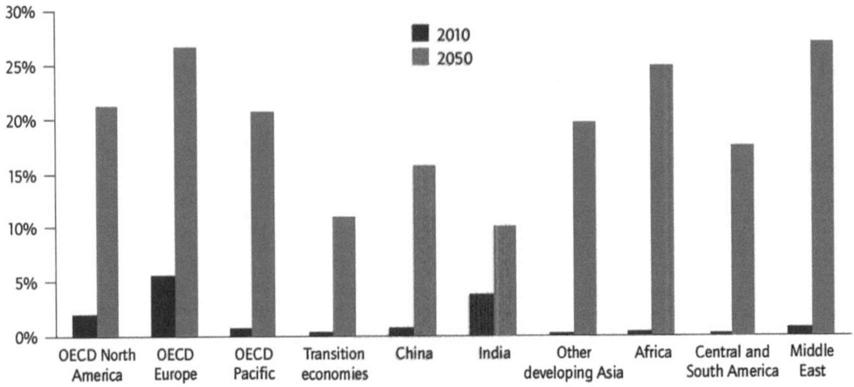

Fig. 1 Portion of variable generation of electricity by region (BLUE Map Scenario) [10]

However, with rapid development and a heavy reliance on fossil fuels to generate electricity globally, greenhouse gas emissions have also brought on new challenges for the global community, namely climate change and global warming. Social norms are favoring the adoption of more environmentally friendly and sustainable development practices to be adopted. Volatile fuel prices and resource constrains have also prompted many countries to place energy security high on their national agendas. For instance, Germany's "Energiewende" (roughly translated as "Energy Transition") favors more renewables instead of the conventional coal or nuclear facilities, which has driven the rapid development of wind and solar generation. Within the last decade, traditional generation mix and associated practices have been rapidly evolving. Figure 1 shows the potential increase of variable generation of the world by 2050, according to IEA's estimates [10].

Meanwhile, the need to upgrade and reinforce the grid is also necessary. Most power grids were built over 50–60 years ago and are aging with different degrees of deterioration. With the continuous needs of meeting load growth, equipment retirement, network reinforcements, and the ever-changing market conditions, the power industry finds itself facing a dilemma. Should the industry continue to invest in conventional means, which only offers the status quo at best, or leap to a more efficient, smarter, interactive, and flexible grid for a low-carbon future?

With maturing Information and Communications Technology (ICT) and their wide applications and deployment in different sectors, a technological pull also provides the industry with different tools and applications to enable the making of a more advanced power grid or as it is commonly known a "smart grid."

Global smart grid developments vary widely in terms of scope and pace of development. In fact, even the definition of smart grid may differ from place to place. In most developed countries, smart grid developments focus on renewables, smart meters (SM), and demand response (DR). The main drivers are on renewables integration, energy efficiency, and provision of more choices to consumers. On the other hand, in emerging economies like China, smart grid developments are focusing on building a "strong and smart" grid. The main driver is to build a reliable and strong transmission and distribution system to meet rapid demand growth. Mega-projects like HVDC and EHV lines stretching thousands of kilometers are being built to bring bulk power from remote sites to major load centers. Digital substation and distribution automation (DA), coupled with different smart meter deployment schemes, are also being trialed in major cities to increase reliability or simply to gain the development experience.

This chapter suggests that a holistic view encompassing four key aspects of smart grid development should be examined concurrently. These aspects include technology, economics, regulation/policy, and social acceptance. As different communities or nations have their own sets of existing conditions, their needs, aspirations, resources, implementation, and results could be all different. Like any other grand problems confronting the global community, each community has its own unique characteristics as well as intricate relationships with each other. The choice of solution would therefore differ, but it is important to understand the inherent and related issues in a holistic manner to constantly refine them. It is hoped that this chapter will help facilitate meaningful discussions and further research and development and ultimately make deployment of smart grids a universal reality—a necessity for this and future generations to build a low-carbon world.

2 Capabilities of Smart Grids and Implications

The notion of a "smart grid" should not be misconstrued to say that the existing power grids are dumb or unsophisticated. Instead, we should appreciate and anticipate what will be required of future grids. In essence, a smart grid is the conventional grid equipped with extensive information-based and technology-based equipment and systems to do or support most, if not all, of the following:

- Integrate a larger share of renewable energy (RE);
- Enable customers to participate in supplying and balancing the grid;
- Accept new load entities such as electric vehicles (EV) and smart appliances;
- Interconnect with different distributed energy resources (DER) and micro-grids;
- Enable market-based power transactions or new business models;
- Recover robustly and more reliably after extreme disturbances;
- Improve grid efficiency and reduce transmission losses; and
- Do the above while upholding or exceeding existing reliability and safety standards, with a sound economic base under the allowable regulatory framework and social acceptance.

With full deployment, smart grids could enable us to reduce carbon dioxide emissions significantly. For instance, it was estimated the potential for carbon dioxide (CO2) reduction in US alone could range from 4 to 27 % based on

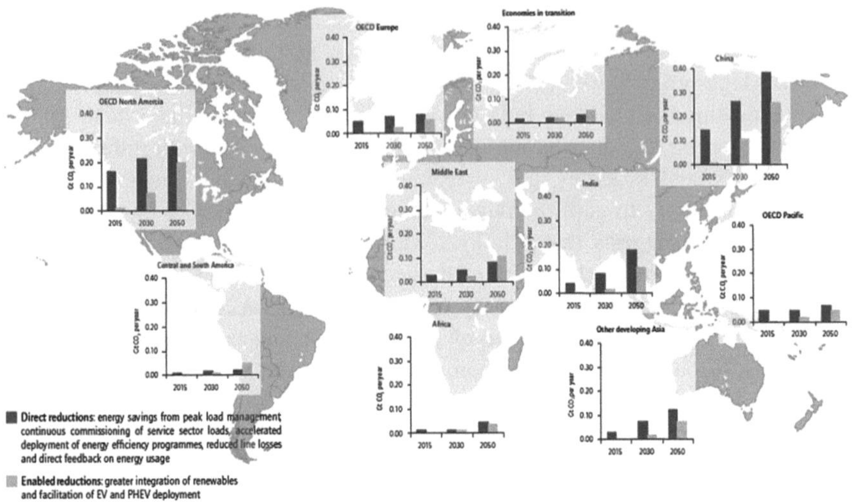

Fig. 2 Regional CO2 emissions reduction through smart grid deployment [10]

different studies [23]. Figure 2 shows an estimate of regional CO2 emissions reduction through the deployment of smart grids according to [10].

However, like all infrastructure establishments, any fundamental change of this magnitude would involve many aspects. Based on the main smart grid capabilities, Table 1 highlights the key implications with respect to four major aspects, namely technical, economic, regulatory, and societal. The following sections will elaborate on each of the key aspects.

3 Key Technologies and Their Current Stage of Development

If one were to take a quick look over the development pace of the ICT infrastructure buildup in the last decade, the growth is nothing but astounding. For instance, fiber-optic cables (the backbone of today's communication networks) have grown from 95 billion km in 2000 to 180 billion km in 2011 [11]. Mobile cellular subscribers have increased almost nine times to 6.3 billion in 2012 since 2000 [12]. Meanwhile, global Internet users grew from 0.361 billion (5.8 % of world population) to more than 2.75 billion (38.8 % of world population) in 2013 [13]. Most smart grid technologies are also built on this ubiquitous and extensive communication technology and infrastructure. However, beyond the ICT, what are the key technologies that are required in order to fulfill the prescribed objectives of a smart grid? This, of course, depends on the local environment [6, 7, 15, 22]. Instead of being pulled by any political agenda, fashionable rhetoric or technology

Table 1 Overview of smart grid capabilities and implications

Smart grid capabilities	Technical	Economic	Regulatory/Policy	Societal
Integration of renewables	Frequency control; dispatch coordination; system and weather forecasting; resource quality and availability; grid codes; loop flows; power and voltage controls; and backup facility controls	Levelized cost of energy for RE; cost of development; and cost of backup facilities	National policy; RE content and target; feed-in tariff; and government incentives	Willingness to purse a higher RE contents; affordability of the local communities; NIMBY resentment; social welfare redistribution; and poor subsidizing the rich
Customers engagement (smart meters and AMI)	Forecasting; response time; reliability; coordination; and cooperation	Investments; ownership; price of "megawatt"; and elasticity of demand	Approval of smart meters and AMI investments and cost recovery method; ownership of metering data; and privacy	Public acceptance of smart meters; perceived health and privacy issues
Distributed energy resources	Sizing; dispatching coordination; power flow controls; voltage controls; maturity of energy storage devices; grid codes; interconnection; and coordination	Cost and availability of alternative fuels; PV systems and storage devices; Stranded grid assets; and innovative business models	Energy and environmental policies, energy security, roles of grid backup, emissions monitoring and control policies	Public sentiment and acceptance to opt for DER (especially after major interruptions experienced)
Electric vehicles supports (e.g., charging and V2G)	Connection and interface standards; mobile load shifts; grid reinforcements to sustain charging peaks and locations; storage devices; and vehicle to grid	Local tariffs; generation mix; cost of EVs; and incentive for charging	Incentives to push charging stations and EV purchases; RE developments	Acceptance and popularity of EV or plug-ins
Distribution automation	Equipment standards; protection philosophy, scheme, and coordination	Asset management practice; investment deferral; and costs of loss of load or interruptions	Policies to deal with cost recovery; RE integration policy	Public awareness and education of the needs and benefits of DA

lobbyists, it is useful to first examine the nature of different "smart grid" technologies and understand their stage of development and deployment today.

3.1 Smart Meters and Automated Metering Infrastructure

Today, hundreds of millions of SM have been deployed globally. ENEL of Italy was a first mover on SM, and over 30 million, SM have already been installed. The fastest growth today is in China. According to Navigant Research [19], there were 139 million SM installed in 2012 and this figure will likely grow to 377 million by 2020. SM are electronic meters which are more intelligent and versatile than the conventional electromagnetic meter. They offer a wide range of functions including the abilities to capture and store detailed consumption data, communicate by wire or wireless means, remotely connect and disconnect, enable net metering, support dynamic pricing or prepaid, and monitor power quality and outage conditions. Although not all the above-mentioned functionalities are present in these SM deployments, almost all are capable of supporting automatic meter reading, dynamic pricing scheme, and detailed consumption data recording.

Automated metering infrastructure (AMI) refers to the bidirectional communication system linking hundreds and thousands of SMs (the customers) with the backend systems (the utilities/distributors). This is usually accomplished by linking the SMs with collectors and concentrators using one or more of mesh networks, power line carrier, optical fiber, Wi-fi, or Wimax technologies. AMI can carry out multiple functions such as automatic meter reading; meter data management system (e.g., integrity check and lost data retrieval); interfacing with customer supports, billing, and outage management systems; and supporting in-home energy displays and/or Web-based customers' consumption portal.

With SMs and AMI, the detailed usage information, bidirectional communication, and near-real-time access of load entities' condition open up new opportunities for both consumers and utilities [10]. With the versatility and convenience of these technologies, cyber security and privacy issues are also present. The new infrastructure therefore also needs to have high security features built into the design so that customer's privacy is reasonably protected and supply reliability is ensured against any cyber attacks.

3.2 Demand Response and Automated Demand Response

DR refers to the active customer engagement requested by the grid operator to reduce demand when there is insufficient generation or transmission capacity to supply all load entities under normal or abnormal conditions (e.g., severe weather or loss of supply). The customers are generally rewarded by compensation or entitlement to special tariff schemes. DR has been demonstrated to be quite successful in USA. The recruitment of participants and management of DR executions are generally performed semiautomatically with lead time in terms of days and hours ahead. Automated demand response (ADR) incorporates special hardware to control major appliances such as air conditioners and hot water heaters to reduce demand directly upon request from the utility. The lead time is reduced to almost real time and is more certain to the operators. In theory, having the means like DR and ADR to control the demand side is desirable as it enhances the flexibility of the network. In practice, the implementation also depends on the nature of the customer fabric and their corresponding effects on the system when DR/ADR is called upon. The level of compensation also depends on the elasticity of demand, tariffs, program flexibility, and general acceptance to the notion of participating in grid control.

3.3 Phase Monitoring Unit and Wide Area Monitoring System

PMU is a high-precision sensor designed to capture subsecond-level information of the voltage and current conditions on the grid. They can be deployed in key or strategic locations (e.g., interconnection points or major substations) of the grid to form a WAMS. PMUs are synchronized to a common and precise clock such that they can monitor the power flows and voltage profiles together at a much higher sampling frequency compared to conventional means. As a result, PMUs and WAMS offer system operators in a control region (as well as across different control regions) the ability to react in a timely manner to avoid or mitigate the damage caused by any disturbances happening far away. It should be noted that PMUs and WAMS are most beneficial to long-distance bulk power transfers and large networks stretching across a wide area. The integration and usage of them are only gradually being introduced to the industry in the last decade, so many are yet to be convinced of its effectiveness in improving the reliability supply.

3.4 Power Electronics, HVDC, and FACTS Devices

Power electronics have been used for grid applications since the 1960s. Many of the inverter and converter stations in HVDC lines or back-to-back HVDC systems were built on the thyristor-based technology [1]. HVDC lines stretching hundreds and thousands of kilometers can now bring in remote hydro power in bulk and lower costs. The same facilities can also be used to harness mega-size wind and solar facilities. This technology is mature and stable after years of development and deployment experience. Today, thyristors are being gradually replaced by a more efficient and powerful technology using integrated gate bipolar transistors

(IGBT). The main difference is that IGBT can be freely switched on and off, while thyristors can only be switched on. This allows faster and more complicated switching schemes to be implemented (e.g., multi-level modular converter). FACTS[1] devices such as STATCOM (or called SVG in China) or SVC are used to alter the natural flow of AC power such that the voltage profile can be maintained and/or real power transfer capability increased. In short, the rapid deployment of power electronic equipment in the grid today is offering more flexibility and security to grid operations. The benefit is wide as the technology can be applied to both bulk power flow controls and on smaller scales, using distributed and variable interconnections with micro-grids and distributed energy resources.

3.5 Distribution Automation

Technologies used in DA are quite mature. To a large extent, similar monitoring and control features used in SCADA and EMS[2] systems for generation and transmission systems are now downloaded to the distribution level. Through the extensive deployment of Remote terminal unit (RTUs) in substations or supply feeders, DA will normally cover the following functions [21]:

- Monitoring and control of distribution circuits and voltage profile;
- Outage location and automated restoration;
- Semi- or automatic reconfiguration;
- Voltage and reactive power controls (via transformer tap changing or capacitor bank switching);
- Load flow calculation and contingency analysis; and
- Equipment health monitoring.

Together with SMs and AMI, DA can significantly enhance service reliability and robustness.

3.6 Energy Storage

To date, energy storage remains the holy grail for the power industry. However, the future looks more and more encouraging due to the rapid concurrent development of various storage technologies, renewables, and EV. With energy storage, we can harness the full potential of uncertain and intermittent RE to give a more

---

[1] *FACTS* Flexible AC transmission system; *STATCOM* Static Synchronous Compensator; *SVG* Static Var Generator.

[2] SCADA—Supervisory Control and Data Acquisition System; EMS—Energy Management System.

controllable interface with the grid, hence providing a better utilization of low-carbon energy. Traditionally, pump hydro was the reliable and most widely used utility-scale means of energy storage around the world. Part of the reason may be because of the storage means used—water is relatively safe to store in bulk and stable enough without the needs for expensive monitoring or maintenance setups. Sodium sulfur (NaS) or Lithium–ion (Li ion)-based battery systems are both commercially available today, and trials are taking place around the world to test their applicability for different uses. Although their unit costs are still high, the deployment rate and applications are growing, particularly in strengthening the supply consistency of renewables. Figure 3 shows an overview of different storage technologies and their stage of development.

3.7 Micro-grids

Micro-grids are self-contained local electrical networks which have one or more generation units feeding the local loads through a small network. The network usually is low-to-medium voltage (1,000 V or below) with limited nodes connecting DER like solar PV, wind turbines, combined heat and power (CHP), fuel cells, and energy storage devices. They typically deal with loads like community buildings, households, and EVs. Figure 4 shows a schematic comparison between today's distribution grid and a micro-grid. In theory, with the distributed generation and sophisticated control and coordination, a micro-grid would be more resilient to disturbances and capable of operating even if the connection with the main grid is disconnected.

3.8 New Paradigms, Systems, and Technologies for a Flexible Grid

Multiple factors are driving the shift toward a low-carbon future, and power grids have to be more flexible [10] to accommodate higher share of renewables (e.g., >30 %), more AC/DC system interactions, more emission-constrained dispatching in accordance with market activities (power markets and/or carbon markets), more demanding controls over the interconnections, and sophisticated coordination of distributed systems (e.g., micro-grids). It is impossible to list all the new paradigms and associated technological needs in this context as some of them have yet to be discovered. However, it is worthwhile to note that different combinations of the above-mentioned technologies, which increase the grid observability and controllability, are just the basic building blocks. They have to be linked by a robust, reliable, and ubiquitous communication framework and overlaid with a suite of elaborated ICT-based monitoring and control applications.

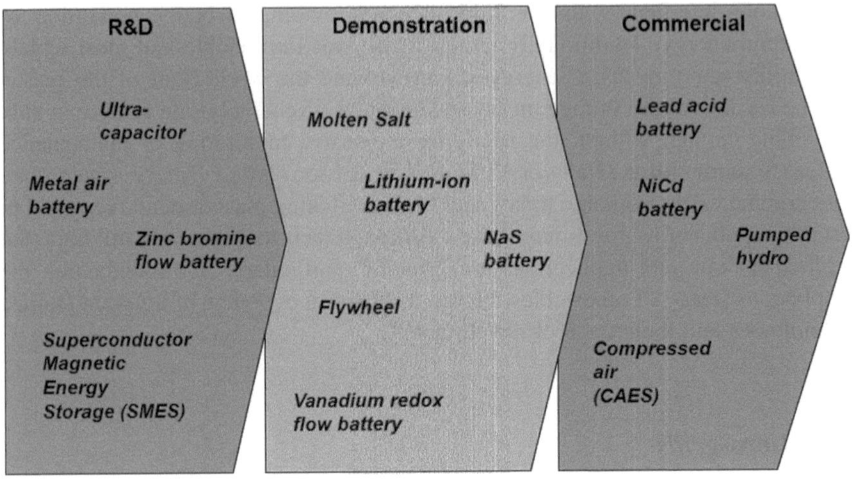

Fig. 3 Storage technologies and stage of development

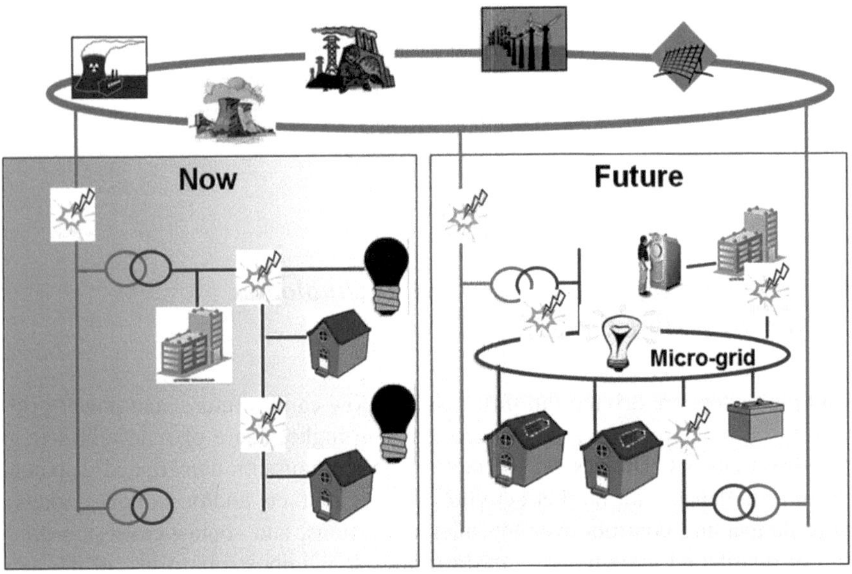

Fig. 4 Comparison of conventional grids with micro-grids

3.9 *Enabling Customer Interactions and Dynamic Pricing Scheme*

Today, the most popular value chain of a smart grid is composed of SM–AMI–Metering Data Repository System–Enterprise Application Systems. This new value chain provides multiple functions to stakeholders including meter reading, energy audits, customer interactions, net metering of distributed renewables, dynamic pricing support and auditing, outage monitoring, remote connection, DR, direct load control, and detailed billing information.

3.10 *Providing Flexibility and Intelligence to Grid Operations*

Grid operators and system planners both understand very well that the power grid is a highly complex machine with many different characteristics and unique practices derived from historical, market, and technical developments. With new technologies, innovative models and algorithms are needed to make the grid more flexible, robust, and reliable. For instance,

- Intelligent tools using remote sensing images to deduce accurate wind and solar generation, optimized with coordinated interconnection control and generation dispatch;
- Decentralized control incorporating DR, DER, and micro-grids instead of the conventional hierarchical approach;
- Data mining on usage data to derive new business opportunities and applications [5];
- Equipment health monitoring diagnostics and monitoring.

4 Economic Values, Benefits, and Market Forces

A power grid is implicitly a natural monopoly because it requires huge investment and long lead time to build and is also costly to maintain. The barrier to entry is so high that grid companies are usually owned by governments, local communities, large corporations, or highly regulated grid operators. To better understand what the different values and/or economic justifications of upgrading to a smart grid are, one must look at it from three different levels, i.e., transmission, distribution, and metering.

4.1 *Transmission*

Smart grid investments on the transmission level are usually driven by conventional needs such as

- Increasing monitoring and control capabilities through PMUs and WAMS for large-scale networks;
- Harnessing remote energy resources (e.g., hydro) using long-distance EHV lines and/or HVDC systems;
- Reinforcing the networks to increase transfer capacity using STATCOM (SVG) or FACTS devices;
- Upgrading aging infrastructure with new facilities (e.g., reconductoring and voltage upgrading) or intelligent software (e.g., dynamic line rating); and
- Resource sharing by building interconnections among neighbors (e.g., back-to-back HVDC tie lines).

The economic benefits would include revenue generation from the import/export of energy, reserve sharing, investment deferral, system reliability, and security improvements (avoiding blackouts or interruptions).

4.2 Distribution

Traditionally, investments in distribution networks are built on increasing service reliability measured by reliability indices such as SAIFI, SAIDI in the USA, and CML and CI in the UK [2, 24]. These indices are used to quantify the reliability needed in order to justify investment. The costs to meet the prescribed level of reliability will then be compared to the cost of interruption/unreliability. This conventional practice is usually guided by regulation in most developed countries.

With increasing numbers of DER such as solar PVs, CHP and fuel cells are gradually being added to the distribution grid on the customer's side. Business values for the distribution grid reinforcements and changes become murkier and more challenging to justify because the new technologies also bring in new players and business paradigm shifts. Management of the distribution networks becomes much more complex. For instance, the return on investment (ROI) for network reinforcements and DA may be diluted, while some new players may benefit more than others without bearing any costs or risks.

More distributed generation would also mean less revenue to the grid company, but the existing feeders and interconnectors remain as assets of the utility. The costs to maintain these assets are unlikely to shrink in the future, but the number of customers may vary. Some of these grid assets may then become a sunk cost as well. As such, innovation and new paradigms to build the business case for distribution-level investments and cost recovery are urgently needed as distribution generation grows [16].

However, charging stations for EVs and the associated infrastructure reinforcements may have a clearer picture with a win–win–win business case. For instance, if the generation mix is reasonably low carbon such as powered by renewables, nuclear, or natural gas and the distribution grid is strong, the added new EV loads could increase the revenue for the utilities. They would reduce roadside pollution and provide a lower-cost transportation means to consumers. Unfortunately, development and popularity of EVs is only taking off slowly globally [14].

4.3 Metering

As previously mentioned, the first mover on utility-scale deployment of SM is ENEL of Italy. ENEL installed 30+ millions SM costing €2.1 billion in 2001. However, the annual benefits through revenue protection, dynamic pricing, and remote connection management were up to €500 million annually. In other words, the payback period was only 5 years [4].

Electricity meters are the cash registers for a power utility company. The evolution from electromagnetic meters to electronic matters is in fact imminent under today's pace of technological progress. Depending on individual utilities, the business case to bring in SM and AMI needs to incorporate existing and future policies on dynamic pricing, renewables targets, net metering, DR, and energy storage facilities (including EVs).

As the industry moves toward widespread deployment of SM and AMI, the utility and consumers are also offered new opportunities such as the provision of energy audits, Internet access, or related applications (e.g., home security monitoring). The key values would include cost savings in automatic meter reading, usage optimization through DR and dynamic pricing, prepaid services, remote disconnect, energy theft reduction, and outage management. These capabilities could add new revenue-generating services for utilities that were not possible or available in the past. However, it should be noted that the deployment of SM is a long-term and capital intensive investment. Without a clear commitment and mandate from the regulatory bodies and policy support, the business case of full deployment (trials are different) will take time to build. Stakeholder education and engagement as well as public supports are key to a successful deployment [8, 9].

5 Regulation and Policies

The planning and operation of power grids are heavily regulated by local, if not national-level authorities. This section first focuses on the primary objectives and current practices that guide electric utilities today in general, followed by the growing challenges in regulation and policy. We then review some global experiences for reference.

5.1 Different Industry Structure

Globally, electric utilities operate in different forms of organizational structure; hence, the regulation and policies applied to them may vary. For simplicity, the following briefly describes the major structures:

- Vertically integrated, investor-owned utilities (IOUs) that own and operate generation, transmission, and distribution and are regulated by the local government;
- Independent system operators (ISOs) are usually set up in a deregulated environment. Their responsibility is to look after the delivery system, while the generation and distribution functions are given to separate business entities. Competitive markets on the wholesale and/or retail levels are set up to interact with the ISO. While the these GENCO and DISCO are more or less market driven, the ISO is always regulated by the local government; and
- Publicly owned utilities (POUs) like the State Grid in China, TVA in US, or BC Hydro in Canada are generally exempt from regulation as they are expected to have the consumer's best interests in mind and hence will offer the best tariff and services to the public.

5.2 Common Regulation Objectives

Today, depending on historical, geographical, and political reasons, heterogenic (e.g., USA), semiheterogenic (e.g., China), and homogenic (e.g., Japan) mixes of IOUs, ISOs, and POUs are present. Despite their differences, a set of common overseeing objectives for the regulators/governments include the following:

- Customer values and rate calculation rules
- Investment cost recovery principle
- Socioeconomic developments
- Efficiency incentives and measures
- Reliability standards
- Environmental considerations
- Other policies (e.g., pollution controls)

5.3 New Objectives and Regulation

With the aspiration of building and deploying smart grids to achieve a low-carbon future, regulators are confronted with some new challenges [16, 17, 25]:

- Technologies are changing the roles of generators and loads;
- Incentives to promote renewables also imply higher rates combined with conventional generation;

- DR may contradict traditional rate structure;
- New entities such as distributed generation, storage devices, and EV are changing the role of the grid and hence their values;
- New players may discourage traditional investment decisions due to higher risks of reasonable rate of return; and
- Evolving and constantly changing public opinions.

As such, government and regulators have to incorporate new objectives and design appropriate measures to nurture a healthy development of the power grids. These objectives may include a combination of the following, depending on their suitability of meeting local needs:

- National policy and/or international commitments on climate, pollution control, etc;
- Target of generation mix for a low-carbon future, including RE target;
- Demand-side management and energy efficiency policies;
- Service reliability standards for distributed and centralized generation;
- Grid charge calculation methodology;
- Guidelines to set up smart grid trials and corresponding evaluation framework;
- Incentives or subsidies for distributed generation, storage, and EV development;
- Electricity market design, monitoring, assessment rules, and dynamic pricing policy;
- Cost recovery principle and means (e.g., for stranded costs);
- Smart grid standards: equipment interoperability for SM, AMI, digital substations, etc;
- Utility's own performance measures: grid efficiency and performance metrics;
- Cyber security requirements: data security and intrusion prevention;
- Privacy protection policy;
- Supports for innovation and R&D;
- Stakeholders' education and coordination; and
- Workforce training policy and programs.

According to a recent report [17], global smart grid development projects have different focuses. In North America, the main objectives are on peak load reduction and dynamic pricing schemes, automatic metering reading, revenue protection, and outage management. In Europe, the focus is more on efficiency improvement through distributed resources and emission reduction means. In Asia Pacific, the objectives vary from country to country. For instance, building a strong transmission grid and automating distribution has been and will remain a strong driver to the rapidly growing infrastructure buildup in China. Australia, however, is closer to the American approach with more emphasis on load management.

Regulator's tasks are increasingly difficult today as new generation and load entities are rapidly entering the arena driven by rapid technological advancements and new business paradigms. Aspirations, decisions, and commitments of the government policymakers and/or the regulators have a long and crucial impact to individual smart grid developments universally. As such, demonstration projects are important. It is imperative for local regulators to provide sufficient funding and flexibility to allow utilities and different stakeholders to conduct demonstration projects such that the policymakers can identify the key aspects to concentrate, mistakes to avoid, skills to acquire, and the pace to undertake.

6 Societal Benefits, Impacts, and Acceptance

Last but not least, the social aspects of what smart grids will bring must be understood and properly addressed for any developments to take place. As different communities have their own unique history, culture, and economic, political, and regulatory environment, their corresponding social issues are different. These are some of the key considerations:

6.1 Economic Growth, Competitiveness, and Jobs

Smart grid development could bring employment to local communities on infrastructure construction, technology development and exports, energy independence, and new markets for products and services. By reducing costs on conventional fuels and dependency, economic developments could increase national competitiveness and economic growth [20].

6.2 Emissions and Pollutant Reductions

Smart grids increase the capability to absorb more renewables, demand-side management, and storage devices. In others words, the dependency on existing fuels such as coal or gas can be reduced and hence less emissions and pollution. As such, the health impacts to local communities and the nation at large could be long term and significant.

6.3 Customer Engagements

Despite what technologies may offer, the new paradigm of offering active participation (e.g., DR, ADR) and financial incentives to load entities to participate in balancing the grid operation requires a committed mass. The aspiration to use energy wisely for a better world, the willingness to relinquish traditional convenience and opt for a better "common good" habitually and continuously, may not be universally accepted nor easily achieved overnight.

6.4 Reactions to the New Prosumers

As more DERs are allowed and developed, the new prosumers' paradigm will undoubtedly go through stages of progressive evolution. It is quite obvious that the more privileged or large load entities will most likely be able to afford the changes and hence enjoy the benefits ahead of others. The subsequent impacts in terms of potential tariff increase, different service reliability, social fairness, and perceived winners may render an unfair judgment of smart grid developments.

6.5 Value of Reliability

System reliability is a complicated notion which encompasses two key aspects, i.e., adequacy and security [2]. Adequacy is a probabilistic measure to gauge the level of resources and redundancy a system should have to ensure the expected load is met. Security, on the other hand, is a deterministic measure to gauge a system's ability to response and recover from a prescribed list of contingencies. Smart grids can offer new or additional reliability benefits through technologies such as DR, DER, and micro-grids. To date, such cost and benefits are most appealing to the niche market such as the military, heavy and big industries, and data centers because interruption or "blackout" is not an option for them. However, the task of convincing policy-makers and the general public on the value of reliability (or the cost of unreliability) is a challenge sometimes because it is difficult to prove or it may be taken for granted. The value of reliability has to be an integral part of the low-carbon future and hence must be understood and accepted by most if not all stakeholders.

6.6 Privacy

Smart meter deployments in the USA have taught us a valuable lesson when the media and some customers question or even totally reject SM as they feel their most sacred guarded place, namely their home, has been intruded. The reality of many security breaches, personal privacy, and financial losses certainly does not help. As such, cyber security of SM and AMI has become one of the biggest concerns as well as requiring substantially expensive development [3]. The value of using the collected data in helping the community to a more efficient and intelligent use of resources must be explained to the stakeholders (from regulator to customers) along with addressing privacy issues. More importantly, choices (including opting out) should be present simultaneously.

6.7 EV Adaption Rate

The adaptation of EVs in different communities varies, which will also affect the pace of smart grid development. Private vehicles in today's societies are no longer just a transportation means, but also a symbol of social status. Unfortunately, the choice of EVs on the market today is still limited. In the meantime, other hybrids (e.g., diesel electric or gasoline electric) or even diesel-based or CNG-based competition exists. How the consumers and local communities open up to the use of EVs will not only take time to come up to speed, but also depends on EV markets globally.

Ironically, social norm changes help fuel the development of renewables, smart grids, sustainable practices, and the notion of a low-carbon futuristic aspiration. However, it is also the general public that represents one of the biggest hurdles. Public inertia is slow to adapt some behavioral changes and paradigm shifts brought on by smart grids. Education would help, but the cultural background, openness, and flexibility of the local communities will also affect the course of local smart grid development.

7 Neither a Silver Bullet Nor One Size Fits All

The conventional grid is designed based on a well-defined set of end points so that the corresponding power flows are generally well understood. The distance, capacity, and associated protection systems of individual connecting lines/cables are also carefully designed and utilized based on prescribed generation and load profiles. The implicit responsibilities of the grid are to interconnect the power systems in a reliable, safe, and sustainable way in an economic manner. With the arrival of more renewables, customer interactions, DER, and increasing mobile loads (e.g., EVs), the grid needs new technologies, regulatory changes as well as a sound business model to ensure its implicit responsibilities and performance remain unchanged. The challenges and the scope of involvement are huge.

Undoubtedly, smart grid development will be evolutionary instead of revolutionary. It will not be one size fits all either. To ascertain a stable and healthy smart grid development for a low-carbon future, different communities will have their own way and choices [16, 25]. The following is only one of many views on how key processes should be included:

7.1 Take Inventory

Review historical and existing conditions including natural resources available, fuel sources, grid structure and extent, geography, regulatory structure, generation mix, interconnections, national policy, political infrastructure, long-term contracts, and social fabric.

7.2 Set Targets

Establish a targeted generation mix, renewables ratio, energy efficiency performance, reliability standards, interconnection commitments, reserves, minimum level of conventional plants needed, climate policy, and pollution control targets.

7.3 Assemble a Road Map

Envision and design a road map with intermediate targets to allow adjustments of development pace and possible technology changes.

7.4 Stage for Flexibility

Staging can help mitigate risks or burdens in financial commitments, complexity of transitioning from legacy systems, technology obsolescence, policy changes, and/or societal reactions.

7.5 Trial Different Options

Identify clear trial objectives and select appropriate assessment methodology. The objectives may include testing the capabilities of selected technologies or the reactions of customers subjected to different schemes. Trial design and execution should be tailored to fit individual communities' needs and encompassing conditions. Learning from other utilities' experiences and sharing one's own findings could accelerate learning.

7.6 Standardize Equipment and Practices

Use of open architecture can ensure interoperability and expedite implementation with less "special" or "exceptional" cases. Following international standards is an option, but using national standards is also acceptable as far as consistency, scalability, equipment interoperability, and cyber security requirements are also met.

7.7 Retrain Workforce

New skills and experiences in operating and planning DER, SM, DR, EV, etc. are needed to enable the existing workforce to manage and deal with the massive implementation and subsequent demand.

7.8 Educate the Public

Educating all levels of the community in terms of the needs and benefits of building a low-carbon future is important. Stakeholder's engagement and public campaigns to promote how smart grids can facilitate a low-carbon future not only strengthen the commitment, but also help stimulate innovation and accelerate the adoption rate.

7.9 Review and Adjust Periodically

Regular reviews and assessments are necessary to ensure the trials or implementation processes are on track and if any adjustments are needed. However, frequent major changes could be detrimental so firm guidelines and strong commitment to reach the goal post are important.

7.10 Research and Development

Technology plays a key role in smart grid developments. Many technologies and associated practices are new in commercial and large-scale applications. R&D is important to help identify risks, pitfalls, and opportunities of new business opportunities.

8 Research Needs

With the gradual shift of our generation mix to favor more renewables and distributed energy resources, the grid naturally needs to adjust not just for the physics and engineering needs, but also the changing business environment. New opportunities and challenges are arising. In particular, the development of distributed renewables, storage devices, CHP, DR, and micro-grids will affect conventional utility operation paradigms to different degrees in different communities.

A healthy research and development program is considered important in this crossroad of the industry. Some of research directions may include the following:

- Economic values of DER for different technologies, markets and regulatory structure;
- Renewable resource assessment, monitoring, and utilization;
- Integration of renewables, DSM, and micro-grid;
- Cost allocation model of smart grid taking into consideration various stakeholders, e.g., prosumers;
- Value of reliability under extreme contingencies or severe weather conditions; and
- Power electronics applications in micro-grids, storage, and bulk power transfer.

9 Conclusion

A low-carbon world is a necessity not only to mitigate climate change, but also to achieve a more sustainable future for our own and future generation. Transformation toward a low-carbon world will need to incorporate more renewables, distributed energy resources, efficient delivery systems, and intelligent use of energy. A smart grid powered by advanced technologies, coupled with sufficient economic conditions and regulatory environment, is an inevitable and crucial component interconnecting the new forms of generation and load entities as well as making the existing infrastructure work more intelligently and efficiently.

Globally, smart grids will have different flavors because of different local needs, available resources, historic and social backgrounds, economies, and regulatory frameworks. Grid changes in terms of more customer engagements, wider integration of renewables, and DER (including storage) will be the main thrust in developed countries such as the USA, EU, and Japan. On the other hand, grid changes in bulk power transmission system developments, network reinforcements, and DA will be the main focus in developing countries and emerging economies such as China, India, and Brazil. In due course, both main streams will converge in terms of technologies which are built on the ubiquitous and versatile Internet, sensors, sophisticated controls, wireless communications, data analytics, and smart decision tools.

However, technology will only be one of the fundamentals for smart grid development and deployment. Healthy economics and appropriate regulatory push-and-pull elements are considered crucial, and they need to be in-synch with technological advances to make any smart grid deployment usable, beneficial, and acceptable to the local communities. Policies to incentivize more renewable contents in the generation profile must also provide sufficient support to reinforce (and smarten up) the associated network to handle intermittency and complex flows of electricity. Policies to encourage trialing/implementing new technologies to allow optimization of grid operations through efficiency improvement, outage management, and asset life extension are welcome. Smart meter deployment must also consider a win–win business model for both utility and customers.

For individual communities, policymakers must set out an inspirational and yet practical vision toward a low-carbon future. A road map and periodic reviews of it will help, but a strong commitment from policymakers and the public is equally important. To decarbonize the electric power generation and increase efficiency and customer engagement, we must also include the efforts and costs of smartening/reinforcing the grid to meet the needs of new players and new challenges. These challenges are not just technical, but also can be economic, social, and environmental. The processes and endeavors are unique to each community. They are neither trivial nor short-lived. Smart grids are the conventional power grids being remade. Long-term commitments toward a low-carbon future, robust technologies, sound business models, and consistent and timely regulatory changes will provide the wind. The industry in turn will set the sail by making our grids stronger, smarter, and more flexible such that we and our future generation can confidently depend on.

References

1. ABB (2013) Retrieved 1 Aug 2013, from http://new.abb.com/about/hvdc-grid
2. Billinton R, Allan R (1984) Reliability evaluation of power systems. Plenum Press, NY
3. Black and Veatch (2012) Strategic directions in the US electric utility industry. Retrieved 2 Aug 2013, from http://bv.com/docs/management-consulting-brochures/2012-electric-utility-report-web.pdf
4. Brattle (2013) Retrieved 1 Aug 2013, from http://brattle.com/_documents/UploadLibrary/Upload805.pdf
5. Cheng JWM et al (2012) An alternative use of power quality information—load signature studies and applications. Paper presented at the proceedings of 15th IEEE international conference on harmonics and quality of power, Hong Kong
6. Chinese Government's Official Web Portal (2013) Full text: China's energy policy 2012 Retrieved 1 Aug 2013, from http://www.gov.cn/english/official/2012-10/24/content_2250497.htm
7. Executive Office of the President (2011) A policy framework for the 21st century grid: a progress report. National Science and Technology Council, from http://www.whitehouse.gov/sites/default/files/microsites/ostp/nstc-smart-grid-june2011.pdf
8. Greentechgrid (2013) Smart meters must better integrate into utility operations. Retrieved 29 Jul 2013, from http://www.greentechmedia.com/articles/read/Smart-Meters-Must-Better-Integrate-Into-Utility-Operations?utm_source=SmartGrid&utm_medium=Headline&utm_campaign=GTMDaily
9. Greentechgrid (2013) The true ROI of smart meter deployments. Retrieved 29 Jul 2013, from http://www.greentechmedia.com/articles/read/the-true-roi-of-smart-meter-deployments
10. IEA (2013) Technology roadmap—smart grids. Retrieved 1 Aug 2013, from http://www.iea.org/publications/freepublications/publication/smartgrids_roadmap.pdf
11. Integer Research (2011) Retrieved 1 Aug 2013, from https://www.integer-research.com/wp-content/uploads/downloads/2011/06/fiberopticgrowth-june2011.pdf
12. International Telecommunication Union (2013) Retrieved 1 Aug 2013, from http://www.itu.int/en/ITU-D/Statistics/Documents/statistics/2013/Mobile_cellular_2000-2012.xls

13. Internet World Stats (2013) Retrieved 1 Aug 2013, from www.internetworldstats.com/emarketing.htm
14. Jlester (2013) Tesla leads, but other EV start-ups stumble as industry continues slow march *Cleantech Finance*
15. Kaplan SM et al (2009) Smart grid—modernizing electric power transmission and distribution. Government Series. Capital.Net
16. Kassakian JG, Schmalensee R (2011) The future of the electric grid an interdisciplinary MIT Study. Retrieved 18 Jul 2013, from http://mitei.mit.edu/system/files/Electric_Grid_Full_Report.pdf%20dated%2018%20July%202013
17. Mulder W, Kumpavat K, Fassen C, Verheij F, Vaessen P (2012) Global inventory and analysis of smart grid demonstration projects. DNV KEMA Energy Sustain. The Netherlands. from http://www.dnvkema.com/Images/DNV%20KEMA%20Report%20Global%20Inventory%20and%20Analysis%20of%20Smart%20Grid%20Demonstration%20Projects.pdf
18. National Academy of Engineering (2013) Greatest engineering achievements of the 20th century. Retrieved 1 Aug 2013, from http://www.greatachievements.org/?id=2949
19. Navigant Research (2013) Retrieved 1 Aug 2013, from http://www.navigantresearch.com/newsroom/installed-base-of-smart-meters-in-china-to-reach-nearly-380-million-by-2020
20. Ristori D (2013) Keynote speech. In: CIGRE symposium. 22–24 Apr 2013), Lisbon
21. Smart Distribution Wiki (2013) Power distribution research. Retrieved 1 Aug 2013, from http://wiki.powerdistributionresearch.com/index.php?title=IEEE/PES_Distribution_Automation_Tutorial_2007/2008
22. Smart Grids (2013) Retrieved 1 Aug 2013, from http://www.smartgrids.eu/mission
23. Succar S (2012) The promise of the smart grid: goals, policies, and measurement must support sustainability benefits. NRDC Issue Brief, from http://www.nrdc.org/energy/smart-grid/files/smart-grid-IB.pdf
24. Sullivan M, Schellenberg J (2011) Smart grid economics: the cost-benefit analysis. FSC Group, from http://fscgroup.com/reports/2011-renew-grid-article.pdf
25. World Energy Council (2013) World energy perspective smart grids: best practice fundamentals for a modern energy system. Retrieved 18 Jul 2013, from http://www.worldenergy.org/publications/3940.asp%20dated%2018%20July%202013

Part II
Technical Characteristics of Smart Grid

Status and Prospects of European Renewable-Based Energy Systems Facilitated by Smart Grid Technologies

Yi Ding, Jacob Østergaard, Poul Ejnar Sørensen, Peter Meibom and Qiuwei Wu

Abstract Renewable energy plays an important role in the future energy framework of the European Union. The European Union will reach a 20 % share of renewable energy in total energy consumption and increase energy efficiency by 20 % by 2020. Smart grids will be the backbone for facilitating the integration of renewable energy resources into future energy systems. The plans and status of renewable energy resources development and energy policy in Europe are introduced in this chapter. The development of smart grid technologies for facilitating the renewable-based energy systems in the European Union is also discussed. The role of Denmark, one of the leading countries for developing smart grid technologies and using renewable energy resources, has been emphasized in this chapter.

Keywords Renewable energy · Wind power · Smart grid · European union · Denmark

1 Background

Two of the major challenges facing the world are climate change caused by global warming and security of energy supply. The European Union (EU) has realized that its future economic growth and jobs are relying on the efficient and sustainable use of natural resources [1]. The EU2020 agenda deliverers a clear European framework for promoting the use of renewable energy resources and reducing

Y. Ding (✉) · J. Østergaard · Q. Wu
Department of Electrical Engineering, Centre for Electric Power and Energy, Technical University of Denmark, Copenhagen, Denmark
e-mail: yding@elektro.dtu.dk

P. E. Sørensen
Department of Wind Energy, Technical University of Denmark, Copenhagen, Denmark

P. Meibom
Danish Energy Association, Rosenørns allé 9, 1970 Frederiksberg, Denmark

D. Mah et al. (eds.), *Smart Grid Applications and Developments*,
Green Energy and Technology, DOI: 10.1007/978-1-4471-6281-0_3,

greenhouse gas emissions for 2020: The EU will reach a 20 % share of renewable energy in total energy consumption and increase energy efficiency by 20 %. In the future energy framework of EU, renewable energy plays an important role tackling the challenge of the climate change as it is an ideal energy source to decarbonize the generation of energy, as well as a means of improving the security of our energy supply by drawing upon inexhaustible sources of energy [2, 3].

However, the existing EU's energy infrastructure cannot cope with the high penetration of renewable energy efficiently and effectively. Without serious upgrading of existing electricity network, monitoring and metering systems, and operation and control strategies, the use of renewable energy generation will develop at a much slower pace and system security may be jeopardized. Opportunities for energy saving and energy efficiency will be missed [1].

Smart grids will be the backbone of the future electricity network of EU, which will facilitate the integration of the high penetration of renewable energy and vast amounts of electric vehicles (EV), heat pumps, and other distributed energy resources with a considerable potential to provide balancing capacity and other flexible services to the power system. Moreover, smart grids provide a possible solution for improving the reliability and security of electricity networks. The Commission of EU is encouraging the implementation of smart grids because it provides an opportunity to boost the future competitiveness and worldwide technological leadership of EU technology [1].

Denmark is also a driver in the way in energy-efficient and energy-friendly economy in EU. In 2012, 30 % of the total electricity consumption came from wind power (Energinet.dk Market Data) [17]. The Danish Energy Agreement of March 2012 (Accelerating green energy towards 2020) [7] sets ambitious targets for 2020, including that approximately 50 % of electricity consumption should be supplied by wind power, and more than 35 % of total energy consumption (including electricity, heat and transportation) should be supplied from renewable energy sources. Danish experience illustrates that dependency of fossil fuel can be reduced without adverse effects on the economic development through positive energy policy and technology development for enhancing energy efficiency.

This chapter discusses the plans and status of renewable energy resources development and energy policy in Europe. The challenges faced by future power systems with high penetration of renewable energy resources are briefly described. Solutions for developing smart grid technologies for catering high requirements of balancing power in the future are discussed.

2 Plans and Status for Renewable Energy Resource Development and Energy Efficiency Improvement

The EU commission has endorsed the European strategic energy technology plan (SET-Plan) to develop and deploy low-carbon technologies during the next 10 years in Europe. The corresponding technology roadmaps are proposed as a basis for strategic planning and decision making on a transition to a low-carbon economy, which includes the following main sectoral targets by 2020 [3]:

- Up to 20 % of electricity in EU will be produced by wind energy resources.
- Up to 15 % of electricity in EU will be produced by solar energy resources.
- The electricity grid in Europe can accommodate the integration of 35 % renewable electricity in a seamless way and effectively maintain the reliable and secure operation.
- Cost-competitive and sustainable bioenergy will occupy at least 14 % of the EU energy mix.
- The framework of CO_2 emission allowance will be effectively developed with carbon capture and storage technologies.
- Around 25–30 European cities will be pioneer of the transition to a low-carbon economy.

For achieving these objectives, new generation of more efficient and more reliable wind turbines and photovoltaics/concentrating solar power plants will be implemented. The offshore resources and deepwater potential of wind power will be exploited. The technology competitiveness for large-scale penetration of solar power generation will be improved.

Meanwhile, an unprecedented European research, development, and demonstration program will be conducted in the next 10 years, which will cost between 58.5 and 71.5 billion euros. Table 1 illustrates the cost estimation of renewable energy development and smart cities initiative [3].

Among EU countries, Denmark is a leading country for developing renewable energy technologies and using renewable energy resources [4]: In 1957, an experimental wind turbine was installed in Denmark and connected to the electricity grid. In the 1980s, the first generation of wind turbines was mass-produced by machine manufacturers. Since then, the Danish Government has encouraged the large power companies to construct wind farms for providing electricity. In the 1990s, the government provided attractive incentives for wind power development, and more and more private individuals join the investment in constructing wind farms and producing wind power because of favorable profits. In the past three decades, wind power generation capacity has grown from nearly 0 to almost 3,500 MW. The Danish electricity grid has also been operated effectively and securely with the high penetration of wind power.

Figure 1 illustrates hourly wind power generation and demand in October 2010 in the western Denmark (DK1) provided by the Danish transmission system operator [5]. The annual wind generation was 27.1 % of the annual demand in DK1 in 2010. Still, the figure indicates that the hourly wind generation exceeded the hourly demand in the nights of and October 3 and October 13. This was the case in 71 of the 8,760 h in 2010, corresponding to 0.8 % of the time.

Besides increasingly using renewable energy, a wide range of energy-saving initiatives and energy efficiency improvement have also been included in the Danish action plan [6]: Denmark is among the countries in the EU with highest energy efficiency. Since 1980, the Danish economy has increased by 78 %.

Table 1 Cost estimate of renewable energy development and smart cities initiative

European industrial initiatives	Total (b€)
Wind energy	6
Solar energy (PV and CSP)	16
Bioenergy	9
Carbon capture and storage	10.5–16.5
Electricity grid	2
Sustainable nuclear energy	5–10
Smart cities	10–12
Total	58.5–71.5

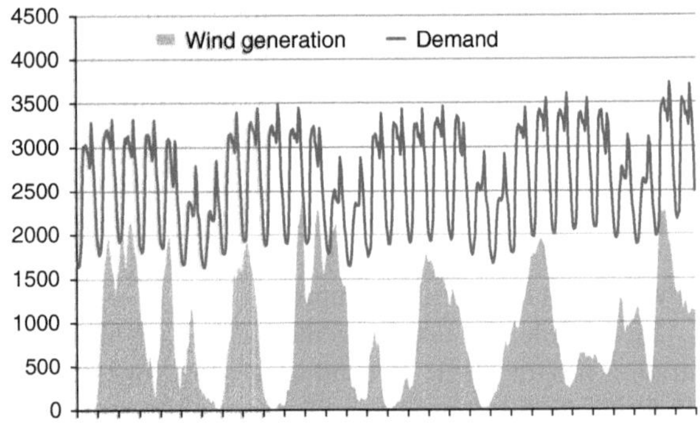

Fig. 1 Wind power generation and demand in November 2010 in the western Denmark

However, energy consumption has remained approximately constant and CO_2 emissions have been reduced with 24 % [18].

The Danish gross energy consumption has been decreased by 2 % from 2006 to 2011 based on the target of Danish energy policy and still further by 2020. Energy consumption from buildings and vehicles is primary target to be reduced. Comparing with current standard, energy efficiency requirements for new buildings will be much stricter. Electric vehicles (EVs) will cover 10 % of the road transportation, and 15 % of energy consumption from road transportation will be supplied by electricity [8]. EVs will also constitute a flexible demand resource enabling smoothing of the power fluctuations from renewable energy resources.

Allowance and tax schemes also work as lever for achieving the goal of energy savings. In Denmark, gasoline-based vehicles have received very high taxes, which is up to 180 % [8]. The total price of a medium-sized car is two times than that in the USA. The gasoline price in Denmark is also approximately double of that in the USA due to heavy taxes. Comparing with that, EVs have not received tax yet. From 2013, most generation companies in EU will have to purchase CO_2 emission allowance from the market, which is expected to increase the electricity prices by approximately 20 % [4]. It will encourage the electricity production from renewable energy sources, which is allowance free. Thus, the emission allowance will spur more investment in renewable energy field and EVs.

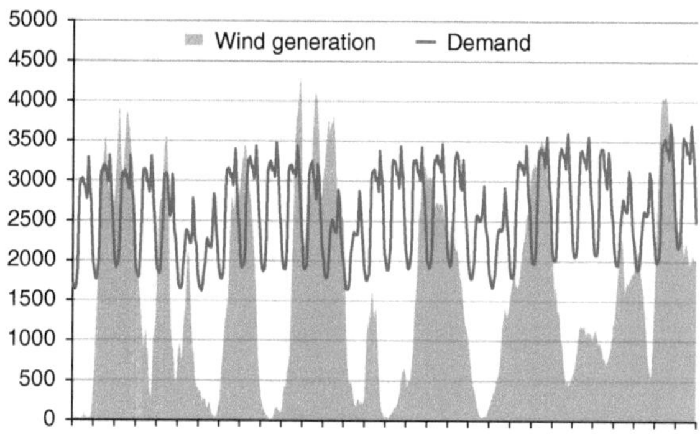

Fig. 2 Estimated wind power generation and demand in November in the western Denmark with 50 % wind power penetration

3 Challenges and Solutions

The high penetration of renewable energy resources will also challenge the power grid of EU for two reasons [9]. Firstly, the share of fluctuating and less predictable electric power production will increase significantly. The Danish balancing challenge with 50 % wind power generation is illustrated in Fig. 2, where the wind generation has been estimated by a simple scaling of the 2010 data so that wind covers 50 % of the demand.

Subtracting wind power leaves a residual demand and an overflow in November as shown in Fig. 3. With the applied simple scaling of the wind power generation, the hourly wind generation will exceed the hourly demand in 1,254 of the 8,760 h in 2010, corresponding to 14.3 % of the time. In reality, the overflow periods will probably be even longer with 50 % wind generation, because most of the new wind power development will be built as large offshore wind power plants. Experience with the operation [10] and modeling [11] of large offshore wind plants has shown that the power fluctuates relatively more than the power from the land-based wind turbines, first of all because there is a much higher geographical concentration of the offshore wind plants compared to the land-based wind turbines.

Obviously, higher renewable energy penetration in Europe will increase competition and costs for balancing power because the conventional energy resources are reduced and the need for balancing resource will increase. Also both current energy markets (spot, regulating power) and ancillary services are designed for conventional large generation, which may be too rigid as the practicability of the new-generation technologies.

Secondly, the electrification of transport and heat will put a pressure on the distribution grids in the power system; hence, the operators of the distribution grids may face two options in order to manage the subsequent grid congestions, either increase grid capacity or implement control signals, e.g., real-time price signals that reflect the variation in demand for capacity during real-time operation. However, these two options are not entirely a choice of one over the other. Some

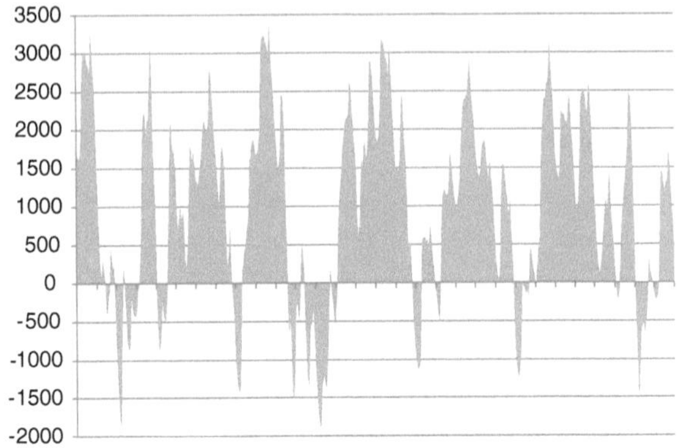

Fig. 3 Estimated residual demand and overflow with 50 % wind power penetration

additional grid capacity will be needed; however, implementing control signals will improve the utilization of existing distribution capacity; hence, the need for new capacity will decrease.

If the control signals are set as locational real-time price signals it will adapt to the local supply and demand conditions, taking grid congestions into consideration during system operation.

To address the future needs and challenges, an efficient and effective market-based tool must be developed, which will attract distributed energy resources (DER) and active end user to provide balancing power for the transmission system operator. The EcoGrid EU—A Prototype for European Smart Grids is a large-scale smart gird demonstration project [9, 12], where a new real-time electricity market is proposed to allow the participation of small-scale DERs and small end consumers into the existing electricity markets. Today, most of the DERs and end users face barriers to supply balancing services to provide balancing services, e.g., requirements on bidding size, complex bidding strategies in the markets, complying with schedules, etc. The objective of the new real-time electricity market is to remove these barriers [9, 12].

While the EcoGrid EU emphasizes on the participation of small-scale DERs and small end consumers for providing system-wide balancing services, another Danish national project iPower—a Strategic Platform for Innovation and Research within Intelligent Electricity is more focused on the aggregation of DERs providing flexibility services, trading services for the efficient and reliable operation of distribution grids. An aggregator-based Flex market [13] has been proposed for promoting small-scale DERs to provide the flexibility services for distribution system operator such as overload and voltage management. Three possible trading setups including bilateral contracts, auctions, and supermarkets have been identified in the Flex market of iPower.

4 EU Smart Grid Initiatives

With the technology advancements, the future EU and Danish power grid have the following design objectives [14]:

- Sustainable to reduce carbon emission.
- Secure and reliable to provide electricity supply.
- Economic to reduce electricity prices and costs for other services.
- Flexible to promote customer participation.

The vision of Smart grid is defined by the European Commission Smart Grids Advisory Council as "an electricity network that grants connection access to all network users and provides efficient and reliable services in order to fulfill customers' needs whilst responding to the changes and challenges ahead" [15].

Key components of smart grid include the following [14]:

- Smart meters: Smart meters are key enabling component for smart grids, which allow the end user to monitor and control consumption in real time. Smart meters also enable two-way communication and interaction between system operator and end users. It will in turn drive development and demand for smart controllers of electrical appliances.
- Intelligent grid management: Intelligent grid management allows flexible power generation, transmission, and active distribution. The bidirectional power flow between transmission and distribution is also possible, which allows prosumers to sell their own power surpluses in the future.
- Software-based added-value services and applications: These services allow DERs and end consumers to economically optimize their electricity production/consumption.
- System-wide oversight and coordination—the "glue" in the system: A uniform communication infrastructure will enable real-time and system-wide communication, which integrate both producers and consumers. The uniform communication infrastructure will support PMU-based wide-area measurement system (WAMS) for system security supervision and control.

There are several smart grid projects implemented in EU. Table 2 [12] summarizes the contribution and key deliverables from previous and ongoing EU smart grid projects. The assessment of technical feasibility is the focus of the majority of smart grid projects. The EcoGrid EU and iPower projects focus on market-based system operation to a higher degree, so that both DERs and end users can provide flexibility services on both transmission and distribution levels.

Table 2 Behnke and Nielsen [12]: EU smart grid project

Project title	Description	Technical R&D	Market design	Technology demonstration	Market demonstration
FENIX (EU)	Identification of the technical capabilities of DER to provide system services through aggregation using the concept of the Virtual Power Plant. Small-scale demonstration	***	*	***	
More MicroGrids (EU)	Integration of small-scale distributed energy resources through the microgrid approach design of alternative control strategy to enable autonomous operation. Small-scale demonstration	***		***	
EU DEEP (EU)	Development of innovative business models for integration of DER into current system and market operation	**	**		**
DISPOWER (EU)	Survey analyzing the present power supply system and its components including information and communication technologies. Laboratory facilities for developing, testing, and demonstration	**	*	**	
ADDRESS (EU)	New Active Distribution Networks (ADN) able to balance in real-time power generation and demand allowing all players to benefit from the increased flexibility of the entire system. Three test fields planned	***	**	**	*
EcoGrid EU (EU)	Demonstration of a large-scale real-time market participation for DER and flexible demand	**	***	**	***
TWENTIES (EU)	Large-scale demonstration of smart grid solutions until 2014 encompassing virtual power plants, HVDC offshore grids, wide-area monitoring systems, and dynamic line ratings	***	*	***	

*** indicate a significant contribution, and * indicates a limited contribution

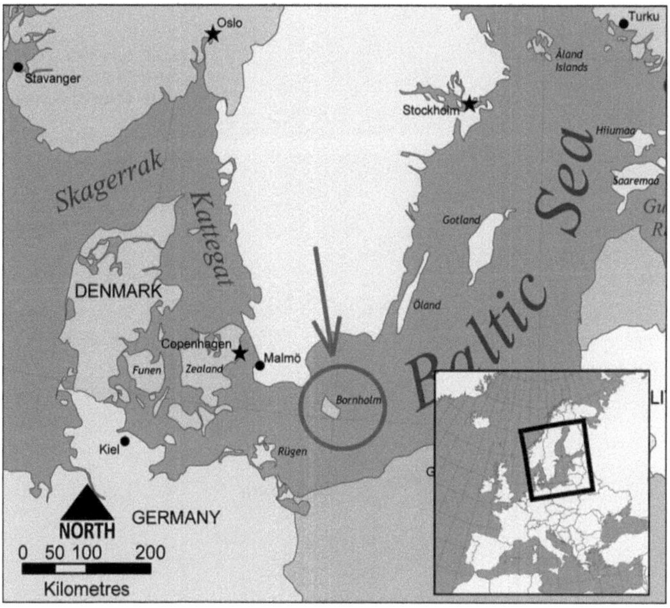

Fig. 4 The Danish Island of Bornholm [16]

5 Case Studies: Bornholm—"Fast Track" for EU Smart Grid

The Danish island of Bornholm is located in the Baltic Sea as shown in Fig. 4 [16], which is a unique test site for smart grid projects. The Bornholm distribution system is integrated in the electricity market of western Denmark (DK2) and is interconnected with Nordic power system as shown in Fig. 5 [12, 16]: The system has a peak load of 55 MW and very high penetration of renewable energy resources. The capacities of wind power and CHP units are 30 and 16 MW, respectively. In the future, the amount of renewable energy resources and energy storage devices such as PV, micro-CHP, fuel cell, and heat storage will be established for smart grid demonstration purpose.

The Bornholm distribution grid is a meshed 60-kV network, which also includes 60/10-kV substations, 10-kV feeders, and 10/0.4-kV substations. The Bornholm grid is also connected to the Nordic transmission grid through a 60-kV AC cable.

The Bornholm power system is a pioneer of future distribution systems of EU, which has very high penetration of renewable energy. The public and municipality of Bornholm support and encourage the "Bright Green Island" strategy. The goal of renewable energy resource penetration in the Bornholm power system will reach 100 %. Therefore, the Bornholm system is well suitable for testing the feasibility and robustness of smart projects of EU. The Bornholm power system will provide a very good test bed for the smart grid projects.

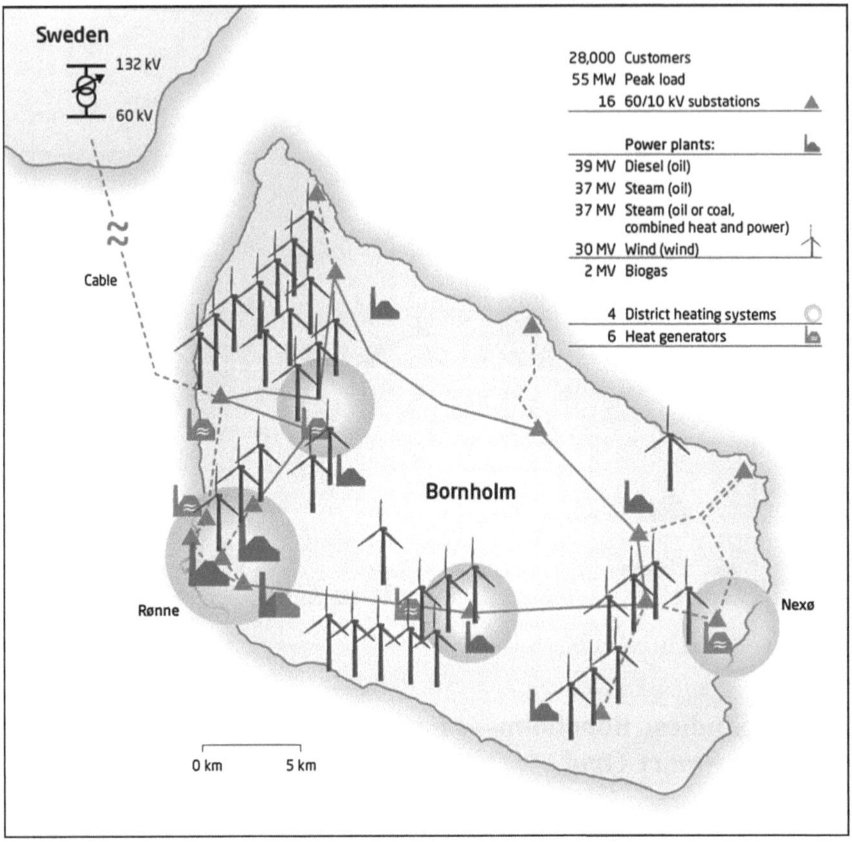

Fig. 5 The Bornholm distribution system [16]

6 Conclusions

The future EU power system will integrate high penetration of renewable energy resources, which will significantly challenge the system operation and control. Smart grid provides an efficient and effective tool for addressing the future needs. This chapter synthesizes challenges faced by future power systems with high penetration of renewable energy resources. Energy policy in EU is also summarized. Several smart grid projects implemented in EU, e.g., Ecogrid EU and iPower, are introduced.

Acknowledgments This chapter is an updated republication of the paper "Towards a European renewable-based energy system enabled by smart grid: status and prospects" by Y. Ding, J. Ostergaard, P. E. Sorensen, P. Meibom, Q. Wu, and Dianli Xitong Zidonghua (Automation of Electric Power Systems) 35(22), 12–17. The copyright permission for reusing the paper has been granted by Dianli Xitong Zidonghua.

The chapter is partially based on a large-scale smart gird demonstration project of EU—EcoGrid EU. The Center for Electric Technology, Technical University of Denmark, is responsible for the new real-time electricity market design in the project.

References

1. Communication from the Commission to the European Parliament (2011) The Council, the European Economic and Social Committee and the Committee of the Regions, smart grids: from innovation to deployment. Brussels
2. Communication from the Commission to the European Parliament (2011) The Council, Renewable Energy: Progressing towards the 2020 target. Commission Staff Working Document, Brussels
3. Communication from the Commission to the European Parliament (2009) The Council, the European Economic and Social Committee and the Committee of the Regions, on investing in the development of low carbon technologies (SET-Plan). Commission Staff Working Document, Brussels
4. Energinet.dk (2009) Green thinking in Denmark, wind power to combat climate change—How to integrate wind energy into the power system. Denmark
5. Energinet.dk (2010) Download of market data. Available at http://www.energinet.dk/EN/El/The-wholesale-market/Download-of-market-data/Sider/default.aspx
6. ENERGI (2009) "The Danish example"—the way to an energy efficient and energy friendly economy. Available at http://www.ens.dk/en-US/policy/danish-climate-and-energy-policy/behind-the-policies/Thedanishexample/Documents/The%20Danish%20Example%20Engelsk%20FINAL%20200209%20vs2_final.pdf
7. Danish Energy Agency (2012) Accelerating green energy towards 2020. The Danish energy agreement of March 2012
8. Østergaard J et al (2009) Electric vehicles in power systems with 50 % wind power penetration: the Danish case and the EDISON programme. European conference electricity and mobility, Würzburg, Germany
9. Energinet.dk (2011) EcoGrid EU—A prototype for European Smart Grids. Available at http://energinet.dk/en/forskning/EcoGrid-EU/sider/EU-EcoGrid-net.aspx/
10. Akhmatov V et al (2005) Integration of offshore wind power into the western Danish power system. In: Proceedings of Copenhagen offshore wind. Denmark
11. Sørensen P et al (2008) Modelling of power fluctuations from large offshore wind farms. Wind Energy 11:29–43
12. Behnke K, Nielsen AH (2009) EcoGrid Europe—A demonstration project for the future. Available at http://ec.europa.eu/research/conferences/2009/smart_networks/pdf/behnke.pdf
13. Zhang CY et al (2013) A flex-market design for flexibility services through DERs. ISGT Europe, Copenhagen
14. SAP AG (2011) Smart grids for Europe—Benefits, challenges and best practices. The SAP perspective
15. European Commission (2006) European smart grids technology platform. Available at http://ec.europa.eu/research/energy/pdf/smartgrids_en.pdf
16. Østergaard J, Nielsen JE (2011) The Bornholm power system —An overview. Available at http://www.dtu.dk/sites/powerlabdk/English.aspx
17. Energinet.dk (2013) Download of market data. http://www.energinet.dk/EN/El/Engrosmarked/Udtraek-af-markedsdata/Sider/default.aspx
18. Danish Energy Authority (2009) Energy statistics. http://www.ens.dk

Arcturus: An International Repository of Evidence on Dynamic Pricing

Ahmad Faruqui and Sanem Sergici

Abstract This chapter introduces *Arcturus*, an international database of dynamic pricing and time-of-use pricing studies. It contains the demand response impacts of 163 pricing treatments that were offered on an experimental or full-scale basis in 34 projects in seven countries located in four continents. The treatments included various types of dynamic pricing rates and simple time-of-use rates, some of which were offered with enabling technologies such as smart thermostats. The demand response impacts of these treatments vary widely, from 0 % to more than 50 %, and this discrepancy has led some observers to conclude that we still do not know whether customers respond to dynamic pricing. We find that much of the discrepancy in the results goes away when demand response is expressed as a function of the peak-to-off-peak price ratio. We then observe that customers respond to rising prices by lowering their peak demand in a fairly consistent fashion across the studies. The response curve is nonlinear and is shaped in the form of an arc: as the price incentive to reduce peak use is raised, customers respond by lowering peak use, but at a decreasing rate. We also find that the use of enabling technologies boosts the amount of demand response. Overall, we find a significant amount of consistency in the experimental results, especially when the results are disaggregated into two categories of rates: time-of-use rates and dynamic pricing rates. This consistency evokes the consistency that was found in earlier analysis of time-of-use pricing studies that was carried out by EPRI in the early 1980s. Our analysis supports the case for the rollout of dynamic pricing wherever advanced metering infrastructure is in place.

A. Faruqui (✉) · S. Sergici
The Brattle Group, San Francisco, CA, USA
e-mail: ahmad.faruqui@brattle.com

D. Mah et al. (eds.), *Smart Grid Applications and Developments*,
Green Energy and Technology, DOI: 10.1007/978-1-4471-6281-0_4,

1 Introduction

Many utilities have been deploying advanced metering infrastructure (AMI) as part of their grid modernization activities. In addition to yielding operational efficiencies in the distribution system, such as lowering the cost of reading meters, faster detection of outages, and theft detection, AMI is also the enabler of dynamic and time-of-use pricing (called time-varying pricing in the rest of this chapter).[1] Through the use of time-varying rates, utilities can lower their cost of doing business and customer bills by lowering, since such rates lower peak loads and improve utility load factors [8].

As of this writing, about a quarter of US households are on smart meters; the number is projected to approach a hundred percent in a decade. However, only two percent of the households are on any type of time-varying rate and only a very small percentage of these households are on any form of dynamic pricing rate.[2]

Over the past decade, a number of dynamic pricing and time-of-use studies have been conducted around the globe in order to inform policy making. Some of these have been randomized experiments, some have been quasi-experiments, some have been demonstrations, and some have been full-scale deployments. A full-blown meta-analysis would require the analyst to normalize for differences in experimental design, which in turn would require access to detailed information on the experiment design, implementation, and findings. Such a study was carried out by EPRI in the early 1980s by using data from five experiments with time-of-use pricing [1]. Lack of individual customer data prevents us from carrying out such an analysis at this time. However, as a first step in that direction, we have assembled the results from 34 studies in a database called Arcturus.[3]

The 34 studies in our analysis encompass a total of 163 "treatments" where a treatment is defined as a unique combination of a time-varying pricing design and enabling technology. At first glance, there is little consistency in the results: the amount of demand response exhibited across the 163 treatments ranges from 0 to 58 %. This wide range of impacts has led observers such as Joskow [10] and some policy makers to conclude that our understanding of customer behavior is not strong enough to proceed with universal deployment of dynamic pricing and time-of-use pricing, even though smart meters are being deployed. However, this range just represents the raw data, unfiltered by the intensity of the price signal that was conveyed to participants. If the data from those treatments that only featured

---

[1] Time-varying pricing refers to time-of-use (TOU) rates as well as dispatchable rate structures such as critical peak prices (CPP) and real-time prices (RTP). AMI is only a prerequisite for dynamic pricing programs, whereas TOU rates can be implemented with legacy meters.

[2] Federal Energy Regulatory Commission.

[3] Faruqui and Sergici [5] and Flaim et al. [9] summarize the results from some recent studies but do not attempt a meta-analysis of the type reported here. A previous meta-analysis, more limited in scope that this one, is contained in Faruqui and Palmer [4]. A comprehensive bibliography on dynamic pricing can be found in Enright and Faruqui [2].

time-varying prices are plotted separately from those that featured time-varying prices with enabling technology, even sharper results are obtained, as enabling technologies increase demand response even more.

We examine the impact of the price ratio on the magnitude of the reduction in peak demand using a log-linear regression model. Because the amount of demand response varies with the presence or absence of enabling technology, such as a smart thermostat, an energy orb or an in-home display, we include a variable that indicates the use of enabling technologies.[4] We find a statistically significant relationship between the price ratio and the amount of peak reduction. The interaction variable between price and the use of enabling technologies is also statistically significant indicating that there is a boost to the peak reduction when prices are paired with enabling technologies. This relationship is termed the Arc of Price Responsiveness for reasons that will become clear later in this chapter.

2 The Time-Varying Rate Designs

Time-varying rate designs charge different electricity rates at different times of the day and year.[5] These rates reflect the time-varying cost of supplying electricity and incentivize consumers to decrease their electrical usage during peak hours and/or shift consumption to less expensive off-peak hours. This chapter examines the resulting peak demand reductions from four types of time-varying rates: time-of-use (TOU), critical peak pricing (CPP), peak-time rebate (PTR), and variable peak pricing (VPP) rates. The last three options fall under the rubric of dynamic pricing. While real-time pricing (RTP) rates also fall into that rubric and have been offered to customers in some of the published studies, the lack of a clear price ratio inhibits us from using these treatments in this chapter.

A time-of-use (TOU) rate could either be a time-of-day rate, in which the day is divided into time periods with varying rates, or a seasonal rate into which the year is divided into multiple seasons and different rates provided for different seasons. TOU rates are fixed by period and consequently offer certainty as to what the rate will be and when they will occur. In a time-of-day rate, a peak period might be defined as the period from 12 p.m. to 6 p.m. on weekdays, with the remaining hours being off-peak. The price would be higher during the peak period and lower during the off-peak period, mirroring the variation in marginal costs by pricing period. TOU rates with three periods have also been offered. Such rate schemes include a shoulder (or mid-peak) period, where the cost of electricity is lower than

[4] Some studies characterize only smart thermostats as enabling technologies as these devices automatically adjust temperature settings without requiring an action from the customers. For the purposes of this chapter, we characterize smart thermostats, energy orbs, and in-home displays as enabling technologies since these devices either automate actions for customers or equip them with information to act on.

[5] For a detailed discussion of time-varying rates, see Faruqui et al. [7].

peak period rates, but higher than off-peak period rates. Additionally, TOU rates may include two peak periods (such as a morning peak from 8 a.m. to 10 a.m. and an afternoon peak from 2 p.m. to 6 p.m.).

On a critical peak price (CPP) rate, customers pay higher peak period prices during the few days a year when wholesale prices are the highest (typically the top 10–15 days of the year which account for 10 to 20 % of system peak load). This higher peak price reflects both energy and capacity costs and, as a result of being spread over relatively few hours of the year, can be in excess of $1 per kWh. In return, the customers pay a discounted off-peak price that more accurately reflects lower off-peak energy supply costs for the duration of the season (or year). Customers are typically notified of an upcoming "critical peak event" one day in advance, but if enabling technologies are used, these rates can also be activated on a day basis.

Like on a CPP rate, customers on variable peak price (VPP) rates pay higher peak period prices during a few days a year when wholesale prices are highest. The main difference between a critical peak price and a variable peak price is that the variable peak price varies from one event day to the next, as determined by market rates. On-peak prices generally vary each day based on day-ahead market prices. On non-event days, the VPP rate acts like a normal TOU rate, with fixed period prices.

If a CPP tariff cannot be rolled out because of political or regulatory constraints, some parties have suggested the deployment of a peak-time rebate (PTR). Instead of charging a higher rate during critical events, participants are paid for load reductions (estimated relative to a forecast of what the customer otherwise would have consumed). If customers do not wish to participate, they simply buy electricity through at the existing rate. There is no rate discount during non-event hours.

Participants in real-time pricing (RTP) programs pay for energy at a rate that is linked to the hourly market price for electricity. Depending on their size, participants are typically made aware of the hourly prices on either a day-ahead or hour-ahead basis. Typically, only the largest customers—above one megawatt of load—face hour-ahead prices. These programs post prices that most accurately reflect the cost of producing electricity during each hour of the day and thus provide the best price signals to customers, giving them the incentive to reduce consumption at the most expensive times.

Figure 1 presents a visual representation of these rate types.

Enabling technologies such as programmable thermostats and in-home displays (IHDs) can be offered with time-varying rates in order to enhance the effectiveness of the rates by automating response and minimizing customer transaction costs. Programmable communicating thermostats (PCTs) can receive a signal during a critical peak pricing event and automatically reduce air-conditioning usage to a level that is specified by the customer, reducing the need to manually respond to high-priced events. Information-enhancing technologies such as in-home displays (IHDs) can give customer information such as the amount of electricity that they are using, what it is costing them, how that translates into their carbon footprint,

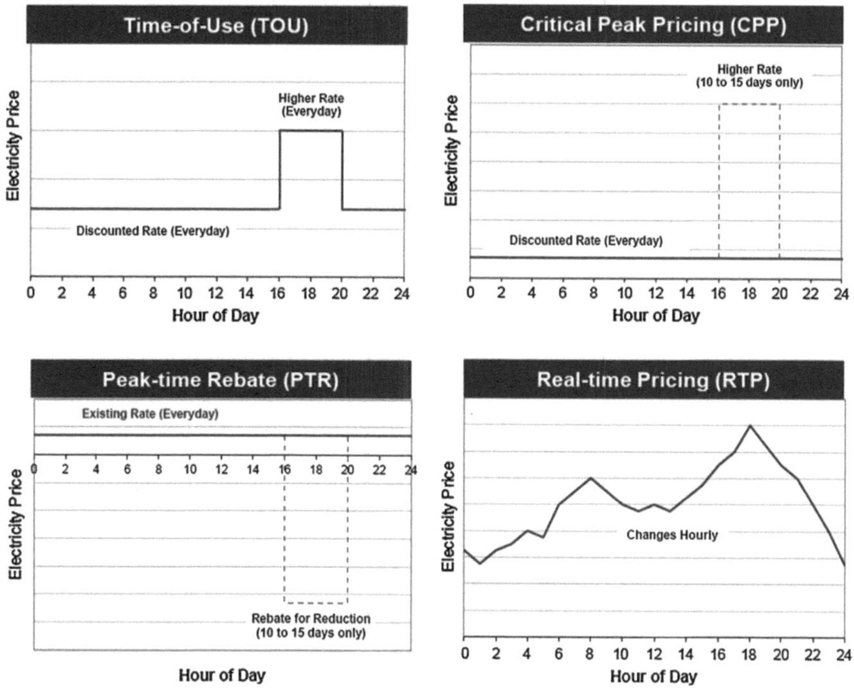

Fig. 1 Common time-varying rate options

how close they are to energy savings goals, and other such data. The information can also be provided online through Web portals or even through a smartphone. Energy orbs provide visual feedback to customers by changing color depending on the price of electricity.

3 The 34 Studies

The 34 studies encompassing 163 experimental treatments in the Arcturus database span four continents and seven countries. Figure 2 sorts the peak reductions for each of the 163 experimental treatments from lowest to highest. At first glance, there is little consistency in the results, for demand response varies from 0 to 58 %. Some of the variation in demand response can be attributed to the different rate types tested, while the rest is potentially due to other factors such as differences in experimental design, socio-demographic characteristics, and climate conditions. Grouping the results by rate type slightly improves the resolution, but not by much. There still remains significant variation among pricing types, as shown in Fig. 3. Due to their tendency to have higher price ratios than TOU rates, we hypothesize

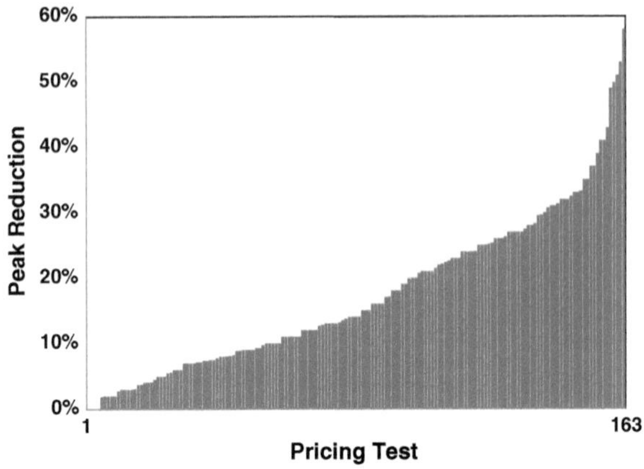

Fig. 2 Impacts from residential time-varying pricing tests, sorted from lowest to highest

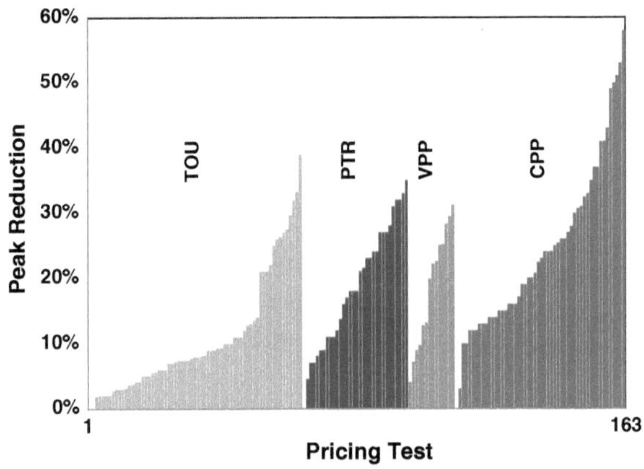

Fig. 3 Impacts from pricing tests, by rate type

that CPP and PTR rates tend to result in higher customer response.[6] We hypothesize that this is primarily due to the use of high price ratios for these rates. By filtering by rate type and the use of enabling technologies, as done in Fig. 4, we can see a clearer picture emerge from the data. The use of enabling technology appears to increase demand response to levels above pricing-only observations for a given price ratio.

Even after sorting the observations by rate type and the use of enabling technology, significant unexplained variation remains. As hypothesized before, the range of results may be due to the variation in the peak-to-off-peak price ratio employed across the studies among other reasons. In order to examine this, we carried out an exploratory data analysis by plotting demand response as a function

[6] For the PTR rate, the effective critical peak price is calculated by adding the peak-time rebate to the rate the customer normally pays during that time period (in the absence of the rebate). This is essentially the opportunity cost of consuming every kWh of electricity.

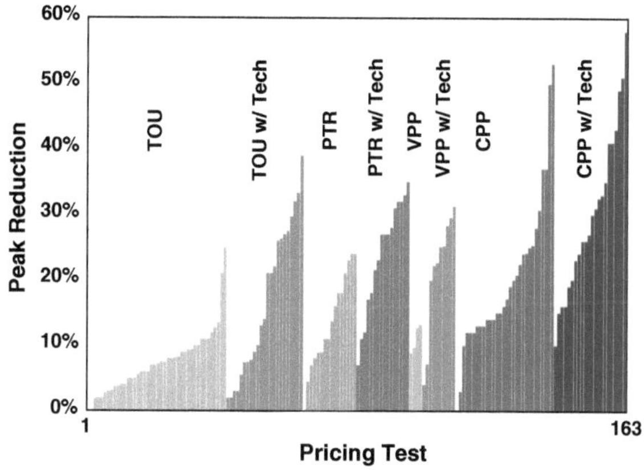

Fig. 4 Impacts from pricing tests by rate type and use of enabling technologies

of the price ratio. The plots initially focus only on pricing treatments that were not accompanied by enabling technology. As seen below in Fig. 5, the 92 price-only treatments fall into a tight pattern. These are then followed by plots that focus on pricing treatments that were also technology-enabled.

The 71 treatments involving price and enabling technologies have a more diffuse pattern, but peak reductions still tend to increase with the peak-to-off-peak price ratio. In addition, for a given price ratio, peak reductions for these technology-enabled treatments tend to be greater than those exhibited by price-only treatments (Fig. 6).

We took this exploratory analysis one step further and estimated a simple regression model for the 163 experimental treatments to quantify the effect of the price ratio and use of enabling technology on demand response. For the purposes of this cursory analysis, we assumed that the main determinant of the variations in the peak reduction is the variations in the peak-to-off-peak price ratio. Using a log-linear specification, we model the amount of demand response, expressed as a percentage, as a function of the log of the price ratio, with and without enabling technology.

$$y = a + b \times \ln(\text{price ratio}) + c \times \ln(\text{price ratio} \times \text{tech})$$

where

y	peak demand reduction expressed in percentages;
ln (price ratio)	natural logarithm of peak-to-off-peak price ratio;
ln (price ratio × tech)	interaction between the ln (price ratio) and tech dummy variable (where tech takes the value of 1 when enabling technology is offered in conjunction with price)

Table 1 presents the regression results.

The results reveal that as the peak-to-off-peak price ratio increases, the peak reduction also increases. In addition, the positive and significant relationship

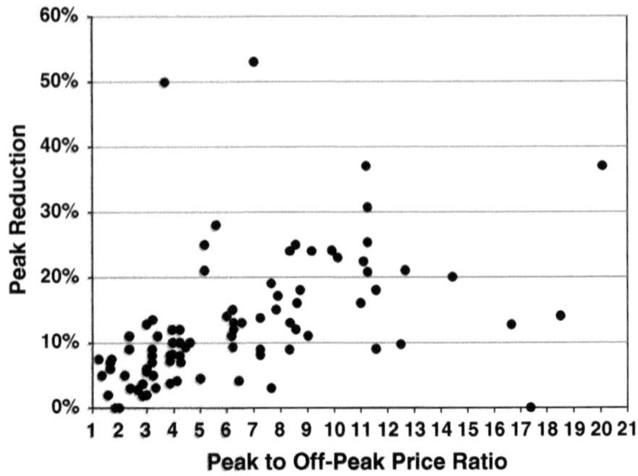

Fig. 5 Price-only treatments

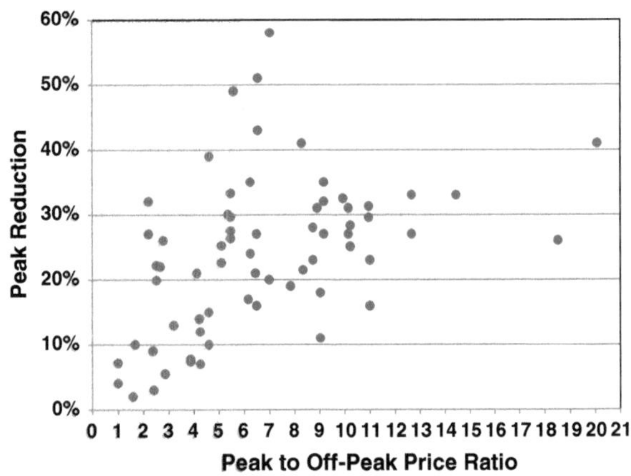

Fig. 6 Price + Technology treatments

Table 1 Regression results

Coefficient	Regression
Ln(Price Ratio)	0.051***
	0.011
Ln(Ratio_EnablingTech)	0.056***
	0.008
Intercept	0.045*
	0.020
Adjusted R-Squared	0.372
F-Statistic	49.02
Observations	163

Standard errors are shown below the estimates
***$p \leq 0.001$
**$p \leq 0.01$
*$p \leq 0.05$

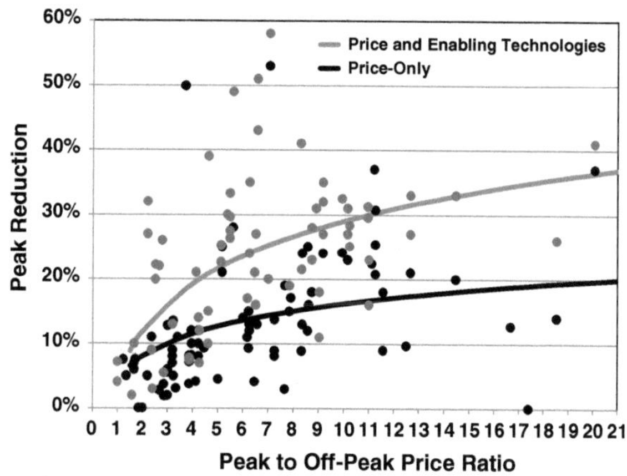

Fig. 7 Area of price responsiveness

between peak reduction and the ln(price ratio × tech) variable indicates that the use of enabling technology further boosts demand response. The R-squared value of 0.37 means that approximately 37 % of the variation in the dependent variable (i.e., peak demand reduction) can be explained by the independent variables.

As seen in Fig. 7, the analysis yields two "arcs of price responsiveness" for pricing-only treatments and technology-enabled pricing treatments. These Arcs of Price Responsiveness can be used to make preliminary assessments about expected customer impacts from various time-varying rates. For example, for a peak-to-off-peak price ratio of 5:1, the expected peak reductions for price-only and price-technology treatments are 12.7 and 21.6 %, respectively. For a price ratio of 10:1, these reductions would increase to 16.2 and 29.0 %, respectively.

Figure 7 also shows that some of the 163 treatments yielded either extremely high or extremely low impacts. In order to minimize the impact of these outliers on the regression estimators, we re-estimated the models using a robust regression technique that utilizes the MM-estimator. This reduces the influence of moderate outliers and completely eliminates the impact of gross outliers by down-weighting their influence in estimating the parameters of the model. "Moderate" and "gross" outliers are defined based upon the scale of the data and such that, were there truly no outliers, all observations would receive the same treatment as in a standard ordinary least squares model.

The regression results from the MM-estimation are presented in Table 2.

As done earlier, the Arcs of Price Responsiveness (shown in Fig. 8) can be used to make preliminary assessments about expected demand response from time-varying rates. For a price ratio of 5:1, the expected peak reduction in price-only and price-tech experimental treatments is 11.5 and 20.1 %, respectively. For a price ratio of 10:1, expected peak period reductions are 15.2 and 27.6 %, respectively.[7] The preliminary results with the 5:1 price ratio are similar to the results from the

[7] By using MM-estimation, our re-estimated arcs predict impacts that are lower than before.

Table 2 Regression results using MM-estimation

Coefficient	Regression
Ln(Price Ratio)	0.054***
	0.011
Ln(Ratio_EnablingTech)	0.054***
	0.008
Intercept	0.027·
	0.016
Number of outliers	3.000
Observations	163

Standard errors are shown below the estimates
*** $p \leq 0.001$
· $p \leq 0.1$

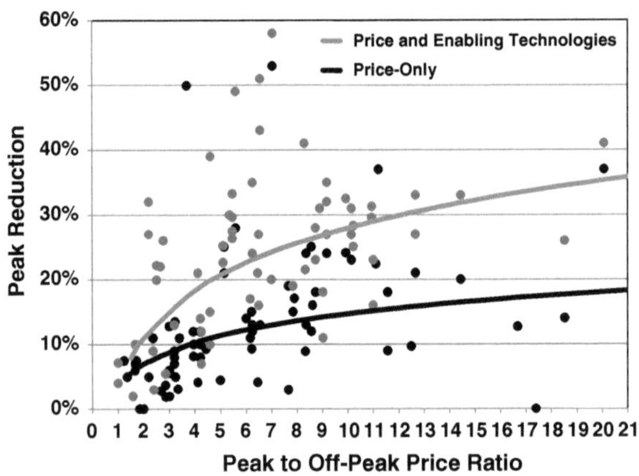

Fig. 8 Arc of price responsiveness (using MM-estimation)

California Statewide Pricing Pilot (SPP) in 2005; this study featured a CPP rate with a price ratio of 6.56 and resulted in a 13 % peak reduction [3].

While this approach led to some interesting and statistically significant results, it became apparent that the analysis could be refined by splitting the Arcturus database and corresponding regressions into two: one for TOU-only treatments and one for dynamic pricing treatments (i.e., CPP, PTR, and VPP). Not only are TOU price ratios consistently lower than other rate types, as can be seen in Figs. 9 and 10, but customer reactions to TOU pricing are probably different. In TOU pricing, the altered rates are applied every day, as opposed to only on discrete event days which is the case in dynamic pricing. This gives TOU customers more opportunity to change their energy consumption habits, leading to greater impacts from TOU. On the other hand, it is easier to change behavior for a few hours per year on dynamic pricing rates.

Table 3 presents the regression results of the TOU-only arc using MM-estimation.

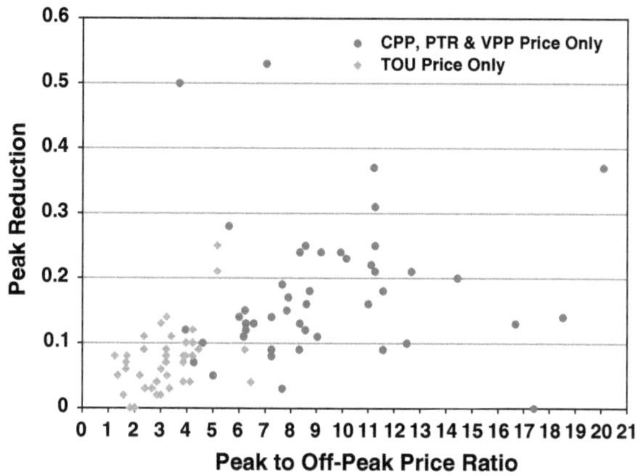

Fig. 9 Price-only treatments split by TOU and non-TOU treatments

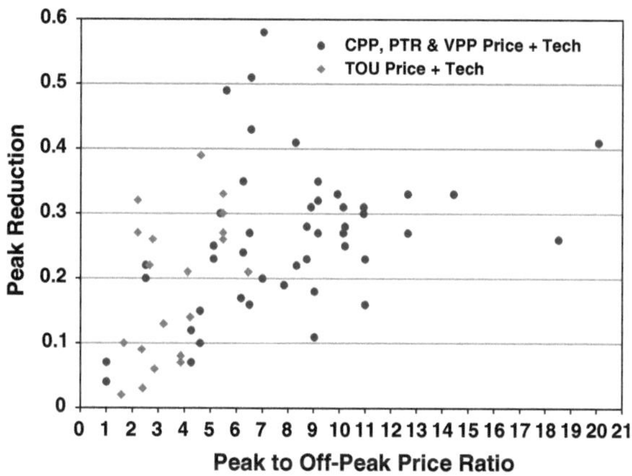

Fig. 10 Price + Technology treatments split by TOU and non-TOU treatments

Table 3 Regression results of the TOU-only arc using MM-Estimation

Coefficient	Regression
Ln(Price Ratio)	0.056**
	0.021
Ln(Ratio_EnablingTech)	0.067*
	0.026
Intercept	0.009
	0.022
Number of outliers	0
Observations	65

Standard errors are shown below the estimates
$^{**}p \leq 0.01$
$^{*}p \leq 0.05$

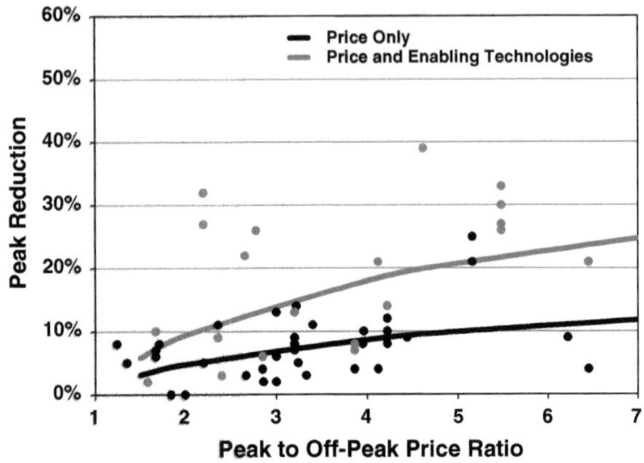

Fig. 11 TOU-only arc of price responsiveness (using MM-estimation)

The results again reveal that as the peak-to-off-peak price ratio increases, the peak reduction also increases. In addition, the positive and significant relationship between peak reduction and the ln(price ratio × tech) variable indicates that the use of enabling technology further boosts demand response.

As done earlier, the Arcs of Price Responsiveness (shown in Fig. 11) can be used to make preliminary assessments about expected demand response from time-varying rates. For a price ratio of 2:1, the expected peak reduction in price-only and price-tech experimental treatments is 4.7 and 9.4 %, respectively. For a price ratio of 5:1, expected peak period reductions are 9.9 and 20.7 %, respectively. The preliminary results with the 2:1 price ratio are similar to the results from the California Statewide Pricing Pilot (SPP) in 2005; this study featured a TOU rate with a price ratio of 2:1 and resulted in a 4–5 % peak reduction [3].

Table 4 presents the regression results of the non-TOU arc (CPP, PTR, and VPP) using MM-estimation.

The results again reveal that as the peak-to-off-peak price ratio increases, the peak reduction also increases, although the relationship is weaker and less significant. However, the positive and significant relationship between peak reduction and the ln(price ratio × tech) variable remains strong and indicates that the use of enabling technology further boosts demand response.

As done earlier, the Arcs of Price Responsiveness (shown in Fig. 12) can be used to make preliminary assessments about expected demand response from time-varying rates. For a price ratio of 5:1, the expected peak reduction in price-only and price-tech experimental treatments is 13.8 and 21.7 %, respectively. For a price ratio of 10:1, expected peak period reductions are 15.9 and 27.2 %, respectively.

Table 4 Regression results of the non-TOU (CPP, PTR, and VPP) arc using MM-estimation

Coefficient	Regression
Ln(Price Ratio)	0.030·
	0.015
Ln(Ratio_EnablingTech)	0.049***
	0.008
Intercept	0.090**
	0.031
Number of outliers	2
Observations	98

Standard errors are shown below the estimates
***$p \leq 0.001$
**$p \leq 0.01$
·$p \leq 0.1$

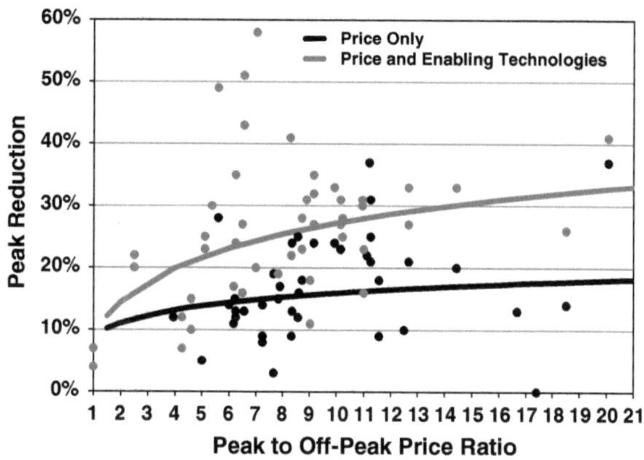

Fig. 12 non-TOU (CPP, PTR, and VPP) arc of price responsiveness (using MM-estimation)

4 Comparison to Earlier Meta-Analysis of TOU Experiments

It is useful to put the results of our analysis in historical perspective. We have done this by comparing them to an earlier meta-analysis of TOU pricing experiments. This was carried out in the early 1980s by Christensen Associates under contract to EPRI and managed by Ahmad Faruqui. In this meta-analysis, data from the five best residential TOU experiments were combined and analyzed. This yielded a variety of elasticities of substitution: (1) for the average household across all five experiments, (2) for households with all major electric appliances living in a hot climate, and (3) for households with no major electric appliances in a cool climate [1]. The elasticity of substitution for this meta-analysis captures a customer's decision to shift usage from higher-priced peak periods to lower-priced off-peak periods. Using Brattle's price impact simulation model (PRISM), which grew out of California's Statewide Pricing Pilot, we have used these elasticities to simulate

Fig. 13 Meta-analysis of 5 TOU experiments

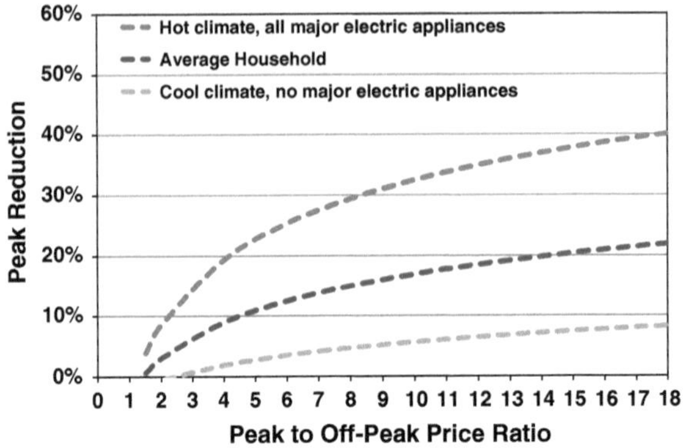

Fig. 14 Price-only arc superimposed

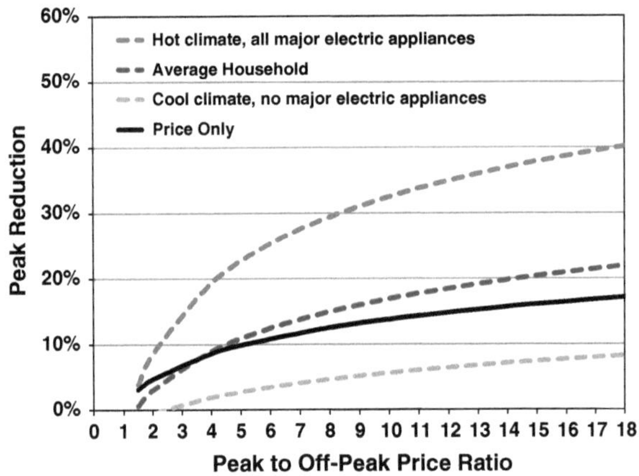

the impact of different price ratios on peak demand [8]. The results are shown in Fig. 13.

And to put these results in perspective, Fig. 14 shows the new Arc of Price Responsiveness for pricing-only TOU treatments superimposed on the previous figure. The results are similar between the average household results from the early 1980s and the price-only result from the recent studies.

5 Elasticity Estimates

In our review of these 34 studies, we find that the impacts also vary widely among the experiments using the same rate design. Other than the differences in the peak-to-off-peak price ratios, this variation is largely attributed to the variation in price elasticities and sample designs. Most studies estimated two types of elasticities: (1) substitution elasticity that captures customers to substitute relatively inexpensive off-peak consumption for relative expensive peak consumption and (2) own price elasticity that captures the change in the level of overall consumption due to the changes in the average daily price. Based on our review, substitution elasticities from the experiments range from −0.07 to −0.40 while the own price elasticities range from −0.02 to −0.10. Availability of the enabling technologies, ownership of central air-conditioning, and the type of the days studies (weekend vs. weekday) are some of the factors that yield variations in the price elasticities.

Price elasticities allow the estimation of peak impacts for a given time-varying rate design. Brattle's PRISM model predicts the changes in electricity usage that are induced by time-varying rates by utilizing a constant elasticity of substitution (CES) demand system. This demand system consists of two equations. The substitution equation predicts the ratio of peak-to-off-peak quantities as a function of the ratio of peak-to-off-peak prices and other factors. The daily energy usage equation predicts the daily electricity usage as a function of daily price and other factors. Once the demand system is estimated, the resulting equations are solved to determine the changes in electricity usage associated with a time-varying rate. PRISM has the capability to predict these changes for peak and off-peak hours for both critical and non-critical peak days. Moreover, PRISM allows predictions to vary by other exogenous factor such as the saturation of central air-conditioning and variations in climate. The model can be set to demonstrate these impacts on different customer types.

6 Conclusion

Our survey of 34 time-varying rate projects across four continents and seven countries has shown convincingly that the amount of demand response increases as the peak-to-off-peak price ratio increases but at a diminishing rate. Enabling technologies, coupled with dynamic pricing, enhance demand response.

Of course, there are many drivers of demand response besides the price ratio. These include the length of the peak period, the number of pricing periods, climate, and appliance ownership. Additionally, the manner in which dynamic pricing rates are marketed to customers has an undeniable impact on customer response, since marketing shapes customer awareness and education. Finally, the non-random selection of customers into some of the time-varying rate experiments can affect the validity of their results.

Because we were unable to control for these factors in this meta-analysis, there are some outliers in our dataset which require further inspection. Even then, the surprising amount of consistency in the results shows that utilities and policy makers can be confident that dynamic pricing and time-of-use pricing will yield significant load reductions.

Acknowledgment This chapter is an updated republication of the paper "Arcturus: International Evidence on Dynamic Pricing" by Ahmad Faruqui and Sanem Sergici, The Electricity Journal 26 (7), 55–65. The copyright permission for reusing the paper has been granted by Elsevier. The authors would like to thank Eric Shultz and Isaac Toussie of Brattle for their excellent research assistance in developing Arcturus and to Toni Enright for assistance in editing and formatting the chapter. We have also benefited from numerous discussions on dynamic pricing with Ryan Hledik and Neil Lessem of Brattle. Finally, Faruqui would like to acknowledge his debt to J. Robert Malko (now a professor at Utah state) for instilling in him a love for time-varying rates when both worked in the Electric Utility Rate Design Study at EPRI.

References

1. Caves D, Christensen L, Herriges J (1984) Consistency of residential customer response in time-of-use electricity pricing experiments. J Econometrics 26:179–203
2. Enright T, Faruqui A (2013) A bibliography on dynamic pricing and time-of-use rates, version 2.0. Retrieved from SSRN http://ssrn.com/abstract=2178674
3. Faruqui A, George SS (2005) Quantifying customer response to dynamic pricing. Electr J 18(4):53–63
4. Faruqui A, Palmer J (2012) The discovery of price responsiveness—A survey of experiments involving dynamic pricing of electricity. EDI Q 4(1):15–18
5. Faruqui A, Sergici S (2010) Household response to dynamic pricing of electricity—A survey of 15 experiments. J Regul Econ 38(2):193–225
6. Faruqui A, Sergici S (2011) Dynamic pricing of electricity in the Mid-Atlantic Region: econometric results from the Baltimore gas and electric company experiment. J Regul Econ 40(1):82–109
7. Faruqui A, Hledik R, Palmer J (2012) Time-varying and dynamic rate design. Prepared for the Regulatory Assistance Project (RAP). http://raponline.org/document/download/id/5131
8. Faruqui A, Hledik R, Tsoukalis J (2009) The power of dynamic pricing. Electr J 22(3):42–56. Available at http://www.sciencedirect.com/science/article/pii/S1040619009000414
9. Flaim T, Neenan B, Robinson J (2013) Pilot paralysis: why dynamic pricing remains overhyped and underachieved. Electr J 26(4):8–21
10. Joskow PL (2012) Creating a smarter U.S. electricity grid. J Econ Perspect 26(1):29–48

Microgrids and Distributed Energy Future

Jin Zhong and Yuqian Song

Abstract Microgrid appears with the development of distributed generations (DGs) and distributed renewable energy resources. It is usually located in low voltage networks and connected to the power grid through switches. In remote or isolated areas, microgrid is an alternative way of power supply instead of installing expensive long-distance transmission lines. A microgrid combines DGs, loads, distributed energy storages, and cogenerations to form a mini power system. The system could be operated as grid-connected mode or islanded mode. The autonomous operation is one of the features of microgrid. Distributed renewable energy resources and small-scale clean energy generating units are the major generation resources in microgrids. The development of microgrids and distributed clean energy generations will be one of the solutions to carbon emissions and global warming. Microgrid is a transition step from conventional power systems to smart grid. The demand management and generation control in a microgrid could significantly improve energy efficiency and system operation. In this chapter, the concept and development of microgrid are introduced. The challenges and operation issues are discussed for microgrids. In the end, the operating microgrids all over the world are introduced.

1 Introduction

Electricity is a necessity for people in a modern society. Power systems traditionally are classified as generation, transmission, and distribution. Due to economies of scale, power systems have been operated as centralized large systems since generators were first developed in 1890s. Centralized large-scale thermal power plants and

J. Zhong (✉) · Y. Song
Department of Electrical and Electronic Engineering, The University of Hong Kong, Pokfulam, Hong Kong
e-mail: jzhong@eee.hku.hk

D. Mah et al. (eds.), *Smart Grid Applications and Developments*, Green Energy and Technology, DOI: 10.1007/978-1-4471-6281-0_5,

hydropower stations are usually located far from load centers. High-voltage transmission lines transmit electricity to urban and suburb areas. Then, electricity is distributed to customers through distribution lines, which have lower voltage levels than transmission lines. Advantages of high-voltage (220–800 kV) transmissions are: larger transmission capacity; ability of transmitting power for longer distance; and less transmission losses. Due to high investment costs of power generation and transmission facilities and the complexity of real-time power system operation, a vertically integrated power company owning generation, transmission, and distribution facilities had been the most common business model for power industry for many years until some countries started power system deregulations in 1990s. Vertically integrated power companies operate power plants, transmission, and distribution lines in the region and dispatch the system based on power generation costs to supply customers in a reliable way. The traditional vertically integrated power system and its operation mechanisms significantly facilitate the development of power technologies and the investment in power industry, especially when power technique development reached its golden era in 1960s and 1970s.

In traditional power systems, customers are connected to the distribution system. Utilities supply electricity to end users with fixed electricity prices. Except some large industry customers that can sign power supply contracts with utilities, most small customers are passive customers. They turn on switches to use electricity and pay the bill based on kWh consumptions. They do not know or care about which generators generated electricity for them and have no choices of power suppliers or electricity prices.

However, this situation has changed since 1990s from two aspects: generation technology development and energy policies.

Due to economics of scale, traditional power generations are usually centralized large power plants. Thermal conversion efficiencies of these thermal power plants are around 30–35 %. With the development of new generation technologies, new types of generating units with higher efficiencies (40–50 %) are commercialized and become common in supplying electricity, for example, gas turbines, microturbine, combined-cycle gas turbine, cogeneration, etc. Most of them use gas or oil as fuel, which is cleaner than coal-fired power plants. However, the sizes of these units are relatively small compared to traditional centralized generating units. Due to higher fuel costs of oil and gas, their operating costs/marginal costs are higher than coal-fired units. On the other hand, they can start up/response in a very short time and have a lower startup cost. The small-sized fast-response units with new generation technologies could be used as peak generators in traditional power systems. However, these types of generators are commonly installed at customer premises using dispersed energy resources. Compare to centralized power generations, they are called *distributed generation* (DG). Distributed generations (DGs) are usually small-sized generators with high efficiencies and can be easily installed. For fossil-fuel-based DGs, gas or oil is often used. Wind turbines and photovoltaic panels using dispersed renewable energy are also classified as DGs. Some DGs are installed at customer buildings applying cogeneration technologies to improve the total thermal efficiency by providing thermal energy in addition to electrical energy to customers.

Energy polices and power system deregulations provide chances for DG technologies. The new generation technologies of DGs became popular and commercialized in 1980s–1990s. During this period, power industry was under restructuring all over the world. Vertically integrated power systems were deregulated, and centralized power generations were unbundled from transmission systems. Independent power providers (IPPs) invest in and own generating units. Power companies are not the sole owner of generating units. IPPs and customers could buy their own generators and provide electricity from the customer side. Power system deregulations provide opportunities and market environments for DG technologies and their applications at the demand side. DGs could be used as the supplementary generator to supply electricity to residential buildings and industry loads. DGs operating during system peak hours have the effect of peak shaving. From planning aspects, installing DGs in distribution systems could relieve transmission congestion and lower the requirement of transmission capacity, hence reducing the investment in transmission expansion planning. In the deregulated power system, one may invest in small generators distributed at the customer side, instead of investing in centralized large generators and long-distance transmission lines. This makes it easier for small investors (or large customers) to invest in power generations. This also postpones the transmission expansion problem, which requires large capital investment.

DGs installed in a distribution network could coordinate to provide electricity to local loads. The DGs, small renewable energy sources, energy storage devices, and loads could form a small power system at the low voltage distribution levels. They are called *microgrid*. Microgrids are usually connected to the main power grid. It has two operation modes: grid-connected operation and islanded operation. When a microgrid is connected to the grid, there is power exchange between the microgrid and the main grid. During emergency situations or other conditions, the microgrid is disconnected from the main grid and be operated autonomously as islanded operation. In a remote/independent area, the microgrid is a promising practice for electrical energy supply to customers, and it can make best use of dispersed renewable energy.

Virtual power plant (VPP) is another concept for dispatching DGs. The concept of VPP is to use a novel software system to control a cluster of small-scale DGs to act as a single large-scale power plant. VPP is not a physical power plant. It aggregates the outputs from DGs, like wind power, photovoltaic panels, biomass generations, combined heat and power (CHP), and energy storage, by using sophisticated software to control their total power output to follow a preset power generation curve. The power output of all DGs looks like a traditional controllable power plant. VPPs were initially developed mainly in European countries and operated by municipal utilities. A VPP requires complicated software system for optimization, control, and communication.

Microgrid and VPP operation technologies were developed to manage DGs dispersed in the distribution network. In a traditional power system, few generators are located in the distribution network. Power flows from generators to customers through transmission and distribution networks. There is no reverse power from customer sides to transmission grid. The system operator controls generating units at the generation level. Small generators located in the distribution networks are transparent to transmission system operators. Forming microgrids or VPPs could manage the large penetration of DGs at low-voltage networks. A well-designed management system could improve the overall energy efficiency of distributed renewable energy sources.

Microgrids may be considered as a low-voltage mini-grid connected to distribution networks. The mini-grid connects loads and small generating resources, which are geographically close. Typical microgrids are usually connecting loads and generators in a remote or independent area, for example, a village, university campus, and residential areas. CHP is a promising generation technology in microgrid. The overall electrical efficiency and thermal efficiency of a CHP is very high, as it could recover waste heat of power generation to thermal usage for residential customers, which is energy efficient for residential customers. Distributed energy storage devices, such as batteries, ultra-capacitors, and flywheels, are also very important components of microgrids. The energy storage devices could be coordinated with renewable energy resources (wind turbines and photovoltaic panels) to store the electrical energy when electricity demand is low while wind or sunshine is strong and to discharge the electrical energy when demand is high and wind or sunshine is weak.

The future of microgrids is to serve as a controllable component in the distribution system. By controlling distributed energy storage, CHP and small gas turbines, microgrid should be able to optimize energy consumption and to improve the efficiency of renewable energy utilization, hence reducing carbon emissions. In smart grid, microgrid will serve an important role in demand response and could significantly shave the peak and lower the capacity requirement of peak generating units (usually fossil-fuel-based generators).

In this chapter, we will introduce the basic concepts of microgrids, their operation modes, technical and economic issues of microgrids, and practical projects of microgrids.

2 Concept of Microgrid

One of the challenges to reliable power system operation is the growing number of DGs, which have smaller sizes than conventional generating units and in proximity to end users. Most DG technologies are based on energy sources with low/zero carbon emissions, which are helpful to global warming issues. On the other hand, as distributed generators are close to customers, their waste heats could be further utilized to heat up or cool down commercial/residential buildings, thus improving the total energy efficiency. The high penetration of small generators in distribution networks raises some technical issues and operation issues to conventional power systems. The concept of microgrid offers a transition step to an autonomous, low emission, smart power system. A well-designed control and management system for microgrid should allow advanced control, which is compatible with current power system operations, and the microgrid should be visible and controllable to higher-level system operators.

The concept of microgrid is to group distributed energy resources and loads, so that they can act as a single generator connected to the power grid through a connecting point. The control strategies and techniques developed for microgrid is to have the least impact on the existing electricity infrastructure and current system operation modes.

A microgrid might be connected to the main power grid, or operated as an islanded system. An isolated microgrid is defined as one with no connecting point to the utility grid. This kind of microgrid is usually located in the remote rural areas or on an island. These areas can be supplied by a large number of clean energy generations managed with a good load control strategy. From this aspect, microgrid is a new method of powering remote areas, eliminating the need of installing long-distance transmission lines to the area, which is not economical if the electricity demand and number of users are small. A grid-connected microgrid represents a transition of the traditional power system to smart grid for accommodating growing numbers of DGs. This type of microgrid groups a number of AC generators, renewable energy generations and loads. If the total power generation from the generators is greater than the total load in the microgrid, the excess power can be fed back to the power grid through the point of common coupling (PCC). On the other hand, if generations in the microgrid are not sufficient to supply the load, the microgrid will withdraw electricity from the power grid through PCC. Control strategies are the key techniques for the microgrid to control the load, battery charging/discharging, and the switching between grid-connected operation mode and islanded operation mode. The energy management system of a microgrid needs to optimize and control the battery charging and discharging based on its state of charge (SOC) to compensate the fluctuation of renewable energy generations. For a microgrid with cogenerations/combined heat and power generators, the coordination of power supply and thermal supply is needed. Another important function of microgrids is dynamic load control, which is also the key technology of demand response and smart meter applications in smart grid. The key features of a grid-connected microgrid are its ability of islanding the system from the main grid during emergency events, and its ability of recovering the connection autonomously. The switching between islanded operation and grid-connected operation is usually accomplished through a bidirectional relay at PCC.

3 Motivations and Challenges of Microgrid

The benefits of microgrids are evaluated from various aspects. By adopting distributed renewable energy for electricity supply, microgrids have significant effects on greenhouse gas emission reductions and other environmental issues. However, there are significant challenges in setting up the micro-power systems in the existing mature power grid. The challenges are due to the integration of new technologies, regulation and policy incentives, and compliance to international standards. Despite these challenges, microgrids have been set up all over the world. Emission and environmental issues are not the only incentives for microgrid development. Other motivations are that utilities could postpone installation of new generation capacities and transmission capacity upgrades by connecting distributed energy resources. Some microgrid owners have similar financial incentives of developing microgrids instead of installing new line capacities to the main grid.

In the area of microgrid, the most discussed topics include: control strategies, reliability and power quality issues, islanded operation, microgrid design, and demand response.

4 Operation and Control of Microgrids

Microgrids may be divided into two categories:

1. Isolated microgrids, which operate independently, and are usually not connected to the main grid
2. Grid-connected microgrids are usually connected to the main grid and have power exchanges

For microgrids of the first category, they are usually located in isolated remote areas. The system itself is like an independently operated scale-down power system. The technical requirements are the same. However, they have more intermittent renewable energy sources. Frequency regulation and voltage control are important for microgrid reliable operation. Isolated microgrids are usually equipped with energy storage devices to compensate for generation fluctuations and system disturbances.

For grid-connected microgrids, the main power grid is a major power supply. They could rely on the power grid for power balance and energy backup. From the perspective of the system operator of the power grid, the microgrid is operated as a predictable load (generation lower than demand) or a controllable generator (generation larger than demand) at the connecting point. Grid-connected microgrid usually has higher reliability than isolated microgrid due to its connection to the power grid, which could import/export power from/to microgrid. Local customers of grid-connected microgrid could be supplied by the DGs in the microgrid if the power supply from the main grid is lost or if there is a fault. In emergency situations, grid-connected microgrids could provide black-start services to the main grid.

Grid-connected microgrids have two fundamental operation modes: grid-connected mode and islanded mode. There are two transition modes: transition mode from grid-connected to islanded, and connection restoration mode.

In grid-connected mode, the switch at the point of connection is closed. The microgrid is connected and performed as part of the main grid. Due to DGs, there is reverse power flow from the microgrid to the main grid. In grid-connected mode, system operation information needs to be exchanged for system control. The information includes generation outputs, demands, voltages, and protection relay states. The information serves as control parameter to schedule a transition to islanded operation mode. In islanded mode, some loads might be shut down to guarantee that power supply is enough for important loads. To balance load with limited generation sources, load shedding and load management are needed. To restore the connection of the microgrid to the power grid, the frequency, voltages, and phase angles should be within limits and must be synchronized.

Control of microgrids is one of the most frequently studied topics in the research areas of microgrids. Microgrid control is to ensure that the frequency and voltage parameters are operated within limits. The major control schemes of microgrids include autonomous control, and agent-based decentralized control, etc. With autonomous control, the microgrid is expected to run automatically while maintaining reliable system operation. Decentralized control is applied to microgrids due to its characteristics of distributed energy sources.

In a microgrid, it is important that a generation resource could be flexibly added, removed, or adjusted, more or less like a 'plug-and-play' resource. The control of microgrid should be flexible. During a fault situation in the main grid, the microgrid is expected to be disconnected from the system. This is called intentional islanding. When a fault is detected, the components in the microgrid need to be coordinated to take actions. At this moment, all control schemes must be able to support different operation modes, and some loads are switched off to match existing power supply capacity. Once the fault in the main grid is cleared, the microgrid should be able to recover its connection to the main grid and restore normal grid-connected operation mode.

The operation and control of microgrids could be different from traditional power systems. From the customer side, it is a transition state from conventional system to smart grid. Demand response and demand management are the first stage of smart grid. With economic incentives and proper control strategies, microgrids could participate as a component of smart grid to adjust its performance to improve energy and economic efficiency.

5 Economics of Microgrids

To establish a microgrid, the investment includes the capital costs of generation devices, energy storages, control equipment, management systems, communication systems, and protection schemes. Operation costs include fuel costs for distributed generators and cost of purchasing energy from the main grid.

Costs of distributed energy generations, in most cases, are higher than conventional centralized generating units. This is one of the drawbacks of microgrid. However, considering the high cost of installing transmission lines in a remote area, installing microgrids could be more economical.

With the applications of demand response and smart meters, the economic benefits of microgrids are amplified. As a system with generators and loads, a microgrid could adjust its energy consumption or power generation in response to electricity price signals. During peak hours, the microgrid could discharge its energy storage and export electricity to the power grid at a higher price. During off-peak hours, microgrid could purchase electricity from the power grid at a lower price and charge its energy storage. With a well-tuned management and control system, a microgrid could perform well in demand response and achieve significant financial benefits. In a microgrid with distributed generators, CHP and renewable energy generations, the system could be operated in a cost and energy efficient way. When renewable energy is excessive, it could be used to charge energy storage devices or feed loads that are not time critical, for example, space heating/cooling, laundry machine, etc. The energy storage could be discharged when electricity price is high. Management and control strategies can be designed for different types of loads to minimize the electricity generations from fossil-fuel-based DGs, as well as to minimize the total cost of supplying local loads. For the microgrid with CHP, the economic benefit could be enhanced by coordinating electricity loads and thermal loads. The total energy efficiency can be improved accordingly.

6 Existing Microgrid Projects

Currently, the research efforts and constructions on microgrids have been accelerating in many regions and countries such as North America, the European Union, and Japan. Based on the regional energy policies and actual operation conditions of power systems, relevant industry and research organizations have achieved remarkable accomplishments in microgrid operation, control, and protection, as well as in energy management systems and the impact analysis on power system [13]. Control strategy is one of the core issues of microgrid technology. As the platform of testing and verifying the relevant control strategies and theoretical techniques, the construction of microgrid experimental system has received more attention by the governments of many countries in recent years. So far, a number of laboratory and demonstration projects have been conducted in the European Union, America, Canada, Japan, etc. [1].

6.1 Typical Microgrid Systems in Europe

Microgrid has been considered as one of the most important solutions for enhancing the reliability of power systems and has gained much attention by members of the E.U. Themed as 'Energy, Environment and Sustainable Development,' the

European fifth Framework Programme (1998–2002) [6] and sixth Framework Programme (2002–2006) [7] have promoted a number of research projects on DG control strategies and operation. The green book published by the European Commission 'European Smart grids Technology Platform: Vision and strategy for Europe's Electricity networks for the Future' in 2006 elaborated the concept of smart grid [8] and pointed out the perspective long-term plan, which is to build the smart grid consisting of centralized power plants and microgrids, in order to achieve higher reliability, less environmental pollutions as well as higher economic benefits. This plan was included in the core issue of the seventh framework programme (2007–2013) [24].

6.1.1 Microgrid Laboratories in the E.U.

NTUA Microgrid

National Technical University of Athens (NTUA), as one of the leaders of European microgrid research work, established the NTUA Microgrid [12], which is a 230-V, single-phase, and 50-Hz system. Using a 60-V, 110-W PV generation as the distributed power source, this system is integrated to the power system. Besides, a 60-V, 230-Ah battery is used to maintain the transient power balance. In order to present the multiformity of the microgrid system, a 2.5-kW wind generator is planned to be added. The purpose of the NTUA experimental microgrid is to analyze different control strategies for the underlying level PV and energy storage on both grid-connected mode and islanded mode and to estimate the impact of microgrid on economic benefit and the environmental pollution. However, the NTUA Microgrid is a single-phase system, which does not have general significance for other network structure.

Demotec Microgrid

Demotec Microgrid is located in the Institute for Solar Energy Supply Technology (ISET), the University of Kassel, Germany, which is one of the earliest established microgrid laboratories in Europe [3]. This experimental grid is a three-phase 400-V, 50-Hz system and is connected to the power grid with 175 and 400-kVA transformers. There are simulators for traditional thermal generation units and DGs in the power supply side, as well as different kinds of loads in the demand side. One noticeable feature of the Demotec Microgrid is that the system is composed of sub-microgrids including a single-phase PV-battery system and a three-phase PV-battery diesel generator system. By managing the upper-level controller, the structure of the microgrid can be optimized and re-configured in order to achieve better power quality and reliability. In addition, the microgrid system will feed the power back to the power system when the DG output is higher than the load. Demotec Microgrid is used to conduct islanded operation test based on power invertors, and it is also considered as an analytical tool for studying the impacts on system transient states due to variations of resistive, conductive, and unbalanced loads, as well as the influences of DG fluctuations on system stability. It has been playing an important role in the theoretical development of E.U. microgrids.

Armines Microgrid

Armines Microgrid is a single-phase microgrid located in Paris, France, including PV (3.1 kW), fuel cell (1.2 kW), and diesel generators (3.2 kW), etc. [27]. It uses 48-V, 18.7-kWh battery sets as energy storage devices. The demand side contains controllable resistive, inductive, conductive, and non-linear loads and can be operated on both connected and autonomous islanded mode by computer-controlled disconnecters. Superior-level management system is built based on AGILENT VEE7 and Matlab, which realize the function of sending commands and collecting data, and manage the microgrid in real time.

Labein Microgrid

Labein Microgrid is one of the demonstration platforms of the European 'multi-microgrid' project [29]. It is connected to the 30-kV power network via two 1,250-kVA transformers in Bilbao, Spain.

The framework of Labein Microgrid includes:

- Regular DGs (single-phase 0.6 and 1.6 kW PVs, three-phase 3.6-kW PVs and 6-kW wind turbines).
- Energy storage devices (48 V/1,925 Ah and 24 V/1,080 Ah battery sets, a 250-kVA flywheel energy storage and a 48 V/4,500 F super-capacitor).
- 150-kW resistive and two sets of 50-kW conductive loads.

The purpose of Labein Microgrid is to design the control strategies and demand-side management under grid-connected mode and to study frequency regulation and power quality issues of microgrid. In addition, this system contains a DC bus, in order to promote further researches on DC microgrid technology.

CESI Microgrid

CESI Microgrid is a 400-V, 50-Hz system, with a capacity of 350 kW, it is connected to 23-kV system via a 800-kVA transformer in Milan, Italy [5]. It was first established to test DG-related technologies. Afterward, it takes part in the European 'multi-microgrid' project and is used to analyze operation features of different types of microgrid structures. Besides, control strategies after disturbances, power quality analysis and communication technologies are also studied base on this platform.

CESI Microgrid manages the lower-level microgrid structures via Ethernet, wireless communication, and power carrier. Based on the existing microgrid framework, a DC microgrid is being planned.

6.1.2 Microgrid Demonstration Projects in the E.U.

Kythnos Microgrid

Kythnos Microgrid is co-founded by the company SMA Technology AG and National Technology University of Athens in 1980s on an island of Greece [31]. It does not really qualify as a microgrid since it will only operate on islanded mode. However, it is of great significance for the development of European microgrid theory and technology.

Based on different research objects, Kythnos Microgrid can be configured as both single-phase and three-phase systems. According to the requirements of European 'multi-microgrid' project, it was set up as a three-phase, 400-V, 50-Hz system, supporting research on topics of microgrid operation, control strategies, and system reliability.

Kythnos Microgrid contains two subsystems so far, a three-phase system (PV, battery sets, and diesel generators) can cover the local load, and a single-phase system including 2-kW PVC and 32-kWh battery sets is operated in order to support the communication devices of the entire system. In the future, a 5-kW wind turbine will also be introduced to limit the consumption of diesel as well as to enhance the diversity of the system.

Since the Kythnos Microgrid only operates on islanded mode, it will drop part of the 'non-critical' load when power supply is less than demand, and, on the other hand, smart load will absorb the redundant power when the power supply is excessive.

Continuon Microgrid

Continuon Microgrid is the first civil microgrid demonstration project in Netherlands [21]. It was established in Zutohen in 2006. The system is a three-phase, 400 kVA, 50 Hz grid, which connects to the 20-kV power system via a 400-kVA transformer, and is allowed to feed power back to the power system. The capacity is 335 kW, supports 200 houses through 4 feeder lines, and the peak load is 150 kW.

By managing the charging and discharging states of the battery banks, the system is able to operate stably in islanded mode for 24 h and has the ability to black-start when the power system fails.

EDP Microgrid

EDP Microgrid is a three-phase, 400-V, 50-Hz system, one 80-kW Capstone gas turbine is operated as the major generation [25]. It is connected to the 10-kV medium-voltage system by a 160-kVA transformer.

EDP is one of the demonstration platforms of the European 'multi-microgrid' project. It is able to operate in both grid-connected and islanded modes, and the research efforts mainly focus on two different scenarios:

- Micro-turbines will cover the local load and the microgrid will feed extra energy to the power system.

- Except supporting local load, the micro-turbine will also supply electricity power for civil, commercial, and industrial load.

Based on these scenarios, the studies are conducted on micro-turbine operation characteristics, strategies of microgrid operating mode switching, and load management.

MVV Microgrid

MVV Microgrid was constructed in Mannheim, Germany. The system is a 400-V, 50-Hz, three-phase microgrid connected to the 20-kV distribution network through a 400-kVA transformer [4]. Current DG sources include PV and micro-gas turbine. In the future, the microgrid will be enhanced with fuel cells, battery sets, and fly wheels.

Bornholm Microgrid

The aforementioned microgrids focus on the research topics of low-voltage microgrids. Bornholm Microgrid, Denmark, is particularly noteworthy as the only medium-voltage-level grid among European microgrid demonstration projects [28].

Bornholm is one of the islands on the Baltic Sea. The generation section of this system consists of a 39-MW diesel generator, a 39-MW gas turbine, a 37-MW CHP generation, and 30-MW wind turbines, all of which supply for the 28,000 residents on the island (the peak load is 55 MW). The microgrid contains 950 10/0.4 kV substations and 16 60/10 kV substations and is connected to the Sweden power grid via a 132/60 kV transformer. As one of the demonstration projects of the European 'multi-microgrid,' the Bornholm microgrid contributes to research on microgrid black-start, and re-connection to the system major network.

Apart from the aforementioned microgrids, the European microgrid projects also include UMIST laboratory (England), Germanos Microgrid (Greece), and Kozuf marsh-gas-oriented microgrid, etc. [9, 14].

For most of the European microgrids, micro-sources such as PVs, fuel cells, and micro-gas turbines are usually interfaced through power electronic devices. Major energy storage units are installed on AC buses. In addition, these microgrids normally adopt hierarchical control strategies, as DGs and loads are controlled at the lower level. The upper level is in charge of configuring and managing the micro-sources and energy storage devices to optimize operating conditions of the microgrid, as well as feeding energy to the major power system as a unit.

6.2 Typical Microgrids in the US

The global financial crisis is pushing the US government to attach great importance to renewable energy. Transformation and development of the renewable energy industry has become the core issue of President Obama's Economic

Stimulus Package. Because of the huge potentials of microgrids in enhancing the efficiency of energy utilization, power system reliability, and security, the US government is now strengthening the research on microgrid-related technologies by supporting hundreds of research institutes, colleges, companies, and national laboratories, as well as accelerating the constructions of demonstration projects. The US Department of Energy (DOE) considers microgrid technology as one of the three cornerstones of the future power systems and includes it in the US 'Grid2013' Plan.

Research efforts on DG and microgrid technologies in the USA are mostly led by Consortium for Electric Reliability Technology Solutions (CERTS). As one of the most authoritative organizations in America, maybe even the world, CERTS is the absolute pioneer in the microgrid and DG field and has published a series of programmatic documents describing the microgrid concept and control theory [19]. The concept proposed by CERTS includes two core components: static switches and autonomous DG sources. When there is a failure or disturbance in the major power system network, static switches will automatically switch the microgrid to islanded mode in order to improve the power supply quality. When operating on islanded mode, the DG will adopt control strategies of active power-frequency and reactive power-voltage droop so that the transient power balance can be maintained.

6.2.1 CERT Microgrid Demonstration Projects

Wisconsin Experimental Microgrid

Wisconsin microgrid laboratory is established at the University of Wisconsin–Madison to verify the microgrid concept proposed by CERTS Lasseter and Piagi [20], using two DC voltage sources to simulate DGs, and the load is pure resistive. A series of experiments conducted on this system have determined that the micro-system can achieve transient voltage and frequency regulation as well as smooth switching between grid-connected and islanded modes by adopting the voltage droop control strategy.

CERT Microgrid Demonstration Platform

After the concept was verified in the laboratory, CERTS established a demonstration platform in Dolan Technology Centre, Columbus, Ohio, which is financially supported by the company American Electric Power (AEP) [26].

The system includes three power feeder lines, among which line A is a regular power line, line B connects to a 60-kW gas turbine (including energy storage devices) and controllable load, and line C connects to two 60-kW gas turbines and sensitive loads. This system is mainly used to study the parallel operation issues of DGs, as well as power quality for sensitive loads.

6.2.2 NREL Microgrid Projects

National Renewable Energy Laboratory (NREL) is very active in the research field of DG and microgrid.

NREL Laboratory

NREL laboratory is built to promote research on DG technologies. The system is composed by PVs, wind turbines, micro-gas turbines, and energy storage devices [17].

This system is used to test the reliability of DG system and other renewable energy-related microgrid technologies. The IEEE 1,547 protocol is formulated based on some of the experimental results in NREL laboratory.

Sandia National Laboratory

Sandia National Laboratory is also known as the Distributed Energy Technology Laboratory (DETL), and it was first established by DOE to study and demonstrate PV technologies [32]. Now, it has been developed into a research and test center for multiple DGs including PV, gas turbine, and wind turbine.

With control in the upper management level, the system can be configured as a three-phase, 480-V microgrid system and operates on both grid-connected and islanded modes. By monitoring the values of voltage and current on AC and DC sides, researchers can evaluate the efficiency of energy utilization and the impacts on microgrid operation due to changes in both power supply and load sides.

DUIT Microgrid

The Distributed Utility Integration Test (DUIT) is implemented under the cooperation of DOE, California Energy Commission (CEC), and Pacific Gas and Electric Company (PG&E) [16, 24]. It is the first implementation of microgrid project in the USA. The purpose of this project is to conduct studies on the impacts of high penetration of DG. DUIT focuses on the distribution system reaction to different types and amount of DGs. In addition, voltage and frequency regulation, islanded operation, power quality monitoring and analysis, and microgrid protection, etc., are as well the research focuses of DUIT.

Oak Ridge National Laboratory

The CHP system of Oak Ridge National Laboratory is primarily focusing on the fuel consumption and emission reduction issues [10], by monitoring and analyzing real-time emission data and combustion tail gas.

6.2.3 Other Microgrid Projects and Future Plans in the USA

In addition to the projects promoted by CERTS and NREL, some companies and organizations have participated in this field as well, such as the General Electric (GE) Microgrid demonstration platform, Mad River Microgrid constructed by

Northern Power Systems (NPS); NREL and Washington Electric Cooperative (WEC), as well as the Palmdale Microgrid supported by Public Interest Energy Research (PIER) of CEC.

There are also some microgrid projects still in the planning phase [18], most of which are listed in Table 1.

The development of the US microgrid focuses on the plug-and-play (PNP) ability of DGs. Energy storage devices and distributed sources are operated as a unity and connected to the micro-system via power electronic interfaces. The microgrid is then integrated to the power system through the public access point and is not allowed to feed power back to the power system. There is no central control unit in the micro-system, and the power is balanced by adopting local voltage droop control strategies. Therefore, this particular type of microgrid is superior in maintaining system reliability and flexibility. The stability issues, on the other hand, are quite noteworthy [30].

6.3 Typical Microgrids in Japan

Due to the lack of natural resources such as oil, coal, and gas. Japan has paid great attention to the development of new energy resource utilizations and is now in the lead in the microgrid area. In Japan, New Energy and Industrial Technology Department Organization (NEDO) was established to coordinate institutes, enterprises, and national laboratories in new energy technology development and demonstration test bed construction.

6.3.1 Demonstrative Project of Regional Power Grid with Various New Energies

The Demonstrative Project of Regional Power Grid with Various New Energies (2003–2007) is one of the most well-known plans since NEDO was first established. The project is to build a regional power supply system, which contains multiple DGs without affecting the power system. This plan includes three demonstration projects [11, 23].

Archi Microgrid

Archi Microgrid uses fuel cell, gas turbine, and PV as its DGs. The microgrid supplies 5 % of the electricity demand for 2005 World Expo. One of the typical features of Archi Microgrid is that all DG sources are connected to AC buses through power invertors, and using NaS batteries as energy storage, which is quite effective in balancing power supply and demand. From December 2004 to September 2005, during the demonstrative operation period, the gross generation of the system was 3,716 MWh. The system was tested to operate in islanded mode in September 2005. According to the experiment results, the system is able to

Table 1 US microgrid plan

Name	Purpose	Components
City of Fort Collins	Energy efficiency and control	PV, gas turbine, and thermal energy storage
ATK launch	DG autonomous control	PV, wind turbine, and CAES
Chevron USA	Energy storage and renewable energy	Energy storage, PV, and fuel cell
Illinois institute	Multiple power supply	Gas turbine, PV, and energy storage
Allegheny power	Super power system	PV, gas turbine, and energy storage
Consolidated Edison	Upper-level control and management	Wind turbine, diesel generator, and fuel cell
San Diego G&E	Coastal microgrid	PV, energy storage, and demand-side management
University of Hawaii	Peak shifting	Gas turbine, wind turbine, and demand-side management
University of Nevada	Energy efficiency and control	PV, battery, and demand-side management

operate in a stable way with the contribution from NaS batteries. After the World Expo, the Archi Microgrid was then moved to Tokoname and reactivated in August 2006.

Kyoto Microgrid

Kyoto Microgrid contains four 100-kW thermal generators, 250-kW fuel cells, and 100-kW lead–acid batteries. In addition, two sets of PV panels and 50-kW wind turbines are installed in the remote area. System information such as operation conditions and power quality is monitored and managed remotely via ISDN and ADSL communication links.

Hachinohe Microgrid

Hachinohe Microgrid is supplied by a marsh-gas generator as DG, which feeds six terminal users via a 5.4-km, 6.6-kV overhead line [15]. The power exchange is constant, and reverse power flow is not allowed when connecting to the power system. In islanded mode, the generator is operated to support the frequency of the microgrid.

A central energy management system controls the microgrid via optical fibers from following aspects:

1. weekly supply and demand schedule (maximize the economic benefits due to environmental factors and fuel costs, etc.)
2. economic dispatch schedule every 3 min (adjust DG outputs based on the difference of power supply and demand)
3. power flow control in second level (avoid the impacts of generation and load fluctuations)
4. frequency regulation in 10-ms level (voltage and frequency regulation in islanded mode)

6.3.2 Demonstrative Projects on New Power Network System

Demonstrative Projects on New Power Network System aim to promote new power distribution network consisting of DGs, SVCs, and dynamic voltage regulators. One of its subprojects is the Akagi demonstration project, focusing on the impacts of SVCs on the distribution network. Another subproject is the Sendai Microgrid demonstration project, which focuses on power quality study.

6.3.3 Other Microgrid Projects in Japan

Apart from NEDO projects, some other organizations are also committed to the research of microgrid and have established demonstration projects, such as the Shimizu Microgrid and the Tokyo Gas Microgrid. In most of the Japanese microgrid projects, DGs and energy storage devices are managed by an upper-level energy management unit to maintain the transient power balance. DGs in these grids are not necessarily required to have the ability of 'plug and play,' but the power should be maintained constant at the connecting points, to be sure the microgrid will not influence the power quality of power systems.

6.4 Microgrids in Developing Countries

Bulyansungwe Microgrid in Uganda is representative of microgrid technology development in developing countries [2]. The system uses PV as major generation resources. A set of 21.6-kW batteries and a diesel generator perform as emergency power supply. It supplies electricity to two hotels, a school, and a monastery. The success of Bulyansungwe Microgrid is of great significance for power system development in African countries with rich solar radiation resources.

In addition, the overseas department of NEDO has also promoted a series of microgrid projects in Cambodia, Mongolia, and Thailand.

Power industry in China has been greatly developed in recent years. The Chinese government has taken more responsibilities of emission reduction in international conventions. As one of the most effective approaches of DG utilizations, microgrid technology is now drawing more attention in China.

An integrated microgrid laboratory was established in Hefei University of Technology, China [22]. DGs include PV panels, wind turbines, and fuel cells. Batteries and superconductors are used as energy storage devices. This system is controlled by combining a central management unit and base-level control modules.

In December 2007, a PV generation-based microgrid was established at Hangzhou with the cooperation between Shimizu Institute of Technology, Japan, and Hangzhou Dianzi University, China. This is the only microgrid with PV generation proportion higher than 50 %, and it is now supplying electricity to two buildings at the University campus.

7 Conclusions

Smart grid technologies and emission reduction issues have been attracting more attention than ever. On the other hand, renewable energy technology is improving and developing rapidly. Both factors have greatly influenced the development of power system operation and control. The fluctuation caused by high penetration of DGs has become a big challenge to power system operators. In addition, the demand for more economical and environmentally friendly power system has exacerbated the problem. As one of the most effective solutions of renewable energy utilizations and cost reductions, microgrid technologies have been significantly advanced since 1980s, especially in Europe, North America, and Japan. Although there are differences in basic concepts and focuses, the purposes and effects are basically the same. Based on the practical platform and demonstration projects established in these regions and countries, microgrid technologies have made many contributions to power system operation and control techniques to handle the uncertainties and fluctuations of renewable energies and loads, as well as the reliability and security issues. However, much remains to be studied in the relevant fields, such as control strategies, standardization efforts, and financial issues.

Microgrid technology will keep on evolving and contribute to the power industry. It is a very important phase of power system transformation and smart grid technology development. No matter what form it will take eventually, the concepts and achievements in microgrid technologies will form the milestones on the road to smart grid.

Acknowledgments This chapter republishes three figures from the paper 'Communication-oriented smart grid framework' by Wen, M.H.F.; Ka-Cheong Leung; Li, V.O.K., Proc. IEEE International Conference on Smart Grid Communications, Brussels, Belgium, Oct 2011. The copyright permission for reusing the paper has been granted by IEEE.

References

1. Barnes M, Kondoh J, Asano H, Oyarzabal J, Ventakaramanan G, Lasseter R et al (2007) Real—world microgrids-an overview. In: IEEE international conference on system of systems engineering (SoSE'07), 2007 pp 1–8
2. Brandt D (2006) Montreal 2006 symposium on microgrids: microgrids for rural electrification
3. Braun M, Degner T, Vandenbergh M et al (2006) Laboratory DG grid. Dispower D6, 3
4. Buchholz B, Schluecking U (2006) Energy management in distribution grids European cases. Power engineering society general meeting, 2006. IEEE, 2
5. CESI RICERCA DER Test Facility (2013) http://www.microgrids.eu/index.php?page=kythnos&id=6
6. Cordis (1998) European 5th framework programme. http://cordis.europa.eu/home_en.html
7. Cordis (2002) European 6th framework programme

8. European Commission (2006) European SmartGrids technology platform: vision and strategy for Europe's electricity networks of the future. Directorate for research
9. FYROM (2010) Kozuf Microgrid http://www.microgrids.eu/index.php?page=kythnos&id=8
10. Firestone R (2004) Distributed energy resources customer adoption model technology data. Tech. Rep
11. Fukuda H (2008) Overview of the micro-grid related project in NEDO. In: International conference on renewable energy in Asia: a challenge for micro grid concept
12. Georgakis D, Papathanassiou S, Hatziargyriou N, Engler A, Hardt C (2004) Operation of a prototype microgrid system based on micro-sources quipped with fast-acting power electronics interfaces. In: Power electronics specialists conference, 2004 (PESC 04), IEEE 35th Annual 2004. vol. 4, pp 2521–2526, IEEE
13. Hatziargyriou N, Asano H, Iravani R, Marnay C (2007) An overview of ongoing research, development, and demonstration projects. IEEE Power Energy 8:78–94
14. Jayawarna N, Barnes M, Jones C, Jenkins N (2007) Operating microgrid energy storage control during network faults. In: IEEE International conference on System of systems engineering (SoSE'07), pp 1–7, IEEE
15. Kojima Y, Koshio M, Nakamura S, Maejima H, Fujioka Y, Goda T (2007). A demonstration project in hachinohe: microgrid with private distribution line. In: IEEE International conference on paper presented at the system of systems engineering 2007. pp 1–6
16. Kroposki B (2003) DUIT: Distributed utility integration test, CA. Tech. Rep. http://www.nrel.gov/docs/fy03osti/34389.pdf
17. Kroposki B (2004) Interconnection testing and applications. Paper presented at the 2004 power engineering society general meeting special technical session
18. Kythnos MS (2008) Overview of federal R&D on microgrid technologies. 2008 Symposium on microgrids
19. Lasseter R, Akhil A, Marnay C, Stephens J, Dagle J, Guttromson R, et al (2002) Integration of distributed energy resources. In: The CERTS Microgrid Concept
20. Lasseter RH, Piagi P (2006) Control and design of microgrid components. PSERC Publication, 06, 3
21. Loix T (2009) The first micro grid in The Netherlands. Bronsbergen, KU Leuven(1–2)
22. Meiqin M, Ming D, Jianhui S, Chang L, Min S, Guorong Z (2008) Testbed for microgrid with multi-energy generators. In: Paper presented at the Canadian conference on electrical and computer engineering 2008, CCECE 2008. pp 000637–000640, IEEE
23. Morozumi S (2007) Micro-grid demonstration projects in Japan. Paper presented at the power conversion conference-nagoya, 2007. PCC'07
24. Muldur U (2007) A new deal for an effective European research policy (electronic resource): the design and impacts of the 7th framework programme. Springer, Berlin
25. Hatziargyriou N (2007) Overview of microgrid R&D in Europe. 2007 Symposium on microgrids
26. Nichols DK, Stevens J, Lasseter RH, Eto JH, Vollkommer HT (2006) Validation of the CERTS microgrid concept the CEC/CERTS microgrid testbed. Paper presented at the power engineering society general meeting, 2006. IEEE
27. Osika O, Degner T, Hardt C, Lange H, Vandenbergh M, Dimeas A, et al. (2005) Description of the laboratory micro-grids. Microgrids project report
28. Østergaard J, Nielsen JE (2008) The bornholm power system: an overview
29. Sanchez E, Santiago J, Oyarzabal J, et al (2007) Report on field tests for islanded mode part I—Labein's μGrid description
30. Stevens J (2005) Characterization of microgrids in the United States: final whitepaper. Tech. Rep
31. Vandenbergh M, Geipel R, Landau M, Strauss P, Tselepis S (2008) Performance evaluation of the Gaidoroumandra mini-grid with distributed PV generators. In: Proceedings of the 4th European PV-hybrid and mini-grid conference 2008
32. Whitaker C, Newmiller J, Ropp M, Norris B (2008) Renewable systems interconnection study: test and demonstration program definition. Tech. Rep

Communication and Network Security Requirements for Smart Grid

Victor O. K. Li

Abstract To relieve the problems of climate change due to global warming, increased renewable energy generation based on wind and solar has been advocated in the power grid. Due to the intermittent nature of such generation, and the requirement of real-time power balance between generation and consumption, increased renewable energy generation may cause instability in the power grid. The smart grid is an attempt to deploy information and communication technologies (ICT) on the power grid to solve the instability problem due to renewable energy sources. The smart grid also enables customer participation in such applications as differential pricing and demand response and promises improved grid operating efficiencies, self-healing capabilities, and resiliencies against cyber attacks. In this chapter, we discuss the communication and network security requirements for smart grid.

1 Introduction

Concerns with the problems of climate change due to global warming prompted governments throughout the world to pursue policies aiming at the reduction of greenhouse gases, mostly carbon dioxide. One important government policy to achieve carbon reduction is to mandate increased renewable energy penetration in electricity generation. However, such increases may adversely impact power system operations. Due to the intermittent characteristics of renewable energy sources such as wind and solar, the outputs of wind farm and solar generators are difficult to forecast. Any change in wind conditions or a cloudy sky may result in very large changes in power outputs. Since a power system requires real-time

V. O. K. Li (✉)
Department of Electrical and Electronic Engineering, The University of Hong Kong, Hong Kong, People's Republic of China
e-mail: vli@eee.hku.hk

D. Mah et al. (eds.), *Smart Grid Applications and Developments*,
Green Energy and Technology, DOI: 10.1007/978-1-4471-6281-0_6,

power balance between generation and consumption, increasing renewable energy penetration may cause instability in the power grid. The smart grid is an attempt to deploy information and communication technologies (ICT) on the electric grid to solve the instability problem due to renewable energy sources.

Another application of the smart grid is to allow active customer participation. This ranges from allowing customers to choose when to consume electricity (to take advantage of differential pricing) to allowing customers with small-scale renewable energy generators to sell electricity to the grid. To facilitate such participation, real-time two-way communications and smart meters are required.

The smart grid also promises improved operating efficiencies, self-healing capabilities, and resiliencies against cyber attacks.

To summarize, the objectives of the smart grid include (1) accommodating different types of electricity generation, including renewable resources, and storage options; (2) enabling active participation by consumers; (3) optimizing assets and operating efficiency; (4) providing good power quality for electricity supply; (5) providing self-healing capability from power disturbance events; (6) guaranteeing operating resiliency against physical and cyber attacks; and (7) enabling new products, services, and markets [1].

The European Union launched the smart grid project in 2003 [2]. The US Electric Power Research Institute started the IntelliGrid project (EPRI) [3]. The US Department of Energy initiated a Grid 2030 project [4]. Under the Energy Independence and Security Act of 2007, the US National Institute of Standards and Technology (NIST) is tasked with coordinating the development of a framework for information management to achieve interoperability of smart grid devices. The NIST report (Phase I) provides a conceptual reference model for the smart grid [5]. There is an urgent need to establish protocols and standards for the smart grid.

One may claim that the electric power grid with modern energy management system (EMS), consisting of large number of remote terminal units (RTU) sending real-time data every 2 s from substations in the power grid to the computer control center via the supervisory control and data acquisition (SCADA) system, and the advanced application software processing the data in the control center to ensure economic and reliable operation, is a smart grid already. For an introduction to SCADA and EMS, interested readers are referred to Wu et al. [6]. However, the computer and communication technologies employed by EMS and SCADA do not represent state-of-the-art ICT technologies. Moreover, most power systems do not support real-time monitoring and control in the lower-voltage distribution system from substation down. In addition, phasor measurement unit (PMU) [7], which provides global positioning system (GPS) time-stamped measurements in milliseconds, may be utilized to upgrade EMS to a smarter transmission grid. Therefore, it is now possible to have a unified smart grid which covers all parts of the power grid, including the centralized fossil-fuel generators, the distributed renewable energy generators, the transmission and distribution networks, and the smart meters and smart appliances at the consumer premises.

2 Smart Grid Enabling Technologies

According to the US National Energy Technology Laboratory (NETL) [8], we need the following technologies to deploy a smart grid: (1) sensing and measurement, (2) advanced control methods, (3) advanced components, (4) Improved interfaces and decision support (IIDS), and (5) integrated communications. These are summarized as follows:

(1) Sensing and measurement: The goal includes enhancing power management with frequent meter readings, preventing energy theft, enabling consumer choices and demand response, and supporting new control strategies. Two key sensor components will be deployed, including the PMU, and smart meters under the advanced metering infrastructure (AMI). PMUs acquire time-synchronized phasor measurement data for power system operations and have been proved capable of significantly improving the performance of power system monitoring, protection, and control. However, they are quite expensive, and one key problem is to minimize the required number of PMUs, while satisfying the requirement of wide-area monitoring with full coverage. This optimal PMU placement (OPP) problem has attracted much research, and integer linear programming approaches have been proposed. However, due to their computational complexity, such approaches may not scale and more recently, heuristic approaches, such as chemical reaction optimization [9], have been used [10]. Another important sensing infrastructure is the AMI, consisting of smart meters at consumer premises and the two-way communication system which connects the smart meters to the control centers of the service provider. AMI enables such applications as differential pricing and demand response.

(2) Advanced control methods: Traditional power dispatch attempts to always generate enough power to meet the demand. Due to start-up delays in traditional power plants, to guard against under-estimated demands, "spinning reserves" are deployed, i.e., the generator is kept running, so it can be put online at short notice. This does not help in our efforts to reduce carbon dioxide emission. With smart grid, and with an improved estimate of the system state due to advanced sensors such as PMUs, it is possible to have risk-limiting dispatch [11], in which the probability of not meeting the operating constraints is reduced to an acceptable level. Of course, the goal would be to eliminate this risk by making use of energy storage devices and such techniques as demand response. While battery storage is still relatively expensive, there are many studies on utilizing the batteries of electric vehicles hooked up to the grid to provide storage [12]. Demand response gives the utility company the flexibility to reduce the load of selected users, at relatively short notice, in return for reduced electricity rates.

(3) Advanced components: PMU is an important component of smart grid. Another recently developed component is the electric spring [13]. This power electronics device can be used to smooth the fluctuations in power generation

due to solar and wind sources. They may be distributed over the power grid to stabilize the system even when there is substantial wind and solar power generation.

(4) IIDS: This includes the " … essential technologies that must be implemented if grid operators and managers are to have the tools and training they will need to operate a modern grid. IIDS technologies will convert complex power system data into information that can be understood by human operators at a glance. Animation, color contouring, virtual reality, and other data display techniques will prevent 'data overload' and help operators identify, analyze, and act on emerging problems." [8, Appendix B5].

(5) Integrated communications: This is the "… infrastructure for real-time information and power exchange, allowing users to interact with various intelligent electronic devices …" [8, Appendix B1]. Since a typical power grid has an existing communication system, it is important for the new communication infrastructure to be compatible with the existing system. It is also important to consider security and privacy issues.

We note that to achieve the full functionalities of the smart grid enabling technologies, communication technology plays a fundamental role. Therefore, in the rest of this chapter, we will focus on the communication technology of smart grid. We shall first describe a communication-oriented smart grid framework, followed by communication requirements, and finally, security and privacy requirements.

3 Communication-Oriented Smart Grid Framework

To better understand the communication requirements for smart grid, it is necessary to develop a communication framework, on which all aspects of smart grid communication can be analyzed. Most existing work on smart grid communications addressed the communication specifications, but they only focused on some specific parts of the entire smart grid network. For example, Aggarwal et al. [14] presented a communication framework for the distribution network, while [15] focused on the home-area network. The US NIST proposed a framework for smart grids [16]. As shown in Fig. 1, this framework consists of seven domains or entities, namely markets, service providers, bulk generations, transmissions, distribution, operations, and customers. The role of each domain is summarized as follows:

(1) Markets: This domain consists of electricity market participants and operators. The goal is to match the energy production with consumption efficiently, through issuing real-time electricity pricing signals. This domain supports energy bidding, distributed energy resource (DER) aggregation, and energy retailing. In addition, ancillary operations, such as frequency regulation and voltage support, are carried out based on information received from this domain.

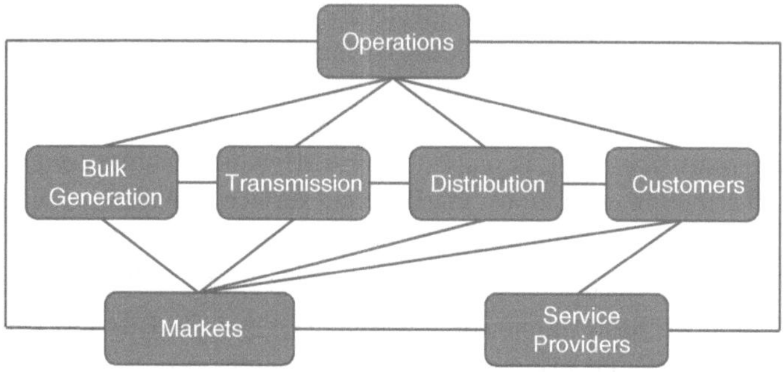

Fig. 1 NIST seven-domain smart grid framework

(2) Operations: This domain consists of electricity service operators. The goal is to ensure reliable and smooth power system operations. Various control and monitoring applications, such as SCADA and EMS, have their control centers in this domain. Moreover, many operational management applications, including fault management, asset maintenance, and operation planning, can be found here.

(3) Service providers: A service provider offers electricity services to customers. This domain acts as an intermediary among operations, markets, and customers. It provides customer management, billing, and other emerging services.

(4) Bulk generation: This domain consists of the electricity generation plants. Bulk generation is typically directly connected to transmission and interfaces with operations, markets, and transmission to support generation control, power flow measurement, plant protection, and other applications. One of the most important functions of bulk generation is to control greenhouse gas emissions, via increased renewable energy sources, and with the deployment of advanced storage devices for smoothing out the imbalance between energy generation and consumption.

(5) Transmission: This domain consists of transmission facilities, such as long-distance overhead lines and transformers. It connects bulk generation and distribution. The most important functions of this domain are transmission stability maintenance and energy loss reduction. This is achieved via voltage monitoring and control carried out at transmission substations. This domain interfaces with markets to procure ancillary services such as frequency regulation services, and with operations for transmission scheduling.

(6) Distribution: This domain consists of distribution facilities, including distribution transformers and underground cables. Distribution interfaces with transmission, customers, markets, and operations. It works with operations to provide real-time management of power flows, and with markets to provide real-time generation and consumption data. It also supports asset and line monitoring and control, distributed energy generation, and bi-directional power flows.

(7) Customers: A customer is an electricity end user, such as a household, a commercial business, or an industrial factory. Since it is increasingly common to have distributed electricity generation and storage facilities, such facilities are also included in this domain. This domain is electrically connected to distribution. It is supported by AMI, which enables communications with distribution, operations, markets, and service providers, facilitating such applications as demand response, building/industrial automation, and the ability by the end user to sell electricity back to the grid.

This high-level NIST framework is complete, but perhaps too complex for those focusing on the underlying communication networks.

In Wen et al. [17], a new communication-oriented framework for smart grid is introduced. This three-entity framework, shown in Fig. 2, is designed to be flexible to accommodate potential new smart grid technologies. It consists of three entities, namely operation network, business network, and consumer network. Operation network manages electricity generation, transmission, and distribution and typically includes automation technologies related to the legacy SCADA systems, wide-area measurement systems (WAMS), and large-scale EMS. Business network is used by the electricity market participants, including the metering service providers and government regulators, to coordinate the electricity market. Consumer network handles the communication for the electricity end users. It includes a home-area network as part of the AMI.

The design principles for this three-entity framework are summarized as follows:

- Simplicity: This framework is designed with only three entities and three types of inter-entity communications, for ease of analyses and further development.
- Completeness: Although simple, this framework represents a complete picture of smart grid communications and is flexible to accommodate all existing and future smart grid applications.
- Compatibility: The proposed framework is compatible with the seven-domain NIST framework. The mapping is as follows. Operation network in the three-entity framework includes the domains of operations, bulk generations, transmissions, and distribution in the NIST framework. Business network in the three-entity framework contains the domains of markets and service providers in the NIST framework. Consumer network in the three-entity framework corresponds to the domain of customers. Due to this compatibility, one can adopt this simple three-entity framework instead of the seven-domain NIST framework to study communication issues defined under the NIST framework.
- Ease of deployment: Three levels of communication requirements have been identified, and system components with the same level of communication requirements are grouped into the same entity. Components in operation network demand the most stringent requirements in terms of cyber security, data availability, and quality of service (QoS). Those components in business network have relatively less stringent requirements, and those components in consumer network have the least stringent requirements. Thus, they are divided

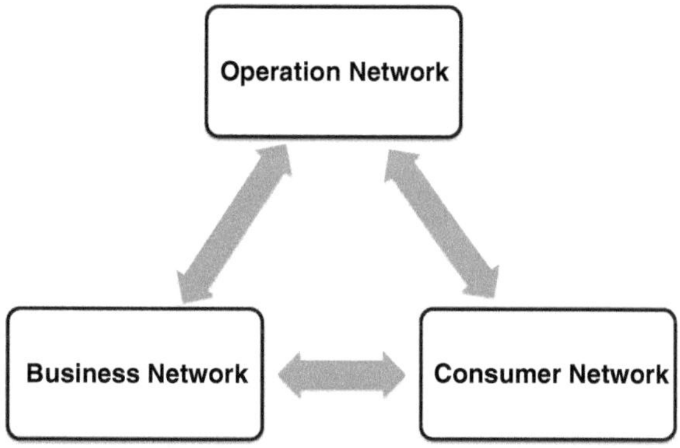

Fig. 2 Three-entity framework for smart grid communication

into three entities for ease of network deployment and inter-entity communication control.

- Ease of evolution: Operation network has been in existence for decades as the core of power system automation, and huge investments have been made by power companies. Due to such investments, future research and development efforts in this area will likely focus on the evolution of the existing network, as opposed to a fundamental change, to meet the requirements for smart grid. Internet technologies have been proposed for supporting business network [18]. Research and development efforts will likely focus on designing new applications and electricity market regulation schemes, such as in Jin and Mechehoul [18] and Rahimi and Ipakchi [16]. Consumer network is relatively new, and research and development on smart metering, demand response, and DER management will prove fruitful.
- Ease of collaboration: The future electricity system requires the expertise of electrical engineers, ICT professionals, business experts, and government officials. The communication-oriented framework must facilitate the collaboration of these experts from various fields.

We shall now describe these three entities in detail.

3.1 Operation Network

Operation network is responsible for electricity generation, transmission, and distribution, and for maintaining the stability and efficiency of the entire power system. As shown in Fig. 3, it consists of eight major components, described as follows:

- Business network gateway and consumer network gateway connect operation network and business network, and operation network and consumer network, respectively.

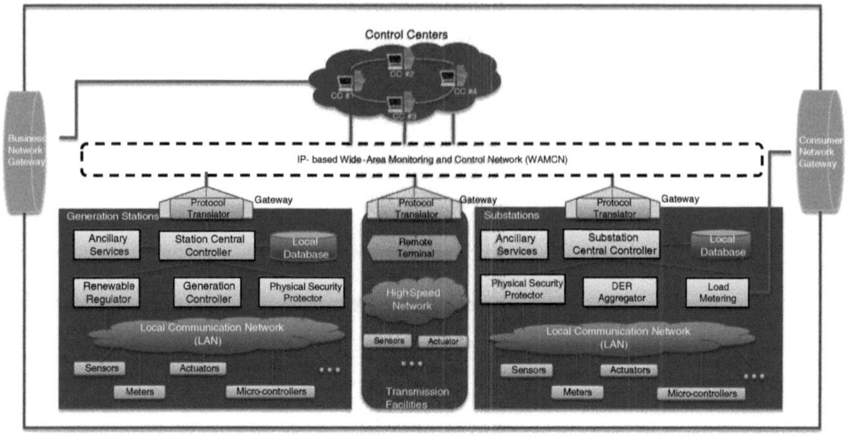

Fig. 3 Operation network [17]

- Control centers gather and process grid operation data. Different control centers communicate with each other to control the whole operation area via a dedicated, secure, high-speed network, managing the various facilities in a smart grid.
- A monitoring and control database is used to store the historical data of the system, including the status parameters of the grid during its operation, and event logs for operators.
- A wide-area monitoring and control network (WAMCN) acquires data from the remote stations or substations and issues control commands. These remote stations can communicate with each other to get a better picture of the system state.
- A generation station is a plant that generates electricity. Each generation station deploys a gateway with a built-in protocol translator to connect the station network to WAMCN.
- Transmission facilities include all the field devices remotely located from the power stations and substations. Most of these devices are monitoring devices (such as PMUs) and control devices (such as actuators). They communicate with the control centers or nearby substations to provide status report of the monitored facilities.
- A substation distributes electricity to the consumers. A gateway is deployed for a substation to access WAMCN.

3.2 Business Network

The business network consists of the components of electricity market regulator, smart meter service provider, demand responder, and electricity market participants. As shown in Fig. 4, they are connected via an IP-based virtual private

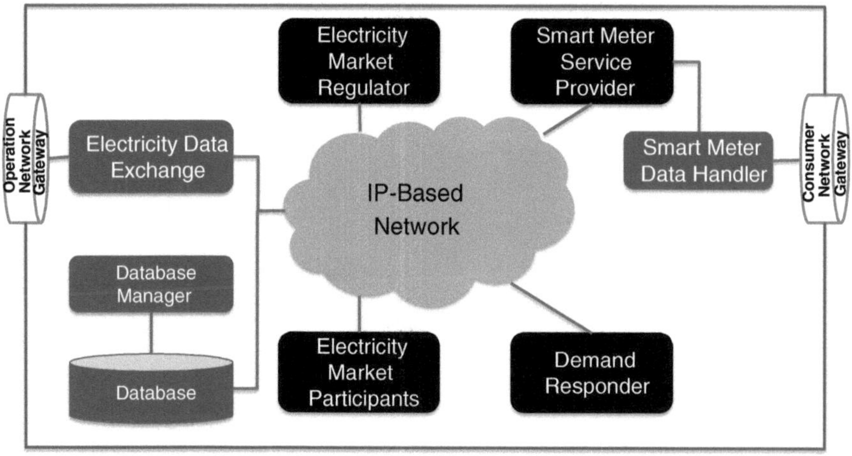

Fig. 4 Business network [17]

network (VPN) and supported by a database manager that manages the electricity market information. They are described as follows.

- An electricity market regulator refers to the government or quasi-government organization which regulates the market.
- Smart meter service providers are the utilities that provide smart metering services to the end users.
- A demand responder refers to an electricity utility that attempts to alter the aggregate electricity consumption by the customers so as to match total consumption with the total power generation. To achieve this, the utility may provide incentives to the end users to either reduce consumption when the total demand is high, via differential pricing, or to agree to switch off some appliances on demand, via reduced utility rates.
- Electricity market participants are those that trade electricity.

3.3 Consumer Network

As shown in Fig. 5, consumer network is a local area network at a consumer premises. It may be located in one residential unit, or may connect multiple units. The major components are as follows:

- A smart controller coordinates the entire consumer network. It is responsible for switching on or off loads automatically according to the current grid operating status based on agreed contracts, analyzing smart meter data, and managing the local energy storage.

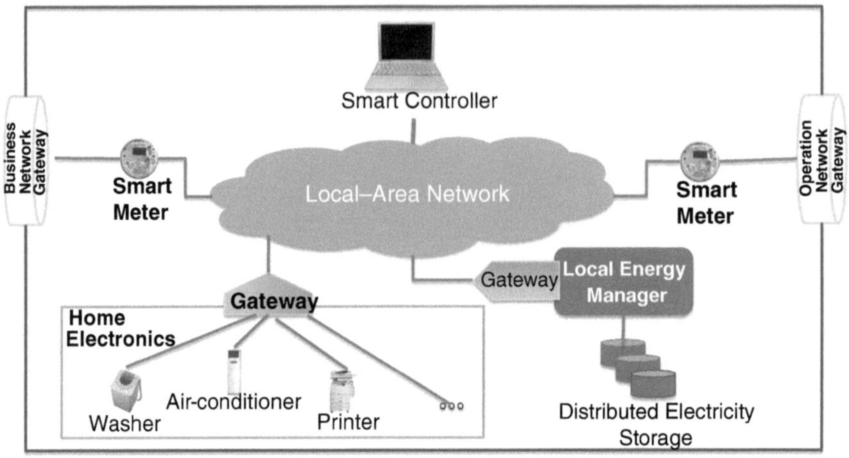

Fig. 5 Consumer network [17]

- A smart meter, which connects to operation network and business network, estimates the electricity usage schedule and sends it to operation network. It also receives real-time electricity rates from business network.
- Home electronics or appliances.
- A local energy manager handles the energy generation and storage at the consumer premises.

Communications among the three entities are required for an effective smart grid.

The communication between operation network and business network requires the highest level of reliability and security, as they form the backbone of the entire smart grid network. The communication between operation network and consumer network also requires high security, and since customer usage schedule will be transmitted to operation network, customer privacy must be ensured. The communication between business network and consumer network, on the other hand, requires high data availability and high reliability, but relatively less stringent security. As will be discussed later, there may be a trade-off between security and QoS requirements.

4 Communication Requirements

It has been advocated [19] that an electricity communication superhighway is required for supporting generation, transmission, substations, consumers, and distribution and delivery controllers. One major challenge is the development of a communication infrastructure to support universal connectivity and system-wide real-time monitoring. Multiple recipients must be able to receive up-to-date system status information, with various latency and rate requirements. In particular, the following communication requirements have been identified [17]:

(1) Capability to handle large volume of data: To enable real-time monitoring and control, deployment of such equipment as PMUs is required. The time resolution of PMU data is in milliseconds, and with a large number of PMUs deployed, the amount of such sensor data will be huge. There will also be large amount of data generated by the power generators, the consumers, and the distribution system. Therefore, the total amount of data generated will be many times the amount of data generated today. The brute force approach to solving these data deluge problem is to install additional transport capacity, probably at prohibitively high costs. But perhaps a more important question is whether we can process and effectively utilize all the data generated. In Li [20], three ways are proposed to handle the large volume of data. Firstly, data redundancy can be reduced by exploiting the spatial and temporal correlation of data. Secondly, adaptive messaging may be used. This calls for different QoS and different levels of security protection for different types of data, i.e., there will be a trade-off between QoS and security protection. Finally, it is desired to transform data to knowledge. After all, even if one can build a communication infrastructure to transmit all data to the operator, he will not be able to handle all of it. Instead, one may transform the deluge of data into specific events, and only send information of the occurrence of certain events to the operator. This will require domain-specific expertise and the development of data-to-knowledge transformation techniques, perhaps based on artificial intelligence.
(2) Extensive coverage: The network must cover the whole power system, including power generation, distribution, and consumption at the customer premises.
(3) QoS support: The system must be flexible to accommodate different reliability, delay, and throughput constraints. As mentioned earlier, there may be a trade-off between QoS and security requirements.
(4) Cyber security: The system must be secure from cyber attacks. This is such an important requirement that we will devote the next section to it.

5 Security and Privacy Requirements

The power grid is perhaps the most important national infrastructure in many countries. Secure and privacy-preserving communications are crucial to smart grid operations. Cyber security experts observed that some types of smart grid sensors can be easily hacked [21]. In addition, worms can spread easily among smart grid sensors [22]. Li et al. [23] has identified several potential attacks on smart grid:

(1) Distributed denial-of-service (DDoS) attack: In a DDoS attack, hacked sensors, which may be geographically distributed, are instructed by the hacker to simultaneously send a large volume of traffic to a victim sensor or to the control center. The victim sensor or control center will be overwhelmed by the huge volume of traffic, and normal data cannot be handled properly. Therefore, it is necessary to develop methodologies to enable a sensor or control center to differentiate between authentic and bogus traffic, and to identify malfunctioning equipment.
(2) Simultaneous shutdown attack: An attacker can pretend to be a control center and send fake control messages to command sensors and other system devices to shut down, thus rendering the smart grid ineffective. Therefore, one must ensure the confidentiality, integrity, and authenticity of control messages.
(3) Fake demand for power: An attacker can intercept and alter the estimated usage schedules from the smart meters to operation network, to dramatically increase or decrease the demand for power. This may lead to power imbalance in the system and cause power failures. Again, it is important to ensure the confidentiality, integrity, and authenticity of the data transmitted.

Other major security and privacy problems have also been identified and studied [24, 25]. For the communication between the control center and the smart meter, it can be proved that the statistical analysis approach cannot protect the system from false data injection attacks [26]. The statistical analysis approach will not work anyway since it requires the system to handle a large volume of data in real time, and the control center of the smart grid will not be able to give a timely response. In addition, the smart meter has been identified as the point of weakness of the AMI and of the smart grid communication system [27]. As the smart meter is located at the customer premises, it is difficult for the control center to protect it from potential attackers. Moreover, a smart meter is an inexpensive device and its memory and processing power are too limited to run complex protective measures. As a result, it is possible for attackers to hack its stored data [28] or even reverse engineer and modify its logic [29]. Therefore, smart meters can be considered as the most vulnerable components in smart grid. In Chim et al. [30], a three-layer architecture, as shown in Fig. 6, is proposed to study the security and privacy problems for the communication between the smart meter and the control center of a smart grid.

At the top layer resides the control center of the power operator. At the middle layer are the substations in the distribution network, and each substation is responsible for the power supply of a service area. The smart meters, placed at the customer premises, are located at the lowest layer. A smart meter communicates

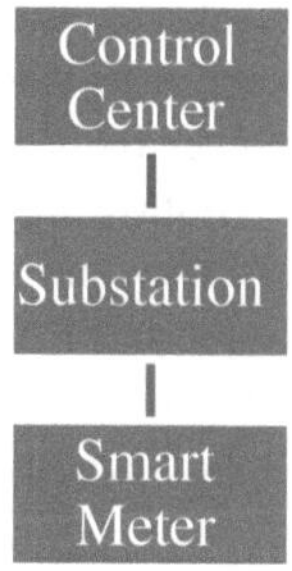

Fig. 6 Three-layer architecture for smart meter communications

with the control center via the substation in its service area. The control center may be a single server located inside the power plant or may be geographically distributed.

Under this three-layer model, the following security and privacy requirements have been identified:

(1) Message authentication: Each message sent by any smart meter should be checked to confirm that it is valid.
(2) Identity privacy preservation: The real identity of the customer should be confidential to everyone, including the power operator, to protect the privacy of customers.
(3) Message confidentiality: The content of the message sent by any smart meter to the control center should be unknown by any third party.
(4) Traceability: The total amount of power used by each customer in a certain service period should be known by the power operator (i.e., the control center), so a proper electricity bill can be prepared.

In Chim et al. [30], a privacy-preserving advance power request paradigm is proposed for smart grid. Presently, there is no need for a customer to inform the power utility how much electricity he/she will require. The power utility estimates the electricity demand (the load) based on historical and other factors and generates enough electricity to satisfy the load. Spinning reserves may be required to cover any shortfall due to load underestimation. Excess generation due to an overestimation of the load and the spinning reserves lead to waste. Under this proposed paradigm of advance power request, power is generated based on explicit customer demands. A customer is required to submit daily electricity usage plans, i.e., the required electricity required at hourly intervals throughout the day. These requests can be sent with different advance notices, such as 1 day ahead or 1 week ahead, but not in real time. This process may be automated by the smart meter, which communicates with all household appliances and stores the usage profile of the customer, and predicts the power usage for the day based on artificial intelligence or other techniques. Upon receiving all the electricity usage plans from all customers in its service area, the power operator schedules its power generation. A similar power request model was described in Li et al. [23]. Since the electricity usage plan includes the usage pattern of the customer, it is important to keep it private. In addition, one must ensure a customer does not make excessive power requests on purpose.

Following the three-layer model as shown in Fig. 6, and targeting the four security and privacy requirements identified, a privacy-preserving advance power request (PPAPR) scheme is proposed [31]. It consists of four modules, namely system start-up, credential request, power usage plan submission, and reconciliation. They are described as follows:

(1) System start-up: In a public key infrastructure (e.g., RSA), each party is assigned a public (known to everyone) and private (known only to itself) key pair. A key is just a string of "0" and "1" bits. When A wants to send a message to B, A encrypts the message using B's public key. B then uses its private key to decode the message. A also uses its private key to generate a digital signature on the message, and B can use A's public key to verify the signature. In the proposed PPAPR scheme, during system start-up, the control center assigns itself an RSA public and private key pair to sign credentials and for customers to encrypt messages to it. Before a new smart meter is put into service, it must be registered and assigned a unique identity.
(2) Credential request: At the beginning of each service period, say each month, customers request a certain number of credentials (or tickets) from the control center for the power usage for that service period. Customers are authenticated using their real identities. Each customer (via his/her smart meter) sends credential-signing requests to the control center. Each credential is of the format <CID, DOI>, where CID is a unique credential identity of each credential and DOI is the issue date that the credential. By presenting a credential, a customer can request V units of power. In PPAPR, all credentials are generated by customers. To make a credential anonymous, a customer first blinds it using a blinding factor (i.e., mixes it with some random components, so none can recognize its original content based on the blind version) and sends it to the control center. The control center signs the credential using its private key and sends it back to the customer. The customer performs some computation to remove the blinding factor in order to obtain the control center's signature on the credential. To prevent a customer from generating an invalid credential (e.g., CID already used or outdated DOI), a customer has to generate n times more credentials using different CIDs and blinding factors, where n is predetermined by the control center. For each n credentials, the control center randomly challenges the customer to open $(n - 1)$ of them and verify the details in them. If the information in all the opened credentials is valid, the control center signs the remaining one using its private key. Otherwise, the control center does not sign the credential and returns an error message to the customer. The control center computes and records the number of credentials N_{total} it has signed for that customer. Recall that the control center's public key is known by everyone, while its private key is only known to itself. Thus, its signature can be

verified by everyone, but can only be generated by it. Also note that although all credentials are known by the customers during generation, it will not cause any security problem because a credential is valid only if it contains the control center's signature.

(3) Power usage plan submission: The smart meter estimates and constructs a power usage schedule and attaches enough credentials of power units to it. The power usage schedule and credentials are encrypted using a randomly generated session key, and the session key is encrypted using the control center's public key, and the whole encrypted block is transmitted to the control center. Upon receiving the message, the control center obtains the session key using its private key and obtains the power usage schedule as well as the credentials using the decrypted session key. It then validates each credential by checking against its own signature and ensuring DOI is not outdated. It also records the credential identities CIDs into its local database so that the same set of credential identities cannot be reused. The control center then schedules appropriate control decisions to adjust the amount of power generated.

(4) Reconciliation: At the end of each service period, reconciliation will be performed and a bill will be generated. Customers need to be authenticated using their real identities. The smart meter of a customer sends all the credentials that it has not used to the control center. The control center then checks the credentials, counts the unused credentials N_{unused}, computes the total used credential Nused as $N_{\text{total}} - N_{\text{unused}}$, and charges the customer accordingly.

6 Conclusions

The smart grid is an attempt to utilize ICT on the power grid to solve the instability problem due to renewable energy sources. The smart grid also enables customer participation in such applications as differential pricing and demand response and promises improved grid operating efficiencies, self-healing capabilities, and resiliencies against cyber attacks. Deployment of smart grid requires technologies on sensing and measurement, advanced control methods, advanced components, improved interfaces and decision support, and integrated communications. To achieve the full functionalities of such smart grid enabling technologies, communication technology plays a fundamental role. In this chapter, we focus on the communication technology of smart grid. We describe a communication-oriented smart grid framework and identify the communication requirements, and the security and privacy requirements.

Acknowledgment This chapter republishes three figures from the paper "Communication-oriented smart grid framework" by Wen, M.H.F.; Ka-Cheong Leung; Li, V.O.K., Proc. IEEE International Conference on Smart Grid Communications, Brussels, Belgium, Oct 2011. Permission for reusing the figures has been granted by IEEE.

References

1. Myles P et al (2011) Electric power system asset optimization. DOE/NETL-430/061110, 7 Mar 2011
2. EC (2006) European smart grids technology platform: vision and strategy for Europe's electricity networks of the future. European Commission, Directorate-General for Research, Sustainable Energy Systems, EUR 22040
3. Electric Power Research Institute (2004). IntelliGrid: at a glance. Available at. http://intelligrid.info/
4. Department of Energy (2003) Grid 2030: a national vision for electricity's second 100 years, US
5. NIST (2010) NIST framework and roadmap for smart grid interoperability standards. Release 1.0, NIST Special Publication 1108
6. Wu FF, Moslehi K, Bose A (2005) Power system control centers: past, present and future. Proc IEEE 91(11):1890–1908
7. Ree JDL, Centeno V, Thorp JS, Phadke AG (2010) Synchronized Phasor Measurement Applications in Power Systems. IEEE Trans Smart Grid 1(1):20–27
8. National Energy Technology Laboratory (2007). U.S. Department of energy, modern grid initiative. https://www.smartgrid.gov/sites/default/files/pdfs/a_systems_view_of_the_modern_grid.pdf
9. Lam AYS, Li VOK (2010) Chemical-reaction-inspired metaheuristic for optimization. IEEE Trans Evol Comput 14(3):381–399
10. Xu J, Wen MHF, Li VOK, Leung KC (2013) Optimal PMU placement for wide-area monitoring using Chemical Reaction Optimization. In: Proceedings of IEEE PES innovative smart grid technologies conference, Washington DC, USA
11. Varaiya PP, Wu FF, Bialek JW (2011) Smart operation of smart grid: risk-limiting dispatch. Proc IEEE 99(1):40–57
12. Kempton W, Tomic J (2005) Vehicle-to-grid power fundamentals: calculating capacity and net revenue. J Power Sources 144(1):268–279
13. Hui SY, Lee CK, Wu FF (2012) Electric springs—a new smart grid technology. IEEE Trans Smart Grid 3(3):1552–1561
14. Aggarwal A, Kunta S, Verma PK (2010) A proposed communication infrastructure for the smart grid. Proc IEEE ISGT 2010:19–21
15. Chen KC, Yeh PC, Hsieh HY, Chang SC (2010) Communication infrastructure of smart grid. Proc IEEE ISCCSP 2010:3–5
16. Rahimi F, Ipakchi A (2010) Demand response as a market resource under the smart grid paradigm. IEEE Trans Smart Grid 1(1):82–88
17. Wen MHF, Leung KC, Li VOK (2011) Communication-oriented smart grid framework. In: Proceedings of IEEE international conference on smart grid communications, Brussels, Belgium
18. Jin T, Mechehoul M (2010) Ordering electricity via internet and its potentials for smart grid systems. IEEE Trans Smart Grid 1(3):302–310
19. Gellings C (2003) Smart power delivery: a vision for the future. EPRI J 9
20. Li VOK (2013) Smart sensing and adaptive messaging in smart grid. In: Workshop on clean energy and environment, University of Hong Kong initiative on clean energy and environment, Hong Kong. http://icee.hku.hk/activities/workshops/ICEEWorkshop_CleanEnergyEnvironment2013.html, 30 Apr 2013
21. Meserve J (2009) Smart grid may be vulnerable to hackers. CNN.com (Available online. http://www.cnn.com/2009/TECH/03/20/smartgrid.vulnerability/index.html), 21 Mar 2009
22. Clemente J (2009) The security vulnerabilities of smart grid. J Energy Secur (Available online. http://www.ensec.org/index.php?option=com_content&view=article&id=198:the-security-vulnerabilities-of-smart-grid&catid=96), 18 June 2009

23. Li VOK, Wu FF, Zhong J (2010) Communication requirements for risk-limiting dispatch in smart grid. In: Proceedings of IEEE workshop on smart grid communications, Cape Town, South Africa
24. Khurana H, Hadley M, Lu N, Frincke DA (2010) Smart-grid security issues. IEEE Secur Priv Mag 8(1):81–85
25. Security (2010) The smart grid interoperability panel cyber security working group second draft NISTIR 7628 smart grid cyber security strategy and requirements
26. Liu Y, Ning P, Reiter MK (2009) False data injection attacks against state estimation in electric power grids. In: Proceedings of the CCS'09, pp 21–32
27. Chan CB, Chim TW, Yiu SM, Li VOK, Hui LCK (2013) Smart meter obfuscation and attestation using non-deterministic execution sequence. Submitted for publication
28. Vigo R, Yüksel E, Ramli C (2012) Smart grid security: a smart meter-centric perspective. In: Telecommunications forum, pp 127–130
29. King C (2012) Smart meter security: the human element. Gigacom. http://gigaom.com/cleantech/smart-meter-security-the-human-element, 13 Apr 2012
30. Chim TW, Yiu SM, Hui LCK, Li VOK (2011) PASS: privacy-preserving authentication scheme for smart grid network. In: Proceedings of the IEEE SmartGridComm, Brussels, Belgium
31. Chim TW, Yiu SM, Hui LCK, Li VOK (2012) Privacy preserving advance power reservation. IEEE Commun 50(8):18–23

Part III
Stakeholders in Perspectives: Interests, Power and Conflict

Smart Grids: The Regulatory Challenges

Daphne Mah, Kaboo Po-yi Leung and Peter Hills

Abstract Smart grids present major potential benefits in terms of economic, environmental, and social considerations. The deployment of smart grids however requires not only technological advancement but also the ability to overcome many regulatory barriers. This chapter brings regulator perspectives—an area that is under-explored—into the field of smart grid studies. We examine why regulators should be concerned about smart grid developments, the nature of the regulatory challenges they may face, and what they can do to address these challenges. We have two major findings. Firstly, we demonstrate that smart grids present new challenges to regulators. Regulators are faced with three major challenges: utility disincentives, pricing inefficiencies, and cybersecurity and privacy. Market liberalisation, decoupling, dynamic pricing, and protocols and standards on cybersecurity are the major mechanisms that regulators can deploy to address these issues. Secondly, our international case studies of countries and cities provide an overview of a variety of actual regulatory initiatives in place. This overview shows how economies have pioneered a variety of regulatory approaches that tend to be more participatory to better respond to the more dynamic stakeholder landscape that is emerging.

1 Introduction

Smart grids are a key to both demand-side (e.g. energy saving and energy efficiency) and supply-side (e.g. renewable energy) management of energy systems. Many countries have elevated smart grid deployment to the status of major

D. Mah (✉) · K. P. Leung
Department of Geography, Hong Kong Baptist University, Kowloon, Hong Kong
e-mail: daphnemah@hkbu.edu.hk

D. Mah · P. Hills
The Kadoorie Institute, The University of Hong Kong, Pokfulam, Hong Kong

D. Mah et al. (eds.), *Smart Grid Applications and Developments*,
Green Energy and Technology, DOI: 10.1007/978-1-4471-6281-0_7,

national economic and energy strategies. The Obama administration has positioned smart grid as a key component of the new energy model for America and announced a national policy framework for smart grids in 2011 [4, 31]. South Korea regards smart grid as a new growth engine for the country [69]. In China, smart grids represent a major component of the current national energy plan [75].

The potential benefits of smart grids could be substantial. For example, it has been estimated that an investment of US$338 billion to US$476 billion for a fully functional smart grid could result in benefits up to US$2 trillion in the USA [28]. Another study has estimated that €67 billion for building and running peak infrastructure could be avoided in the EU if dynamic pricing can be adopted [33]. The nation-wide smart grid demonstration project, *Smart Grid, Smart City*, in Australia is estimated to generate AU$3.4 billion of direct financial savings, and a total of AU$5 billion with reliability and environmental savings included [21].

However, to fully utilise the potential benefits that smart grids may offer, many technological, operational, economical, institutional, and policy challenges have to be overcome [108]. Among them, regulatory challenges are particularly significant because the design and operation of smart grids are fundamentally different from traditional power grids. Traditional systems are predominantly centralised, fossil fuel based with the presence of monopoly market conditions. Important features of smart grids—including the existence of a more decentralised power system, the emergence of new market actors (such as independent power producers), more dynamic two-way utility–consumer relationships and the use of massive amounts of energy usage data [20, 82, 97]—present to regulators new challenges that may include utility disincentives, monopoly power, information asymmetry, consumer inertia, and breach of personal privacy.

To cope with these challenges, new regulatory initiatives in relation to smart grid technologies are increasingly being developed and implemented in many economies. Mandatory hourly pricing for large customers in New York [87], electricity price control regulation and efficiency grants in the UK [5, 56], mandatory smart meter roll-out and time-of-use pricing in Italy [66], and the smart metering privacy rules in the Netherlands and California [8, 9, 18] are some examples.

This chapter aims to identify and examine the key issues that may confront regulators in relation to smart grid deployment, and how these challenges can be addressed. We attempt to answer these specific questions: Why do regulators have to be concerned about smart grid deployment? What are the gaps between existing regulatory practices and the new regulatory requirements for smart grid deployment? How and to what extent can regulators address these gaps? What approaches can regulators deploy? Are there any examples of good practice that we can discern from international experience?

Our analysis is based on a review of published work from academic sources, government documents, and reports. This chapter is organised into five sections. Following the introduction, we examine why regulators should be concerned about smart grid deployment. We then identify and discuss three major regulatory issues that confront regulators and the approaches that can be deployed to address these issues. This is followed by an overview of international cases of countries and cities that have pioneered a broad range of regulatory measures to overcome these challenges. Three specific cases are highlighted to illustrate the features, outcomes, and keys to success of these initiatives. We conclude by highlighting the policy implications.

2 Why Should Regulators Be Concerned About Smart Grid Deployment?

Electricity markets need regulators to oversee the effective functioning of the electricity sector through rule-setting, monitoring, and enforcement [10, 11, 22, 106]. Specifically, regulators have important roles to play in two main areas: market structure and conduct. In terms of market structure, regulators can determine ownership, access to the market (who can enter and who should be restricted from entering), and contractual relationships, market planning as well as the mechanisms of allocation [1]. In terms of conduct, regulators are concerned about the production of electricity. Regulation may influence the fuel mix, production technologies, the environmental impacts of electricity generation, the security of supply, and tariffs [1]. Examples of electricity market regulators include the Office of Gas and Electricity Markets (Ofgem) in the UK, the Federal Energy Regulatory Commission (FERC) in the USA, and the State Electricity Regulatory Commission (SERC) in China [84].

In recent decades, two trends have reshaped the roles of regulators in the electricity sector. The first is the global trend towards electricity market liberalisation. Driven by the aspiration for efficiency, many developed and developing economies have since the mid-1980s introduced electricity market liberation in various forms. These measures include privatisation, the establishment of sector regulators, the introduction of competition into generation, and the unbundling of generation, transmission, distribution, and retail activities [56]. Regulatory functions since then have progressively evolved beyond economic efficiency considerations to encompass the development of market codes and standards, monitoring market behaviour related to the abuse of market power and information asymmetries, and the facilitation of dispute resolution [11, 27, 56].

Another trend is related to the context of rising public awareness on climate change impacts. This trend is noticeable particularly since the early 2000s. Regulations seem to be increasingly required to correct market failures in the energy sector, most notably externalities of emissions [2, 100]. Regulators are therefore expected to place more emphasis on the environmental performance of the regulated utilities as well as their economic performance.

It is in these evolving contexts that smart grid deployment presents new challenges by introducing two major additional changes in electricity systems: changes in hardware and software of power grids, and changes in the stakeholder landscape.

In terms of hardware changes, centralised grid systems need to be upgraded to accommodate the integration of more decentralised electricity generation systems with intermittent renewables. Devices such as smart meters, which are capital intensive with high investment uncertainty, are necessary to support the data exchange between suppliers and end users [51]. In the USA, for example, about US$338 to US$476 billion of investment is required for full smart grid implementation over the next 20 years [28]. Smart grid investment in Europe, China, and South Korea is expected to reach €56.5 billion, €71billion, and €16.8 billion, respectively, in the next decades [58]. Regulators therefore have an important role to play in providing adequate incentives for smart grid investment and ensuring the benefits of smart grid are adequately accounted for in economic terms [30, 98].

Apart from hardware, smart grids require new software—which include dynamic pricing systems to incentivise consumers' participation, more open markets to allow new market players and competition, and a new, efficient and reliable data management system that can manage the massive sets of energy usage data [50, 70, 97]. All these changes require regulators to set up new market rules and protocols to ensure resilience, fairness, security, and effectiveness of modernised grids [23].

Another major change that can be brought about by smart grid deployment is a change in the stakeholder landscape. Unlike the conventional top-down linear systems in which established utilities possess dominating roles and consumers are passive, smart grids provide opportunities to electricity consumers, new market players (such as independent power producers) as well as established utilities to take up new roles in energy management systems.

Smart grids allow electricity consumers to take a proactive role in managing their electricity use. Consumers can be better informed and price responsive, and can contribute to energy saving, energy efficiency, and peak load shifts through responding to real-time electricity information linked with dynamic pricing systems [68]. Consumers can also assume the role of "prosumers": consumers can produce electricity and sell it to utilities through decentralised generation technologies such as roof-top solar panels [44, 86]. Furthermore, consumers can become proactive in choosing their electricity suppliers or which kind of electricity products (such as green electricity) they wish to consume because suppliers and utilities are under pressure to develop a larger variety of products and services. For example, the electricity market reforms in Texas since 2002 allow residential consumers to choose from about 200 retail offers provided by about 40 suppliers, and as a result about 40 % of the consumers have switched away from their original providers [62].

Apart from consumers, new players can enter the market not only in traditional fields such as power generation (e.g. as independent power producers), but also in new energy service and product areas such as energy audit and green power marketing [70]. Existing power utilities may also take up new roles in smart grids. Because of the perceived increase in market competition, these established utilities are under greater pressure to diversify their business by expanding services in such areas as renewable energy and energy efficiency.

Smart grid deployment therefore gives rise to new two-way utility–consumer relationships, the co-existence of existing players and new market entrants, and a redistribution of benefits and costs among these diverse stakeholders in more decentralised energy systems. Such changes need to be accompanied by regulatory oversight in areas that range from consumer protection to grid access and to market transparency. Because of the dynamic and complex nature of the new stakeholder landscape, smart grids also require a more participatory approach to energy decision-making in which new market players and consumers can assume a more proactive role.

These changes give rise to a number of important questions: What are the gaps between existing regulatory practices and the new regulatory requirements for smart grid deployment? What should regulators be concerned about? What can be done by regulators to address these issues? In the next section, we will discuss three major issues confronting regulators and the possible regulatory approaches that can be deployed to overcome these challenges. These issues are utility disincentives, pricing inefficiencies, and cybersecurity and privacy.

3 Major Regulatory Issues on Smart Grids: Utility Disincentives, Pricing Inefficiencies, and Cybersecurity and Privacy

3.1 Regulatory Issue 1: Utility Disincentives

Many current regulatory regimes do not adequately address the issues of lock-in situations, inertia, cost increases, and new risks that deter utilities from investing in smart grid technologies [3, 53, 72]. Many require a link between a utility's revenue and the sales of electricity or capital investments, and have tended to reinforce the lock-in effects of established energy technologies. Sunk investments tend to make rapid diffusion of smart grids difficult to achieve [3, 46]. These mechanisms have therefore created an incentive structure that encourages utilities to supply fossil fuel-based power, but discourage them from investing in distributed generation such as renewable energy and demand-side management projects [46].

Regulators can introduce changes in regulatory regimes to address these disincentives. Such regime changes can be made either through introducing radical electricity market reforms that overhaul the electricity sector or through more moderate approaches such as introducing some programme-based changes, for example decoupling mechanisms.

The important role that electricity market reforms or liberalisation can play to facilitate technology innovations including smart grids and renewable energy has been extensively documented (see, for example [48, 51, 62, 70]). Traditionally, generation, transmission, and distribution of electricity have been carried out by vertically integrated monopolies [12]. Market reforms can foster energy innovation through introducing structural changes that are associated with accessibility, market rules, and incentive systems [62]. Liberalised markets tend to improve market accessibility through lowering the barriers for new entrants [62]. Accessibility is conducive to market competition to a large extent because competition can drive innovation, and the consumer's right to choose between suppliers can pressure utility companies to develop new products and services. Accessibility can therefore reduce lock-in effects. Liberalised markets also tend to provide fair market rules and thus a level playing field for both existing and new market players [62]. This is another way to induce competition-driven innovation [62]. In addition, new incentive mechanisms, most notably *decoupling*, can be established to ensure the extra costs incurred from smart grid deployment can be covered. This can stimulate risk-adverse utilities to invest in asset innovation for smart grids rather than expanding existing assets [62, 72, 73].

It is important to note that electricity market reforms have been introduced in different forms that vary in scope and depth across countries [27]. Some countries such as the UK and some states in the USA have introduced competition into all the generation, transmission, distribution, and retail sectors of the power industry that was once vertically integrated and nationalised (HK EAC, 2003). On the other hand, countries such as China, Japan, and South Korea have introduced partial liberalisation in which competition has been introduced but this has mainly limited to the generation sector, and some state monopolies and distorted pricing systems have remained [69, 71, 107].

As noted earlier, moderate regime changes rather than a complete overhaul of electricity market structures can be considered as a pragmatic approach to facilitating smart grid deployment, and it is particularly so in economies where electricity market reforms are partially implemented. Decoupling mechanisms are one of the most widely discussed approaches that regulators can deploy as a first step in major regime changes.

Decoupling is a regulatory approach that breaks or decouples the linkages between electricity sales from revenues [46]. A regulator can introduce two major forms of decoupling: cost-based and incentive-based. Cost-based decoupling mechanisms such as direct-cost recovery and fixed-cost recovery are regulator-approved mechanisms for recovering costs that usually include administrative costs and transmission costs. These mechanisms can take the form of rate cases, tariff surcharges, revenue caps, and price caps.

Incentive-based mechanisms such as performance incentive programmes, on the other hand, provide financial rewards for utility companies according to their achievements in energy efficiency programmes. For example, utility companies may receive a percentage of the achieved savings, an adjusted rate of return for achieving savings targets, or a penalty for failing to meet energy saving goals [54]. Table 1 highlights the features of the major types of decoupling mechanisms. Table 2 provides some details of an example of performance incentive programme in California.

Table 1 Major types of decoupling mechanisms

	Features	Examples
Direct-cost recovery	Recover administrative costs of energy efficiency programmes through rate cases, system benefits charges, and tariff rider or surcharges	Bill surcharge to recover conservation and DSM programme costs for utilities in Florida [74]
Fixed-cost recovery	Recover revenue loss from fixed operating cost via lost revenue adjustment and revenue decoupling mechanisms	A lost revenue recovery mechanism in Montana allows utilities to factor their revenue loss from DSM into monthly rates
Performance incentive programmes	Reward utilities for savings achieved by energy efficiency programmes through grants and adjusted rate of return incentives	The shared savings mechanism in California has four tiers of earning rate according to the level of savings achieved by energy efficiency programmes (see Table 2 for more details)

Source Institute for Electric Efficiency [54]

Table 2 Earning rate for energy efficiency risk-reward incentive mechanism in California

Savings goal achievement	>100 %	85–100 %	65–85 %	<65 %
Earning rate	12 %	9 %	0 %	Penalty

Source Institute for Electric Efficiency [54] and NAPEE [74]

In consideration of the potential benefits that decoupling mechanisms can offer, it is important to note that these mechanisms have not been widely implemented worldwide. About 70 % of IEA member countries have not implemented any decoupling regulations for energy utilities [52]. Some pioneer measures were implemented by some countries such as the USA, the UK, and Germany. As of July 2012, 46 out of 52 states in the USA have introduced direct-cost recovery systems, 27 states with fixed-cost recovery, and 23 states offering performance incentives, in which all measures were found to be contributing to decoupling [54].

3.2 Regulatory Issue 2: Pricing Inefficiencies

Electricity consumers and demand-side energy management are core elements of smart grid deployment. A major challenge for regulators is therefore to create a favourable environment that can incentivise consumer participation. Another major regulatory issues associated with smart grid deployment is electricity pricing. This has direct impacts on electricity consumption patterns of individuals, including the total amount of consumption, when they consume, and what they consume [102].

In general, two aspects of inefficient pricing have created barriers to energy innovations including smart grids, and these need to be more effectively regulated. The first is that external costs of emissions are often not fully reflected in electricity prices. A European study has estimated that if the environmental costs of fossil fuels are accounted for, the external cost of fossil fuel-based electricity generation could be nearly 27 times higher than that of renewables, with 5.8 c€/kWh for brown coal, 1.5 c€/kWh for natural gas, 0.09–0.12 c€/kWh for wind, and 0.21–0.41 c€/kWh for solar PV [25].

Secondly, inefficiency in tariff systems may undervalue the benefits of demand management practices and renewable energy in many ways. For instance, flat rates generally assume consumers are highly price inelastic and provide no incentives to peak-shaving consumption behaviour [36, 102]. Regressive rate systems tend to discourage customers from saving energy because the more energy they consume, the lower the energy rate per unit is [32]. Another example is that except in places where net-metering is in place [26], households have no economic incentives to produce renewable electricity and sell it to their suppliers even though such a community-based option is technologically available.

Regulators can deploy a range of measures to overcome these pricing issues. These include pricing emissions (e.g. congestion charges and carbon taxes), reducing subsidies for fossil fuels, and net-metering with feed-in tariffs [64, 100]. In particular, regulators can introduce changes in tariff systems, most notably through dynamic pricing, as an effective means to incentivise consumer engagement in the areas of peak shaving, energy saving, and energy efficiency.

In contrast to flat rate systems, dynamic pricing is the charging of different electricity rates at different times of the day and year to reflect the time-dependent cost of supplying electricity [35]. Dynamic pricing differentiates energy prices between peak and non-peak hours and therefore provides financial incentives for customers to shift their consumption pattern and to conserve energy [10, 69]. Dynamic pricing can also facilitate a better match between marginal costs and marginal demand and hence improve the efficiency of electricity systems [102]. There are a wide range of dynamic pricing modes and they include, ranging from the least to most varying, block rates, seasonal rates, time-of-use rates, super peak time-of-use rates, critical-peak pricing, variable peak pricing, and real-time pricing [34]. The most widely discussed and adopted schemes are time-of-use rate (TOU), real-time pricing (RTP), and critical-peak pricing (CPP) [49, 89].

Time-of-use rate can be established at least a day before and at even longer time interval, charging higher rates for peak-hour use. In contrast, with real-time pricing, consumers are provided with simultaneous pricing which is established based on the demand and supply balance at that moment, similar in some respects to those pricing systems in stock markets. A more sophisticated variation of real-time pricing is two-part real-time pricing, in which real-time pricing only applies to usage that deviates from a baseline level and therefore provides consumer protection against price volatility. Critical-peak pricing is a combination of time-of-use rates and real-time pricing in which a base rate is applied unless certain load circumstance occurs to trigger peak load pricing. Peak time rebate (PTR) is a variation of critical-peak pricing that compensates peak-shaving consumers rather than penalising peak-riding consumers [34].

In practice, dynamic pricing systems have recorded mixed results. While some dynamic pricing pilots produced minimal responses [69], some have positive outcomes. A study based on 18 dynamic pricing pilot programmes has found that time-of-use and critical-peak pricing could lead to 5 and 20–30 % peak load reductions, respectively [37]. A number of studies also show that the potential energy savings from consumer feedback triggered by dynamic pricing or other demand response measures ranges from 5 to 15 % [19, 38, 83]. Another study has found that the dynamic pricing pilot programme in Ontario led to 11–25 % peak reduction, 6 % consumption reduction, and CAN$4.17 savings in monthly electricity bills on average [47].

However, introducing dynamic pricing for smart grid deployment is often politically sensitive as consumers are themselves highly sensitive to tariff changes [68, 69]. In the USA, for example, although dynamic pricing has been contemplated for decades, there are only a few established systems. In many cases, regulators have opted for voluntary programmes initiated by specific target groups who are more receptive to such changes. One example of such a voluntary programme on dynamic pricing is in Tennessee. In 2011, the Tennessee Valley Authority approved a tariff structure to include options of time-of-use rate for customers to encourage energy efficiency and peak demand reductions [26]. Other states such as California have been making progress on enforcing dynamic pricing. The California Public Utilities Commission (CPUC) requires Pacific Gas and Electric Company (PG&E) to replace flat rate with mandatory peak day pricing for large consumers by May 2010 and mandatory time-of-use tariff for small and medium business and agricultural consumers by March 2012 [13, 17]. CPUC also approved optional TOU and CPP for residential consumers and mandatory TOU and default CPP for small commercial consumers by 2013 and 2014 for San Diego Gas & Electric [16].

3.3 Regulatory Issue 3: Cybersecurity and Privacy

A defining feature of smart grids is the extensive use of information and communication technology in modernising power systems. In smart grid systems, massive sets of energy consumption data are generated, collected, aggregated, and utilised for various purposes including billing, measuring power quality, updating instant electricity prices as well as providing real-time feedback on household energy consumption to incentivise energy efficient behaviour [40, 83]. These data are provided by smart meters and control devices [83].

The transformation to automatic connection and control systems exposes the grid to three types of risks: grid operation failures, data breaches, and cybercrimes. Disruption of grid services can be caused by intrusions to the two-way communication system of smart grids, which may lead to temporary power outages or blackouts. Consumer privacy is another risk because massive sets of personal information and energy usage data are exchanged through smart meters and networks. These data may expose personal details of consumer's activities and occupancy patterns [97]. Unauthorized disclosure of this information resulting from for example cyberattacks may give rise to home security issues [19, 83, 97]. In addition, automatic metering provides a new gateway for cybercrimes. Hackers could intrude into smart meter systems and alter metering data to reduce energy use resulting in electricity theft [99]. There are also concerns about loss of control over national grids under terrorist attack [29, 39, 83]. All these issues can impose profound impacts on national security, public safety, privacy protection, and cybersecurity.

It is important to note that cybersecurity and privacy is particularly a challenging issue to regulators because data management systems involve various parties including electricity producers, distributors, smart technology developers, and energy efficiency service providers and consumers. It is difficult for regulators to ensure sufficient investment in security be incentivised where many actors have a collective stake but diffused and unclearly defined responsibility in grid reliability [83, 99].

Most of the existing safety standards have a narrow focus on the physical component of the grids [15]. Regulators are therefore required to enhance standards in the digital aspects of the grids. Regulators need to strengthen the regulatory framework in relation to the ownership and sale of information [10]. Specifically, regulators need to ensure the integrity of the data management system, preventing cyberattacks, incentivise investment in cybersecurity as well as to enhance consumers protection [83].

There are three measures that a regulator can deploy. These are privacy regulation, facilitating public–private collaboration, and providing financial incentives for cybersecurity innovation. Privacy regulations could be enforced by clearly defining the roles and responsibilities of each party involved in data processing and management [97]. Privacy regulations could prevent misuse of data. More importantly, they can increase consumers' confidence and acceptance of smart grid deployment and this is especially important in a context in which public distrust of governments and utilities is not uncommon in many countries and cities [68, 69]. For instance, California enforces energy data privacy rules as state legislation, and is the pioneer in energy data privacy regulations and has inspired other states to join the path [15].

Public–private collaboration can also help to ensure that policy measures evolve along with the changing risks as well as the needs of diverse stakeholders. Workshops and working group meetings are crucial to address new challenges and solutions. For example, since 2006, the USA Department of Energy has been working with different energy sector stakeholders on developing grid cybersecurity and has implemented 65 initiatives to develop cybersecurity solutions [79]. The private–public collaboration taskforce, the Energy Sector Control Systems Working Group, collected inputs from various stakeholders such as government leaders, asset owners and operators, chief information officers (CIOs), research and

Table 3 Summary of regulatory issues, approaches, and measures for smart grid development

Regulatory Issues	Approaches	Examples of Measures
Utility disincentives	Decoupling	Direct-cost recovery
		Fixed-cost recovery
		Performance incentives
		Time-of-use
Pricing inefficiency	Dynamic pricing	Real-time pricing
		Critical-peak pricing
		Privacy regulation
Cybersecurity and privacy	New protocols and standards	Public–private collaboration
		Financial incentives for cybersecurity innovation

technical experts, security specialists, and vendors and generated two roadmaps to address these issues [29]. Another example of stakeholder collaboration is the Cyber Security Working Group (CSWG) under the USA National Institute of Standards and Technology (NIST). The group consists of more than 650 stakeholders worldwide including utilities, vendors, academia, regulators, and government [77]. This group aims to develop overall cybersecurity strategies for smart grids, and sub-groups have been set up to focus on different topics such as advanced metering infrastructure, privacy, and design principles [77].

Financial incentives for cybersecurity investment are crucial to motivate both the utility companies and technology developers to move into this new area. For instance, the USA Recovery Act delegated US$3.4 billion to 99 projects under the Smart Grid Investment Grant Programme (SGIG), in which awardees are required to prepare and implement cybersecurity plans specific to their projects [29]. The programme also fosters a peer sharing of experience in implementing cybersecurity plans, persistent security risks, and information gaps through a two-day workshop [24].

To sum up, utility disincentives, pricing inefficiencies, and cybersecurity and privacy are the three major regulatory issues that regulators need to effectively address to facilitate smart grid deployment. The lock-in effect and the coupling between electricity sales and utility revenue have created utility disincentives to smart grid investment. Regulators may introduce market liberalisation and decoupling mechanisms to facilitate smart grid deployment through market accessibility, fair market rules, and incentive systems. To overcome the issues associated with pricing inefficiency, regulators may introduce tariff reforms, most notably through different forms of dynamic pricing with a phase-in approach moving from voluntary to mandatory. Regulators may address cybersecurity and privacy issues through introducing regulations on privacy, enabling public–private collaboration as well as providing financial incentives for cybersecurity investment. Table 3 below provides an overview of the major regulatory issues, approaches and examples of specific regulatory measures.

4 Smart Grid Regulatory Approaches: An Overview of International Experience

We now provide an overview of the diversity of regulatory approaches that have been deployed in various places, examine the achievements of these approaches, and then highlight the unresolved issues that confront regulators.

4.1 A Variety of Regulatory Approaches and the Achievements

Our case examples summarised in Table 4 show that the USA and a number of European countries, including the UK, Germany, Italy, France, and the Netherlands, have introduced a wide range of approaches to address the three major regulatory issues. Decoupling, dynamic pricing, and cybersecurity and privacy protocols and standards are the commonly adopted approaches.

Two observations can be drawn from these case examples. Firstly, countries and cities have adopted a diversity of regulatory measures within each approach. The case examples show that the UK, California, and Germany have adopted different forms of decoupling mechanisms (e.g. price caps and revenue caps). Similarly, our case examples that deployed dynamic pricing measures differed in the forms of pricing systems (e.g. time-of-use and critical-peak pricing) as well as the level of obligation (i.e. voluntary or mandatory basis). In relation to cybersecurity and privacy issues, privacy regulations in these case examples also differ with some focusing on data disclosure and some on data aggregation.

Secondly, the implementation of these regulatory measures involved various stakeholders. Regulators worked closely with utilities and consulted other stakeholders on decoupling to develop effective incentive mechanisms through rate case review, energy savings goal evaluation, and project grants assessment. Dynamic pricing measures also involved consumer engagement through outreach and education about new energy pricing systems and to ensure public acceptance. Similarly, some cybersecurity measures were based on private–public or stakeholder collaboration to develop strategies and plans.

Although a comprehensive analysis of these regulatory approaches is yet to be available, these initiatives have recorded some achievements of facilitating smart grid development. Decoupling, for instance, as a major regulatory measure to address the issue of utility disincentives provides new incentives for some utilities to divert investment into new areas. One of the achievements of decoupling is a reduction in network and electricity costs through efficiency gains. The electricity price cap (RPI-X) in the UK, for instance, has effectively promoted efficiency and led to a 7.7 % decrease in operational costs for distribution network operators as well as 50 and 41 % reduction in electricity distribution and transmission charges for consumers, respectively, over a period of 11 years [5]. Decoupling measures

Table 4 A summary of international experience on the three regulatory approaches

Approaches	Place	Since	Measures	Description	Accomplishment	Key to Success	Challenges	Sources
Decoupling	U.K.	2005	Price cap, cost recovery, performance-based	Pricing regulation (RPI-X) capped price based on 5-year cost forecast and benchmarking, adjusted by service quality and efficiency incentives. Packaged with project grants and cost recovery (i.e. IFI and LCN)	Price reduction Grid stability Encourage investment	Transparent process Legislative clarity Unbundling network Data availability	Large profits for distribution system operators at early stage Limited emphasis on emission	[5, 56, 73]
	California, U.S.	1982	Revenue cap, performance incentives, energy saving goals	Promote energy efficiency via revenue cap with rate adjustment, required energy savings and renewable targets, introduce market competition, etc.	Reduce energy use Decoupling	Political commitment Clear targets and priority Public interest	Reform with the least cost acquisition	[91, 94, 105]
	Germany	2009	Revenue cap	Incentive Regulation Ordinance (ARegV) puts a cap on utilities' revenue to encourage network efficiency and cost reduction	Incentives for network efficiency	Political commitment	CAPEX time shift Encourage replacement more than innovation Supply quality issues	[30, 45, 67, 101]
	Italy	2006	Mandatory smart meters roll-out	Mandatory smart meter roll-out with specific incentives, tariff, and minimum functionality requirements. Promote smart meters for measuring service quality	Baseline measures Output-based High penetration	Mandatory obligations Functionality requirements Financial incentives & penalties	Ensure interoperability of meters	[66, 72, 73, 104]

(continued)

Table 4 (continued)

Approaches	Place	Since	Measures	Description	Accomplishment	Key to Success	Challenges	Sources
Dynamic pricing	New York, U.S.	2006	Mandatory RTP	Mandatory Hourly Pricing for large electricity consumers	Business case for mandatory RTP	Liberalized market Consultation Phase-in	Insufficient price signals	[59–61, 63, 80, 87, 90]
	Ontario, Canada	2006–2007	TOU, CPP and CPR pilot	Tested consumers' response to TOU, TOU-CPP, and TOU-CPR for 1.5 years with Hydro Ottawa	Peak shaving Reduce energy use Bill reduction Consumer acceptance	Customer education Automated peak notifications	Needs of data infrastructure upgrades for large-scale roll-out Lack of impacts during winter critical peaks	[47, 81]
	Italy	2008	Mandatory TOU	Mandatory TOU for residential and small commercial users who have not chosen their suppliers in open markets	Price signals Peak shaving Bill reduction	Increase in use of highly efficient appliances Transition	Insufficient price signals Limited room for peak load shifting	[43, 66, 103]
	France	1995	Optional CPP tariff	A utility company offers CPP tariff with peak, off-peak prices and three types of days	Price reduction Over 400,000 customers joined	Pilot and evaluation Customer response assessment	Competitive disadvantage in open market	[41, 65]

(continued)

Table 4 (continued)

Approaches	Place	Since	Measures	Description	Accomplishment	Key to Success	Challenges	Sources
Protocols and standards on cyber-security and privacy	California, U.S.	2009	Regulation—data non-disclosure	Mandate utilities to address cybersecurity measures in their Smart Grid Deployment Plan (SGDP). Regulation against energy data disclosure to third party without prior consent	Regularly review of SGDP Internal cybersecurity plans	Political commitment	Evolving nature of risks Combine compliance-base with dynamic risk management Define acceptable for risk	[8, 9, 15]
	Netherlands	2011	Regulation—privacy rules	Define terms and conditions of data management by various parties. Provide various smart meter opt-out options for consumers. Require utilities to address data processing in annual reports. Combine billing to minimize data processing and transfers	Simplified data management Privacy rules for smart meter roll-out	Bill amendment Privacy impact assessment Advice from data protection authority	Lack of pre-privacy impact assessment Trade-off between privacy rules and economies of scale	[18]
	U.S.	2006	Public-private collaboration	Establish a working group, consists of various public and private sectors, to collect input from cross-section stakeholders and develop roadmap	Identify barriers, strategies, and timeline to achieve cybersecurity	Clear policy vision Inputs from stakeholders Address constantly changing risks with regular updates	Lack of mutual driving force for implementation among various stakeholders	[29, 79]
	U.S.	2010	Financial incentives	A grant program provided funding for smart grid projects with the requirement of preparing and implementing cybersecurity plans (CSP) tailored to the projects	Pilot implementation and review of CSP Peer sharing of lesson learned, risks, and improvement	Funding resources Mandatory CSP Progress evaluation and experience sharing	Only offer one-time incentives for cybersecurity measures	[24]

also effectively altered utilities' investment strategies on energy efficiency. One example is that the gas conservation programme by Pacific Gas & Electric reduced US$46 million of revenue from natural gas sales, which would not have been implemented without a mandatory decoupling measure [91].

Dynamic pricing, on the other hand, as a key regulatory approach to engaging consumers through the use of pricing signals, has recorded some successes in terms of peak load shifts and consumption reductions. A review on dynamic pricing pilot programmes found that time-of-use and critical-peak pricing with automated responding technology installed could possibly reduce peak loads by 5 and 30 %, respectively [76]. The pilot programme in Ontario also demonstrated 6 % reduction in energy consumption and CAN$4.17 savings in monthly electricity cost [47].

Cybersecurity rules have been regarded as an important component of smart grid deployment plans in many economics including the USA, EU, South Korea, and China. However, the assessment of what privacy rules have accomplished is not readily available due to the mostly preventative nature of the regulations [57].

Table 4 below provides an overview of international experience of implementing the three regulatory approaches, followed by three case studies of each approach to elaborate the key features, outcomes, and lessons learned (Boxes 1–3).

Box 1 Case—Decoupling Electricity in California

What happened

Triggered by the first energy crisis, decoupling mechanism was adopted in California from 1982 to late 1990s, with high tail block rates, revenue adjustment, and performance-based incentives [105]. After 1996, the regulatory strategy was shifted to relying on market force and competition for energy efficiency by restructuring the market [105]. During the 2001 energy crisis, the state re-introduced revenue decoupling mechanism, with an explicit political priority on energy efficiency and energy saving goals specific to each utility [14, 94].

Outcomes

Decoupling measures facilitated intensive development of energy efficiency programmes, which led to about US$1.5 billion net benefits from pre-1998 energy savings [105]. The energy saving goals also appeared to succeed in promoting energy efficiency—all three utilities exceeded their saving goals in electricity use by 27–30 % and demand by 21–31 % [14].

Lessons learned

California's experience shows that decoupling requires a mix of decoupling mechanisms and compliance regulation, such as revenue adjustment and energy saving goals, to provide incentives for energy efficiency on both demand and supply side. Clear political commitment and priorities, explicit targets as well as tailor-made energy saving goals specific to each utility, are the keys to succeed of these decoupling mechanisms.

Box 2 Case—Dynamic Pricing Pilot in Ontario

What happened

The Ontario Energy Board Smart Price Pilot (OSPP) started in August 2006 with 373 Hydro Ottawa consumers for one and a half years, aiming to explore implications of province-wide smart meter roll-out and mandatory time-of-use tariff [47]. This pilot programme tested three types of pricing—the time-of-use, time-of-use with critical-peak price, and time-of-use with critical-peak rebate, in which the two critical-peak time-of-use rates charge 3¢ more for on-peak use with an addition of CAD 30¢/kWh charge for peak-hour or rebate for avoided peak-hour use on critical-peak days [47].

Outcomes

Time-of-use with critical-peak price appears to have the most robust impacts on peak shaving—about 11 % demand reduction on average and 25 % during critical-peak hours [47]. All three pricing types led to lower total energy use (6 % on average) [47]. Taking both load shifting and conservation effects into account, over 93 % of OSPP participants paid CAD$4.17 less on their electricity bills per month—35 % savings from load shifting and 65 % from conservation [47]. Most of the participants (78 %) were satisfied with the TOU pricing [47].

Lessons learned

OSPP provides insights into effectiveness of different time-of-use pricing options and programme implementation approach. A comprehensive set of consumer engagement tools, such as automated peak notification, and education package and flyers, is essential to the responsive consumer behaviour to dynamic pricing. Dynamic pricing with penalty on on-peak use appears to be more effective in peak shaving and conservation [36, 76].

Box 3 Case—Cybersecurity and Data Privacy in California

What happened

Several legislative bills, namely SB 17, SB 1476, and SB 674, contribute to the cybersecurity regulation in California. Utilities are required by California Public Utility Commission to address cybersecurity in their smart grid deployment plans (SGDPs) and are prohibited to disclose energy data without prior consent from customers [7–9].

Outcomes

These regulations facilitate over US$19 million utility investment in cybersecurity [85, 93, 96]. Measures are implemented in the following approaches [55, 85, 92, 95]:

- Network upgrade: threat-detective and threat-adaptive control systems, pilot projects
- Private–public collaboration: professional societies and working groups
- Standard and policy: network standards (NIST, NERC), internal security governance policy
- Privacy frameworks: privacy protocols (i.e. fair information practice principles and privacy by design), privacy impact assessments
- Risk management: cyberincident response team and plan
- Internal measures: awareness training, Chief Customer Privacy Officer and team

Lessons learned

California establishes a pioneer case of how regulatory changes can facilitate security and privacy protection measures by utilities via compliance-based reporting and privacy rules. The annual updates of SGDPs provide insights into regulatory improvement and can better respond to the evolving nature of cyber-risks. Defining roles and authorisation limitation of utilities in data disclosure enhance consumers' privacy protection. However, there is a lack of local information sharing, cyberincident reporting, and independent auditing of implemented measures.

4.2 Unresolved Issues

Although these regulatory approaches have recorded some achievements, there are still a number of unresolved issues remaining. The major challenge of decoupling is to optimise the regulatory regime with the least cost [91]. It is a particular area of concern because financial incentives such as grants that are required for implementing decoupling often involve a large amount of public expenditure. For instance, California has received US$314.5 million from the American Recovery and Reinvestment Act (ARRA) for efficiency grants, state energy programmes on training and financing, and energy assurance planning [6]. Sufficient funding is essential to decouple with financing incentives, and the challenge is therefore to ensure active involvement of the government or the development of viable business models for financing. Another unresolved issue of decoupling is concerned with the trade-offs between the implementation of unbundling to guarantee fair access conditions and the additional burden put on small new entrants in terms of costs and complexity of system integration [12].

In relation to dynamic pricing, the effectiveness of this regulatory measure is closely tied to the demand responses of consumers. However, if prices are excessively volatile or manipulated by utilities through exerting market power, residential consumers may become particularly vulnerable [10]. How to ensure consumers will benefit from dynamic pricing would be an area of major concern to regulators. Furthermore, the political and economic feasibility of introducing dynamic pricing may be highly uncertain and vary across different regulatory regimes as showed in some studies [69]. Consumer response to price signals is often deterred by the lack of awareness, knowledge, and ability to respond. For instance, in the case of New York, minimal differences between market prices and dynamic prices do not pay for customer's cost of responding to price signals, such as time spent on monitoring hourly prices and changes in consumption behaviour. Underutilisation of real-time pricing software and lack of knowledge of load reduction strategies also contribute to the insignificant peak-shaving effort of the programmes [59–61, 78, 80, 90].

In addressing the issues of cybersecurity and privacy, regulators are concerned with the evolving nature of cyber-risks, and the trade-off between acceptable risk and costs [15]. What complicates the matter is that utility companies are often not motivated to invest in cybersecurity, especially in relation to data privacy. One of the reasons is that even if cybersecurity is associated with high impacts, it has a low probability of occurrence but high implementation cost [15]. The limited market for smart grid security technology also leads to the lack of incentives for technology developers to provide responsive technical support and innovation while smart security technology is still in its early development stage.

5 Conclusions

To fully capitalise on the potential benefits of smart grid deployment, it requires not only technological advancement but also a good understanding of the regulatory barriers. This chapter has contributed to the literature on smart grid by bringing the regulator perspectives—an area that is under-explored [102]—into the field of smart grid studies. We examine why regulators should be concerned about smart grid developments, the nature of the regulatory challenges, and what they can do to address these challenges.

We have two major findings. Firstly, we demonstrate that smart grids present new challenges to regulators. We found that regulators have important roles to play in establishing rules and mechanisms to address these regulatory challenges and facilitate smart grid deployment. Specifically, regulators can regulate investment incentives, consumer incentives, and the use of data through introducing market liberalisation, dynamic pricing, and protocols and standards on cybersecurity and privacy issues.

Secondly, our international case studies of countries and cities provide an overview of a variety of actual regulatory initiatives in place. This overview showed how these economies pioneered a variety of regulatory approaches that include different forms of decoupling, dynamic pricing and ways to address cybersecurity and data privacy issues. It is important to note that while these regulatory initiatives have secured some achievements, the unresolved issues are substantial. The trade-offs between the potential benefits of market liberalisation and the new burden put on both existing and new market players, the difficulties in predicting consumer responses to dynamic pricing, and the evolving nature of cyber-risks are some examples of the major issues confronting regulators when attempting to regulate smart grid deployment.

Our findings have some important policy implications. Firstly, rather than relying on a single solution, in practice regulators need to use a combination of regulatory measures to enhance regulatory effectiveness. For instance, various means of decoupling measures including direct-cost recovery, fixed-cost recovery, and performance incentives programmes may reduce the financial barriers for utility companies to invest in energy efficiency efforts. But these incentives may still not be adequate. Mandatory requirements for energy efficiency such as energy efficiency resource standards (EERS)—which are energy efficiency targets for utility companies—could be the ultimate driver to enhance the utilities' willingness to promote energy efficiency [88, 94]. The combination of legal requirements and incentive mechanisms may enhance regulatory effectiveness.

Secondly, regulators need to adopt a more engaging approach to effectively involve stakeholders when formulating regulatory solutions. Smart grid deployment presents challenge to the traditional technocratic way of energy decision-making. For instance, in relation to cybersecurity, the nature of potential risks, forms of attacks, and potential consequences are evolving [15]. A static set of rules will not be able to respond to such challenges. Regulators therefore need to engage more widely to formulate more proactive and adaptive measures. Furthermore, as shown in the case study of Ontario's dynamic pricing scheme (Box 2), consumer engagement through effective automated peak notification and education tools is essential to the smooth implementation of dynamic pricing.

Thirdly, regulating smart grid deployment requires an intelligent match of the choices of regulatory approaches and regulatory capacities. Worldwide, countries and cities adopt different regulatory approaches, a reflection of differing institutional endowments and regulatory capacities [42, 56]. It is a challenge to a regulator to learn how to make use of possible regulatory tools to effectively respond to the local context, and the opportunities and constraints for introducing regulatory changes.

Our analysis has however some limitations. Our overview of regulatory approaches in practice (Table 4) does not constitute a representative sample of the regulatory initiatives in this field. It is at best a brief analysis. Future research may generate fruitful findings through a thorough assessment of the extent to which regulatory measures have been implemented in the countries of reference. Furthermore, regulatory alignment between the regulatory tools and a regulator's abilities is a major research area that is under-explored [42]. A comparative study from this perspective may generate useful findings that can contribute to the literature on regulatory governance.

Acknowledgments We gratefully acknowledge the funding of our research by the Hong Kong Baptist University (through the Faculty Research Grants from the Faculty of Social Sciences,

FRG2/12-13/057 and FRG1/13-14/051) and the University of Hong Kong (through the Initiative on Clean Energy and Environment).

References

1. Arentsen M, Künneke R (1996) Economic organization and liberalization of the electricity industry. Energy Policy 24(6):541–552
2. Ashford N, Ayers C, Stone R (1985) Using regulation to change the market for innovation. Havard Environ Law Rev 9(2):419–466
3. Belyaev LS (2011) Electricity market reforms: economics and policy challenges. Springer, New York
4. Blarke MB, Jenkins BM (2013) SuperGrid or SmartGrid: competing strategies for large-scale integration of intermittent renewables? Energy Policy 58:381–390. doi:10.1016/j.enpol.2013.03.039
5. Buchanan A (2008) Is RPI-X still fit for purpose after 20 years? Office of Gas & Electricity Markets, London. http://www.ofgem.gov.uk/Networks/rpix20/Factsheets/Documents1/AB%20Oct%20speech.pdf. Accessed 13 May 2013
6. California Energy Commission (2010) ARRA Funded Energy Programs: investing in California's Energy Future. California Energy Commission, Sacramento. http://www.energy.ca.gov/2010publications/CEC-180-2010-003/CEC-180-2010-003.PDF. Accessed May 15, 2013
7. California State Legislature (2009) Electricity: smart grid systems. SB 17 (2009–2010 edn, vol SB 17). California State Legislature, USA
8. California State Legislature (2010) Public utilities: customer privacy: advanced metering infrastructure. *SB 1476* (2009–2010 edn, vol SB 1476). California State Legislature, USA
9. California State Legislature (2011) Telecommunications: master-metering: data security. SB 674 (2011–2012 edn). California State Legislature, USA
10. Clastres C (2011) Smart grids: another step towards competition, energy security and climate change objectives. Energy Policy 39(9):5399–5408
11. Córdova-Novion C, Hanlon D (2003) Regulatory governance: improving the institutional basis for sectoral regulators. OECD J Budgeting 2(3):16. doi:10.1787/budget-v2-art16-en
12. Cossent R, Gómez T, Frías P (2009) Towards a future with large penetration of distributed generation: is the current regulation of electricity distribution ready? Regulatory recommendations under a European perspective. Energy Policy 37(3):1145–1155. doi:10.1016/j.enpol.2008.11.011
13. CPUC (2011) Decision granting in part and denying in part petitions for modification of decision 10-02-032. Decision 11-11-008. California Public Utilities Commission, San Francisco
14. CPUC (2012) 2010–2011 energy efficiency annual progress evaluation report. In: C. P. U. Commission (ed) California Public Utilities Commission, San Francisco
15. CPUC (2012) Cybersecurity and the evolving role of state regulation: how it impacts the California Public Utilities Commission. California Public Utilities Commission, San Francisco. http://www.cpuc.ca.gov/NR/rdonlyres/D77BA276-E88A-4C82-AFD2-FC3D3C76A9FC/0/TheEvolvingRoleofStateRegulationinCybersecurity9252012FINAL.pdf. Accessed 4 July 2013
16. CPUC (2012) Decision adopting a dynamic pricing structure for residential and small commercial customers and denying the motion for approval of a settlement agreement. Decision 12-12-004. California Public Utilities Commission, San Francisco
17. CPUC (2012) Decision denying petition for modification of decision 10-02-032 and decision 11-11-008. Decision 12-08-005.California Public Utilities Commission, San Francisco
18. Cuijpers C, Koops B-J (2013) Smart metering and privacy in Europe: lessons from the Dutch case. In: Gutwirth S, Leenes R, de Hert P, Poullet Y (eds) European data protection: coming of age. Springer, Netherlands, pp 269–293

19. Darby S (2006) The effectiveness of feedback on energy consumption: a review for DEFRA of the literature on metering, billing and direct displays. Environmental Change Institute, University of Oxford, Oxford. http://www.eci.ox.ac.uk/research/energy/downloads/smart-metering-report.pdf. Accessed 13 June 2013
20. Devine-Wright P (2007) Energy citizenship: psychological aspects of evolution in sustainable energy technologies. In: Murphy J (ed) Governing technology for sustainability. Earthscan, London, pp 63–89
21. DEWHA, Australia (2009) Smart grid, smart city: a new direction for a new energy era. Department of the Environment, Water, Heritage and the Arts, Australia
22. Dixit S, Dubash N, Nakhooda S, Owen G (2008) Clean energy, good governance and regulation. In forum on clean energy, good governance and regulation, Singapore, March 16–18 2008: World Resources Institute, Prayas Energy Group, Renewable Energy & Energy Efficiency Partnership, Centre on Asia and Globalisation
23. DOE (2008) What the smart grid means to Americans. Department of Energy, Washington DC. http://energy.gov/sites/prod/files/oeprod/DocumentsandMedia/ConsumerAdvocates.pdf. Accessed 4 July 2013
24. DOE (2012) Smart Grid Investment Grant Program–Progress Report July 2011. In U. S. Department of Energy (Ed). Department of Energy, Washington, DC
25. Dones R, Heck T, Bauer C, Hirschberg S, Bickel P, Preiss P, et al (2005) Externalities of energy: extension of accounting framework and policy applications (ExterE-Pol). European Commission, France. http://www.externe.info/externe_2006/expolwp6.pdf. Accessed 5 June 2013
26. EIA (2011) Smart grid legislative and regulatory policies and case studies. U.S. Department of Energy, Washington, DC. http://www.eia.gov/analysis/studies/electricity/pdf/smartggrid.pdf. Accessed 4 July 2013
27. Energy Advisory Committee (2003) Electricity market review: role of regulator. Energy Advisory Committee, Hong Kong. http://www.enb.gov.hk/en/resources_publications/policy_consultation/files/Reference11.pdf. Accessed 28 May 2013
28. EPRI (2011) Estimating the costs and benefits of the smart grid: a preliminary estimate of the investment requirements and the resultant benefits of a fully functioning smart grid. Electric Power Research Institute, Palo Alto. http://www.epri.com/abstracts/Pages/ProductAbstract.aspx?ProductId=000000000001022519. Accessed 2 July 2013
29. ESCSWG (2011) Roadmap to achieve energy delivery systems cybersecurity. Energy Sector Control Systems Working Group. http://energy.gov/sites/prod/files/Energy%20Delivery%20Systems%20Cybersecurity%20Roadmap_finalweb.pdf. Accessed 7 June 2013
30. Eurelectric (2011) Regulation for smart grids. Eurelectric, Brussels. http://www.eurelectric.org/media/25920/eurelectric_report__on_reg_for_sg_final-2011-030-0131-01-e.pdf. Accessed 2 July 2013
31. Office Executive (2011) A policy framework for the 21st century grid: enabling our secure energy future. Executive Office of the President of the United State, Washington DC
32. Faruqui A (2012) The ethics of dynamic pricing. In: Fereidoon PS (ed) Smart grid. Academic Press, Boston, pp 61–83 Chapter 3
33. Faruqui A, Harris D, Hledik R (2010) Unlocking the 53 billion Euro savings from smart meters in the EU: how increasing the adoption of dynamic tariffs could make or break the EU's smart grid investment. Energy Policy 38:6222–6231
34. Faruqui A, Harris D, Hledik R (2010) Unlocking the €53 billion savings from smart meters in the EU: how increasing the adoption of dynamic tariffs could make or break the EU's smart grid investment. Energy Policy 38(10):6222–6231. doi:10.1016/j.enpol.2010.06.010
35. Faruqui A, Palmer J (2011) Dynamic pricing and its discontents. Regul Fall 34:16–22
36. Faruqui A, Palmer J (2012) The discovery of price responsiveness—A survey of experiments involving dynamic pricing of electricity. The Brattle Group, San Francisco. http://papers.ssrn.com/sol3/papers.cfm?abstract_id=2020587. Accessed 23 May 2013
37. Faruqui A, Sergici S, Palmer J (2010) The impact of dynamic pricing on low income customers. Institute for Electric Efficiency, Washington DC

38. Fischer C (2008) Feedback on household electricity consumption: a tool for saving energy? Energ Effi 1(1):79–104. doi:10.1007/s12053-008-9009-7
39. GAO (2012) Cybersecurity: challenges in securing the electricity grid. U.S. Government Accountability Office, Washington DC. http://www.gao.gov/assets/600/592508.pdf. Accessed 8 July 2013
40. Giordano V, Fulli G (2012) A business case for smart grid technologies: a systemic perspective. Energy Policy 40:252–259. doi:10.1016/j.enpol.2011.09.066
41. Giraud D (2004) The tempo tariff. In: EFFLOCOM workshop. EDF, Trondheim, 10 June 2004
42. Glachant J-M, Khalfallah H, Perez Y, Rious V, Saguan M (2012) Implementing incentive regulation and regulatory alignment with resource bounded regulators. European University Institute, Italy. http://cadmus.eui.eu/bitstream/handle/1814/22734/RSCAS%20_2012_31.pdf?sequence=1. Accessed 4 July 2013
43. Grattieri W, Maggiore S (2012) Impact of a mandatory Time-of-Use tariff on the residential customers in Italy. In: Current issues in Demand Side Management, Espoo, Finland, 2012: RSE (Ricerca sul Sistema Eneretico S.p.A.)
44. Grijalva S, Tariq MU (2011). Prosumer-based smart grid architecture enables a flat, sustainable electricity industry. In: Innovative Smart Grid Technologies (ISGT), 2011 IEEE PES, 17–19 Jan 2011, 2011, pp 1–6. doi: 10.1109/isgt.2011.5759167
45. Hadre S (2009) Impact of the German regulatory framework on investment incentives. In: The 20th international conference and exhibition on Electricity Distribution—Part 2, 2009 (CIRED 2009), 8–11 June 2009, 2009, pp 1–9
46. Heins S, Sandersen J (2007) Decoupling: divorcing electricity sales from revenues creates win-win for utilities and customers. In: 2007 ACEEE summer study on energy efficiency in industry: "improving industrial competitiveness: adapting to volatile energy markets, globalization, and environmental constraints". Orion Energy Systems, White Plains, NY, 24 July 2007
47. IBM (2007) Ontario Energy board smart price pilot—final report. IBM Global Business Services and eMeter Strategic Consulting, Ontario. http://www.ontarioenergyboard.ca/documents/cases/EB-2004-0205/smartpricepilot/OSPP%20Final%20Report%20-%20Final070726.pdf. Accessed 21 May 2013
48. IEA (2001) Towards a sustainable energy future. OECD Publishing, Paris. http://dx.doi.org/10.1787/9789264193581-en. Accessed 4 July 2013
49. IEA (2011) Empowering customer choice in electricity markets. International Energy Agency, Paris. http://www.iea.org/publications/freepublications/publication/Empower.pdf
50. IEA (2011) Smart grid-smart customer policy needs. International Energy Agency, Paris. https://www.iea.org/publications/freepublications/publication/sg_cust_pol.pdf. Accessed 4 July 2013
51. IEA (2011) Technology roadmap: smart grids. International Energy Agency, Paris. http://www.iea.org/papers/2011/smartgrids_roadmap.pdf
52. IEA (2012) Progress implementing the IEA 25 energy efficiency policy recommendations—2011 evaluation. International Energy Agency, Paris. http://www.iea.org/publications/insights/progress_implementing_25_ee_recommendations.pdf. Accessed June 13 2013
53. IEA/OECD (2003) The power to choose: demand response in liberalised electricity markets. OECD/IEA, Paris
54. Institute for Electric Efficiency (2012) State electric efficiency regulatory frameworks. Institute for Electric Efficiency, Washington DC. http://www1.eere.energy.gov/buildings/betterbuildings/neighborhoods/pdfs/iee_state_reg_frame.pdf. Accessed 4 July 2013
55. IPC (2012) Applying privacy by design best practices to SDG&E's smart pricing program. Information and Privacy Commissioner, San Diego Gas & Electric Co., Ontario http://www.sdge.com/sites/default/files/documents/pbd-sdge_0.pdf. Accessed 4 July 2013
56. Jamasb T, Pollitt M (2007) Incentive regulation of electricity distribution networks: lessons of experience from Britain. Energy Policy 35(12):6163–6187. doi:10.1016/j.enpol.2007.06.022

57. John JS (2013) Report: US smart grid cybersecurity spending to reach $7.25B by 2020. Greentech Media. http://www.greentechmedia.com/articles/read/report-u.s.-smart-grid-cybersecurity-spending-to-reach-7.25b-by-2020. Accessed 4 July 2013
58. JRC (2011) Smart Grid projects in Europe: lessons learned and current developments. European Commission Joint Research Centre and Institute for Energy, Netherlands. http://ses.jrc.ec.europa.eu/sites/ses/files/documents/smart_grid_projects_in_europe_lessons_learned_and_current_developments.pdf. Accessed 4 June 2013
59. KEMA (2009) Mandatory hourly pricing report of consolidated edison company of New York, Inc. KEMA, Inc., Connecticut. http://documents.dps.ny.gov/public/MatterManagement/CaseMaster.aspx?MatterCaseNo=03-E-0641. Accessed 2 July 2013
60. KEMA (2009) Orange and rockland utilities, Inc. Submits its mandatory hourly pricing report. KEMA, Inc., New York http://documents.dps.ny.gov/public/MatterManagement/CaseMaster.aspx?MatterCaseNo=03-E-0641. Accessed 2 July 2013
61. KEMA (2009) Two year evaluation report of Niagara Mohawk Power Corp d/b/a National Grid. KEMA, Inc., Connecticut. http://documents.dps.ny.gov/public/MatterManagement/CaseMaster.aspx?MatterCaseNo=03-E-0641. Accessed 2 July 2013
62. KEMA (2011) KEMA white paper: innovation in competitive electricity markets. KEMA, Inc., Massachusetts. http://www.competecoalition.com/files/KEMA%20Innovation%20in%20Electricity%20Markets%20White%20Paper.pdf. Accessed 2 July 2013
63. KEMA (2012) Mandatory hourly pricing program evaluation (prepared for Consolidated Edison Company of New York, Inc.). KEMA, Inc., Connecticut. https://www.google.com.hk/url?sa=t&rct=j&q=&esrc=s&source=web&cd=1&cad=rja&ved=0CDMQFjAA&url=http%3A%2F%2Fdocuments.dps.ny.gov%2Fpublic%2FCommon%2FViewDoc.aspx%3FDocRefId%3D%257B1BB38D64-2382-4AF4-8A1D-46622C0A5585%257D&ei=zmFvUYW5F-SWiQeCloCQCA&usg=AFQjCNHHv06uLgQCZpq4MReL9Ub-2wsnwg&sig2=QB78lf0Zr73iiPPXLN03rg. Accessed 2 July 2013
64. King C, Strapp J (2012) Software infrastructure and the smart grid. In: Fereidoon PS (ed) Smart grid. Academic Press, Boston, pp 259–288 Chapter 11
65. LBST, HINICIO, Barquín J (2012) Effect of smart metering on electricity prices. Policy Department Economic and Scientific Policy, European Parliament, Brussels. http://www.europarl.europa.eu/document/activities/cont/201202/20120223ATT39186/20120223ATT39186EN.pdf. Accessed May 2 2013
66. Lo Schiavo L, Delfanti M, Fumagalli E, Olivieri V (2013) Changing the regulation for regulating the change: Innovation-driven regulatory developments for smart grids, smart metering and e-mobility in Italy. Energy Policy 57:506–517. doi:10.1016/j.enpol.2013.02.022
67. Maeding S (2009) Investment incentives in the context of revenue cap regulation. In: 20th international conference and exhibition on electricity distribution—part 1, 2009 (CIRED 2009), 8–11 June 2009, pp 1–4
68. Mah DNY, van der Vleuten JM, Hills P, Tao J (2012) Consumer perceptions of smart grid development: results of a Hong Kong survey and policy implications. Energy Policy 49:204–216. doi:10.1016/j.enpol.2012.05.055
69. Mah DN-Y, van der Vleuten JM, Ip JC-M, Hills PR (2012) Governing the transition of socio-technical systems: a case study of the development of smart grids in Korea. Energy Policy 45:133–141
70. Markard J, Truffer B (2006) Innovation processes in large technical systems: market liberalization as a driver for radical change? Res Policy 35(5):609–625. doi:10.1016/j.respol.2006.02.008
71. Mathews JA, Tan H (2013) The transformation of the electric power sector in China. Energy Policy 52:170–180. doi:10.1016/j.enpol.2012.10.010
72. Meeus L, Saguan M (2010) Smart regulations for smart grids. European University Institute, Italy. http://cadmus.eui.eu/handle/1814/14043. Accessed 4 July 2013
73. Müller C (2011) New regulatory approaches towards investments: a revision of international experiences (IRIN working paper for working package: advancing incentive regulation with

respect to smart grids). WIK GmbH, Bad Honnef. http://www.bremer-energie-institut.de/download/IRIN/pub/discussion/WIK-Diskus_353.pdf. Accessed 4 July 2013
74. NAPEE (2007) Aligning utility incentives with investment in energy efficiency. National Action Plan for Energy Efficiency. http://www.epa.gov/cleanenergy/documents/suca/incentives.pdf. Accessed 4 July 2013
75. NEA (2011) 12th five-year plan for energy technology (2011–2015). In: CNEA (ed) China National Energy Administration, Beijing
76. Newsham GR, Bowker BG (2010) The effect of utility time-varying pricing and load control strategies on residential summer peak electricity use: a review. Energy Policy 38(7):3289–3296. doi:10.1016/j.enpol.2010.01.027
77. NIST (2011) SmartGrid Interoperability Panel (SGIP)—Cyber Security Working Group (CSWG). National Institute of Standards and Technology, Washington DC
78. NMPC (2009) Niagara Mohawk Power Corporation d/b/a national grid: two-year evaluation report on mandatory hourly pricing. Niagara Mohawk Power Corporation, New York. http://documents.dps.ny.gov/public/MatterManagement/CaseMaster.aspx?MatterCaseNo=03-E-0641. Accessed 2 July, 2013
79. NSTC, Executive Office of the President (2011) A Policy framework for the 21st century grid: enabling our secure energy future. In: NSTC (ed) National Science and Technology Council, Washington DC
80. NYSEG (2009) New York State Electric and Gas Corp. Mandatory hourly pricing program report. New York State Electric & Gas Corporation, New York. http://documents.dps.ny.gov/public/MatterManagement/CaseMaster.aspx?MatterCaseNo=03-E-0641. Accessed 17 Apr 2013
81. Ontario Energy Board (2008) Staff discussion paper: regulated price plan—Time-of-Use prices: design and price setting issues. In O. E. Board (ed) Ontario Energy Board, Ontario
82. Parag Y, Darby S (2009) Consumer-supplier-government triangular relations: rethinking the UK policy path for carbon emissions reduction from the UK residential sector. Energy Policy 37(10):3984–3992
83. Pearson I (2011) Smart grid cyber security for Europe. Energy Policy 39(9):5211–5218. doi:10.1016/j.enpol.2011.05.043
84. Pearson M (2005) The business of governing business in China: institutions and norms of the emerging regulatory state. World Polit 57(2):296–322
85. PG&E (2011) Smart grid deployment plan: 2011–2020. Pacific Gas & Electric Co., San Francisco. http://www.pge.com/includes/docs/pdfs/shared/edusafety/electric/SmartGridDeploymentPlan2011_06-30-11.pdf. Accessed 4 July 2013
86. Potter CW, Archambault A, Westrick K (2009) Building a smarter smart grid through better renewable energy information. In: Power systems conference and exposition, 2009. PSCE'09. IEEE/PES, 15–18 Mar 2009, pp 1–5. doi: 10.1109/psce.2009.4840110
87. PSC (2006) Proceeding on motion of the commission regarding expedited implementation of mandatory hourly pricing for commodity service, order denying petitions for rehearing and clarification in part and adopting mandatory hourly pricing requirements. In: PSC (ed) Case 03-E-0641 (vol Case 03-E-0641). Public Service Commission, New York
88. RAP (2011) Electricity regulation in the US: a guide. The Regulatory Assistance Project, Vermont. www.raponline.org/document/download/id/645. Accessed 4 Jul 2013
89. RAP (2012) Time-varying and dynamic rate design. Regulatory Assistance project. http://www.hks.harvard.edu/hepg/Papers/2012/RAP_FaruquiHledikPalmer_TimeVaryingDynamicRateDesign_2012_JUL_23.pdf. Accessed 5 Jun 2013
90. RG&E (2009) Rochester Gas and Electric Corp. Two year mandatory hourly pricing summary report. Rochester Gas and Electric Corporation, New York. http://documents.dps.ny.gov/public/MatterManagement/CaseMaster.aspx?MatterCaseNo=03-E-0641. Accessed 17 Apr 2013
91. Risser R (2006) Decoupling in California: more than two decades of broad support and success. In: Aligning regulatory incentives with demand-side resources. Pacific Gas and Electric Company, San Francisco, 2 Aug 2006

92. SCE (2011) Smart grid deployment plan. Rosemead: Southern California Edison Company. http://www.smartgrid.com/wp-content/uploads/2011/09/SG-Plan-SCE.pdf. Accessed 4 Jul 2013
93. SCE (2012) Annual update—smart grid. Rosemead: Southern California Edison Co. http://www.cpuc.ca.gov/NR/rdonlyres/E11D9761-733C-4BD1-BC1B-4B52E1AF9FB1/0/R0812009SCESmartGridAnnualDeploymentPlanUpdate2012.pdf. Accessed 4 Jul 2013
94. Sciortino M, Nowak S, Witte P, York D, Kushler M (2011) Energy efficiency resource standards: a progress report on state experience. American Council for an Energy-Efficiency Economy, Washington, DC. http://aceee.org/research-report/u112. Accessed 13 May 2013
95. SDG&E (2011) Smart grid deployment plan: 2011–2020. San Diego Gas & Electric Co., San Diego. http://www.sdge.com/sites/default/files/documents/smartgriddeploymentplan.pdf. Accessed 4 July 2013
96. SDG&E (2012) Smart grid deployment plan—2012 annual report. San Diego Gas & Electric Co., San Diego. http://www.sdge.com/sites/default/files/documents/1138900767/SDGE_Annual_Report_Smart_Grid_Deployment.pdf. Accessed 4 July 2013
97. SEE Action (2012) A regulator's privacy guide to third-party data access for energy efficiency. State and Local Energy Efficiency Action Network, Washington, DC. http://www1.eere.energy.gov/seeaction/pdfs/cib_regulator_privacy_guide.pdf. Accessed 2 May 2013
98. Sioshansi FP (2012) Smart grid: integrating renewable, distributed & efficient energy. Elsevier/Academic Press, Amsterdam
99. Sorebo GN, Echols MC (2011) Smart grid security: an end-to-end view of security in the new electrical grid. Taylor & Francis, Florida
100. Stern N (2006) Stern review: the economics of climate change. HM Treasury, London. http://www.hm-treasury.gov.uk/independent_reviews/stern_review_economics_climate_change/sternreview_index.cfm. Accessed 4 July 2013
101. Stronzik M (2011) Incentive Regulation and the Return on Equity. WIK GmbH, Bad Honnef. http://www.wik.org/index.php?id=diskussionsbeitraegedetails&L=1&tx_ttnews%5Btt_news%5D=1335&tx_ttnews%5BbackPid%5D=93&cHash=ff0a74079c96bbde7e899a665c723f30. Accessed 5 July 2013
102. Tabors R, Parker G, Caramanis M (2010) Development of the smart grid: missing elements in the policy process. In: IEEE 43rd Hawaii international conference on system sciences, Honolulu, Hawaii, 5–8 Jan 2010
103. Torriti J (2012) Price-based demand side management: assessing the impacts of time-of-use tariffs on residential electricity demand and peak shifting in Northern Italy. Energy 44(1):576–583. doi:10.1016/j.energy.2012.05.043
104. Vasconcelos J (2008) Survey of regulatory and technological developments concerning smart metering in the european union electricity market. European University Institute, Florence. http://cadmus.eui.eu/bitstream/handle/1814/9267/RSCAS_PP_08_01.pdf?sequence=2. Accessed 28 May 2013
105. Weber T, Besa A, Miller B (2006) Decoupling mechanisms: energy efficiency policy impacts and regulatory implementation in less is more: En route to zero energy buildings. American Council for an Energy-Efficient Economy, Pacific Grove, CA, 13–18 Aug 2006
106. Wolak FA (2005) Lessons from international experience with electricity market monitoring. The World Bank, California. http://elibrary.worldbank.org/docserver/download/3692.pdf?expires=1369795827&id=id&accname=guest&checksum=D0B61F56AD1E619E2216DDDDFBDC8642. Accessed 29 May 2013
107. Zhao X, Wang F, Wang M (2012) Large-scale utilization of wind power in China: obstacles of conflict between market and planning. Energy Policy 48:222–232
108. Zio E, Aven T (2011) Uncertainties in smart grids behavior and modeling: what are the risks and vulnerabilities? How to analyze them? Energy Policy 39(10):6308–6320. doi:10.1016/j.enpol.2011.07.030

i-Energy: Smart Demand-Side Energy Management

Takashi Matsuyama

Abstract Much of the discussion concerning smart grids focuses on the potential benefits of technological advances to power suppliers. However, there are many possible advantages that may benefit consumers as electricity grids exploit more of the opportunities provided by the cyber network society and emerge as a form of supporting social infrastructure. A critical element in the development of advanced consumer-friendly grids is the notion of an integrated system for energy management from the consumer's viewpoint. Here, we present such a concept—*i*-Energy—for smart demand-side energy management. The concept embodies four main elements: Smart Tap Network, Energy on Demand Protocol, Power Flow Coloring, and Smart Community. The key features of each are explained, and the characteristics and effectiveness of the required supporting technologies as demonstrated by ongoing research are presented. The importance of bringing together key stakeholders to achieve effective collaboration to implement the *i*-Energy concept is also emphasized.

1 Introduction

This chapter first introduces the concept of *Cyber-Physical Integration Systems* to develop social infrastructures in the twenty-first century such as smart electric power network systems. Then, we propose the concept of *i*-Energy for smart demand-side energy management. This differs from *the Smart Grid*. The former aims at energy management from the consumer's viewpoint while the latter from the supplier's viewpoint. The latter half of the chapter presents four steps in the realization of the *i*-Energy concept: (1) *Smart Tap Network* for monitoring detailed

T. Matsuyama (✉)
Graduate School of Informatics, Kyoto University, Kyoto, Japan
e-mail: tm@i.kyoto-u.ac.jp

D. Mah et al. (eds.), *Smart Grid Applications and Developments*, Green Energy and Technology, DOI: 10.1007/978-1-4471-6281-0_8,

power consumption patterns of individual appliances and dynamic activities of people in homes, offices, and factories; (2) *Energy on Demand Protocol* to realize the priority-based best effort power supply mechanism as well as the automatic ceiling mechanism of power consumption in both watts (W) and watt hours (Wh); (3) *Power Flow Coloring* to allow versatile power flow controls depending on types and costs of power sources; and (4) *Smart Community* for bidirectional energy trading among households, offices, and factories in a local community. The technologies to embody each of these four ideas are presented with research achievements so far obtained.

2 Integrating the Physical Real World and the Cyber Network Society

Until the twentieth century, our society was structured in such a way that most social and personal activities took place in what we could call the physical real world. For example, appliances, furniture, housing space designs, building architectures, traffic systems, and even sporting environments are designed based on physical laws. The last decade of the twentieth century saw the cyber network society emerge, thanks to the advancement of information and communication technologies (ICT). And now, our social and personal activities are conducted in both the physical real world and the cyber network society.

To design twenty-first-century social infrastructures, therefore, it is crucial to study how we can integrate seamlessly the physical real world and the cyber network society (Fig. 1). One idea to realize this cyber-physical integration is to link and merge dynamical flows in the physical real world and the cyber network society; the former are characterized by the flows of goods, people, and energy and the latter by the flow of information.

We believe that the cyber-physical integration will be significant, not only for creating infrastructures for twenty-first-century society, but also in a more academic sense as well. In other words, the discovery of natural laws in the physical real world and the creation of basic theories which have allowed us to harness those laws for technologies have been extremely successful. Such theories include physical models represented by differential equations like the Newton and Maxwell equations. On the other hand, the foundations supporting the cyber network society can be seen in the computational theory of the Turing machine and Shannon's communication theory, on the basis of which many different kinds of ICT systems have been developed and are supporting our everyday activities. So the question is: What is the relationship between physical models on the one hand and computational and information models on the other? Is a theoretical model which could unify them possible? We believe that only after such a unified theoretical foundation has been established, will the way forward toward the cyber-physical integration be shown, leading to the creation of a sound twenty-first-century society. (Our research group is currently

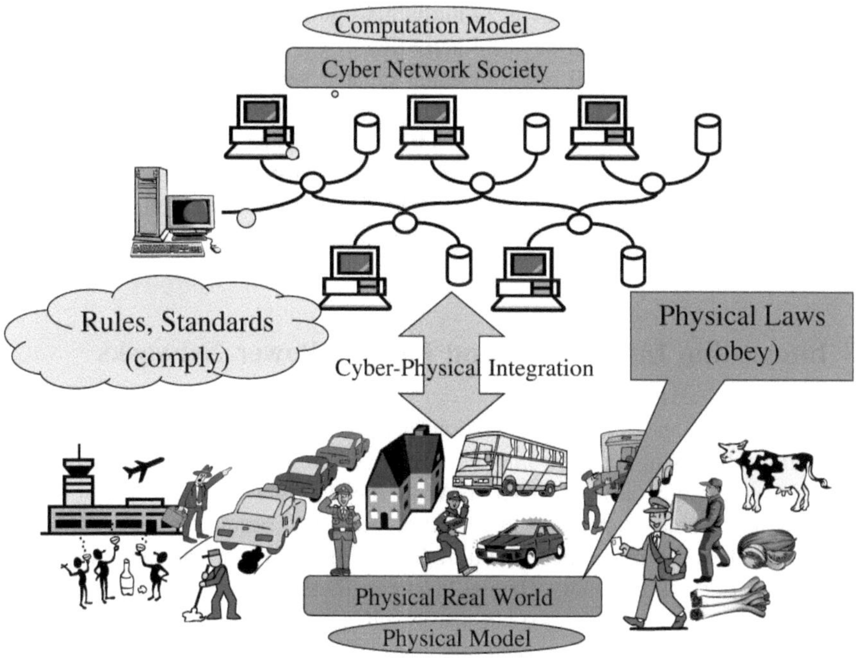

Fig. 1 Cyber-physical integration

involved in research on a hybrid dynamical system [1] as one attempt to set out a unified theory.)

Today, our social and personal activities are being supported by cyber-physical integration systems. The most striking example is *e-money*: the digitization of currency and securities. Value flows in the physical real world in the form of physical things such as precious metals, currencies, and securities, but in the cyber network society, value flows in the form of digital numbers and characters whose values are guaranteed by personal authentication and information security; these functions are critical mechanisms giving economical meaning to the flow of information in the cyber network society.

The second example is the realization of so-called *ubiquitous society*. By attaching bar codes, IC tags, and RF tags as well as microprocessors to objects in the physical real world and recording and managing the positions, types, and processing histories of these objects as information in the cyber network society, it becomes possible to associate flows of goods, traffic, and people with flows of digital data. These ID-ing methods together with location-sensing methods based on GPS and other localization/recognition devices (e.g., car license plate readers) have realized food product tracing systems, electronic highway toll collection systems, car navigation systems, cell phones, and the like. More recently, this trend has spread from objects to people as well, with experiments being performed to monitor the health and behavior of individuals with wristwatch-like devices that record biological states and physical activities.

Thus, the cyber-physical integration has created various infrastructures for the twenty-first-century society. In other words, while many people may be satisfied with currently available ICT services such as smartphones that make things more convenient and amusing, we should continue to develop new ICT to create a new social infrastructure through the cyber-physical integration. So then, we must ask "What is the next stage of the cyber-physical integration?"

3 Integrating Information and Electric Power Networks

Early computer and telecommunication systems were organized as star-shaped networks, with a single large computer or line switching machine in the center and terminals radiating outward from it. With the emergence of workstations and personal computers, however, information networks became more distributed, bidirectional, and personalized, resulting in the creation of the Internet. We should notice that this revolutionary change has been realized not only by advances in ICT (in the physical real world) but also by changes and reforms of societal rules such as the privatization of national telephone departments and the deregulation of telecommunication services (in the cyber network society).

When we envision the future of electric power networks (a major social infrastructure system in the physical real world) from this point of view, their current star-shaped networks, in which energy flows from centralized power plants to factories, offices, and households, can be expected to rapidly undergo the same process of becoming distributed, bidirectional, and personalized, thanks to physical-real-world technological advances in distributed power sources such as wind power plants, photovoltaics, fuel cells, and storage batteries, as well as complementary cyber network society policies designed to stop global warming and incubate new energy services.

Then, the critical question is how we can embody such revolutionary changes in the electric power network. A naïve idea would be to integrate the information and electric power networks by employing ICT to monitor and control electric power flows (Fig. 2). Such a slogan, however, is not enough to develop a new social energy infrastructure. Instead, we should learn from history: As is well known, packet data transmission and the IP protocol are the essential driving forces to implement the Internet. Hence, we should develop innovative technologies to implement distributed, bidirectional, and personalized electric power networks.

Accordingly, for about ten years, we have put forward the idea of *i*-Energy to devise innovative technologies, which will be described later in this chapter, and we believe they will grow as driving forces to implement the integration of the information and electric power networks like the packet data transmission and the IP protocol in the Internet.

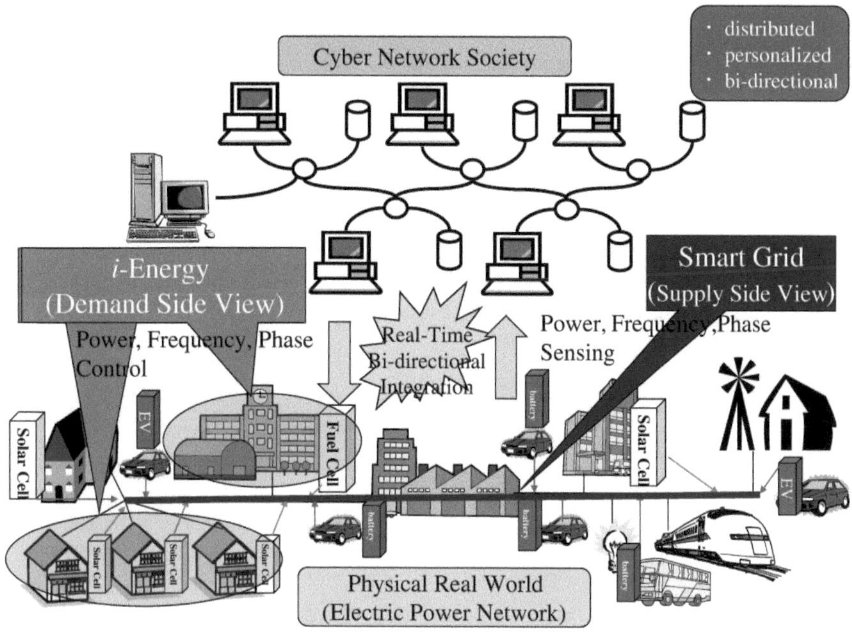

Fig. 2 Integration of information and electric power networks

3.1 i-*Energy Versus Smart Grid*

The idea of using ICT to improve electric power network management is similar to *the Smart Grid*. However, our idea of *i*-Energy differs largely from the smart grid; the former is for the demand (i.e., consumer)-side energy management while the latter for the supplier (i.e., producer)-side one (Fig. 2).

Figure 3 illustrates the scheme of demand response and dynamic pricing, which are the major functionalities supported by current smart grid systems. One of the most important roles of power suppliers is to balance the power generation with the power consumption. If the power generation capability were much larger than the expected maximum power consumption, the balance control would not be so difficult. In reality, however, the power generation capability is limited due to the cost of having excessive power plants, the reduction of greenhouse gas emissions, and the political policy to stop and/or limit nuclear power plants. The aim of smart grid systems is to make the balance even with the limited power generation capability, that is, to suppress the power consumption by asking consumers to reduce their power consumption according to power supplier's requests. In short, the scheme of current smart grid systems is designed from the power supplier's viewpoint: A supplier asks consumers to help the power balancing control by the supplier.

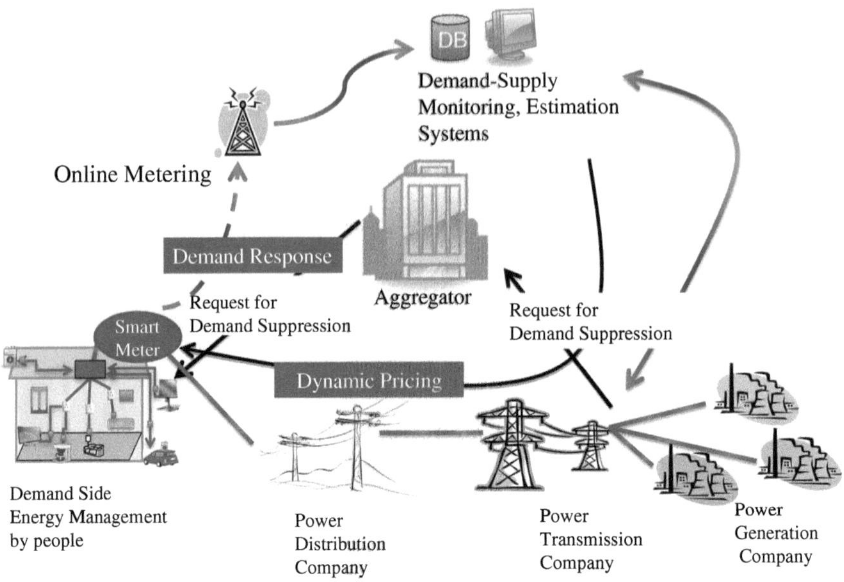

Fig. 3 Scheme of the demand response and the dynamic pricing

i-Energy systems, on the other hand, manage appliances and distributed power generation/storage devices in consumer's living spaces: households, offices, factories, and local communities. Thus, consumers manage all power consumption, generation, and storage by themselves to realize the power balance control as well as the reduction in the energy costs and the greenhouse gas emissions. That is, *i*-Energy systems enable consumers to become *prosumers*, i.e., live as both consumers and producers. Note that the population of prosumers has been rapidly growing in this century; supported by feed-in tariff systems, a large number of households have introduced photovoltaics to produce and sell electric power as well as to consume it.

The reasons why we focus on the demand-side energy management are as follows:

1. R&D balancing between power supply and demand-sides: Since electric power networks function well only when the power generation and consumption are balanced, supplier-oriented smart grid systems are not enough and consumer-oriented systems should be developed at the same time.
2. Cooperation with existing social infrastructures: Since modern countries, cities, and towns have accumulated large-scale complicated electric power networks and huge social and personal activities are conducted everyday on such infrastructures, it would not be reasonable to propose their drastic change whatever merits innovative technologies could bring. In other words, we will be able to implement innovative technologies rather easily into households, offices, factories, and communities if we can prove their merits.

3. Bottom-up creation of clean-slate social infrastructures: In developing countries, on the other hand, many people, perhaps more than half of the human population, live without stable or sufficient electric power network services. Even if their sizes are small, compact demand-side energy management systems can support such people and improve their lives at a reasonable cost, which would be far less expensive than implementing national-wide electric networks as in developed countries.

4 Technologies to Implement the *i*-Energy Concept

We have been developing technologies to implement the *i*-Energy concept and conducting real-life experiments to evaluate their utility for 10 years. The following are the four steps in our R&D road map:

Step (1) *Smart Tap Network* for visualizing power consumption patterns of individual appliances and monitoring human activities based on observed data

Step (2) *Energy on Demand Protocol* to realize the priority-based best effort power supply mechanism and the automatic ceiling mechanism of power consumption in both watts (W) and watt hours (Wh)

Step (3) *Power Flow Coloring* to allow versatile power flow controls depending on types and costs of power sources

Step (4) *Smart Community* for bidirectional energy trading among households, offices, and factories in a local community.

In what follows, overviews of our technologies and experimental results are presented. As for technical details, please refer to the references and the demo videos on the Web.

4.1 Smart Tap Network

The first step to implement the *i*-Energy concept is to develop a smart tap network: attach *Smart Taps*, which consist of voltage and current sensors, a power controller, a wireless (i.e., Zigbee) communication module, and a microprocessor to individual electric devices inside households, offices, and factories to create a sensor network that monitors detailed power consumption patterns of individual devices (Fig. 4).

We developed the following application systems by implementing smart tap networks in an ordinary apartment room, an independent house, an office room in our university, and an actual small factory, respectively:

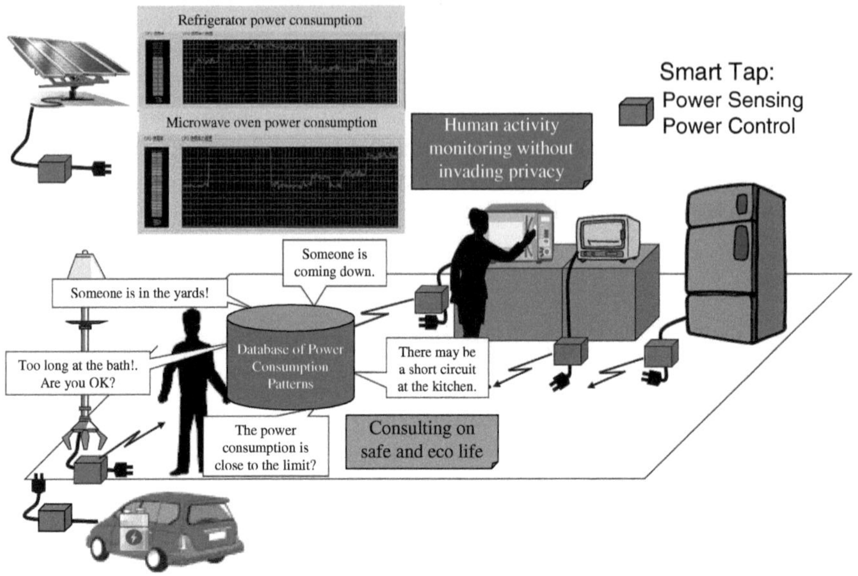

Fig. 4 Smart tap network

1. Real-Time Visualization of Power Consumption Patterns: Fig. 5 illustrates a TV screen displaying power consumptions of all appliances in the apartment room. Since the data are renewed every 0.5 s, we can monitor which appliances are operating in which modes. With this function, energy-saving awareness can be raised to switch off and/or reduce power of useless appliances. Such power control can be conducted just by clicking appliance icons on the screen by a TV game machine or a smartphone; power control commands are issued to the smart taps attached to the selected appliances. Real-life experiments at the smart apartment room showed that at best, about 20 % power reduction in Wh a day can be attained with this real-time visualization and control system.
2. The smart tap we developed is designed to measure the voltage and current at 20 kHz sampling rate, which enables us to identify types of appliances by analyzing the current waveform over one AC cycle (see Fig. 6). We developed a novel method of appliance identification, by which 16 different types of home appliances can be identified with 99 % accuracy [2]. This means that an appliance can be identified simply by plugging it into a socket. Our recent study showed that aging phenomena and malfunctions of appliances can be detected by analyzing current waveforms, which is useful for early detection of problems with devices, thereby contributing to a safe and secure life.
3. The power control of some appliances is done manually by a person, while for others, it is done automatically. If we can recognize which power control events are by a person, we can estimate his/her location since the location of each smart tap is known a priori. In fact, the newest version of the smart tap is implemented

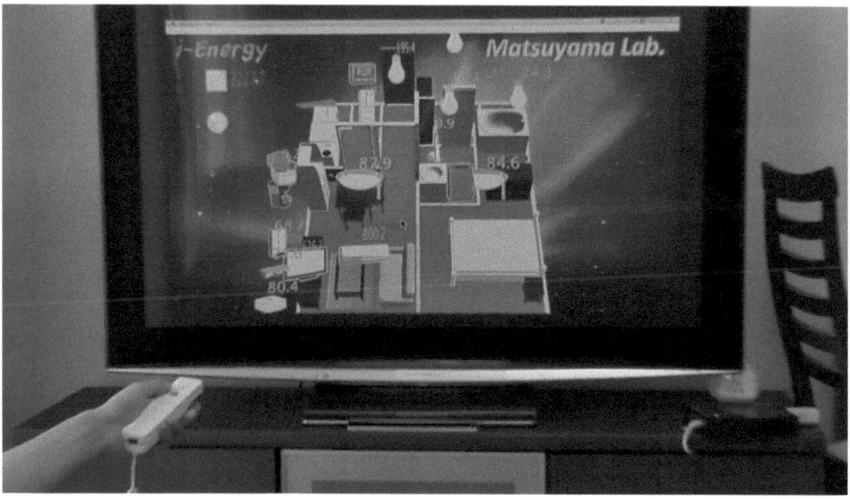

Fig. 5 Real-time visualization of individual appliance power consumption: Every 0.5 s, the consumed power is displayed on *top* of each appliance icon. By clicking an appliance icon, we can monitor its power consumption profile as well as control its supplied power. The overall power consumption pattern in the apartment room can be monitored to analyze which appliances consume how many watts

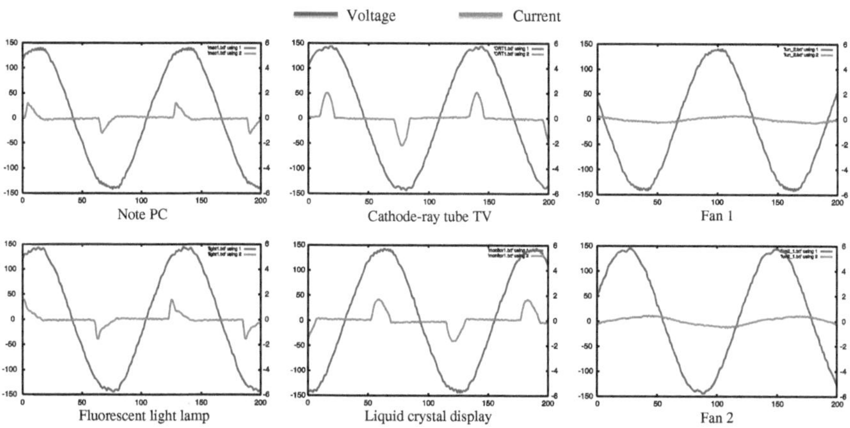

Fig. 6 Voltage and current waveforms of appliances. While the former is maintained constant, the latter varies a lot from appliance to appliance

in a wall socket. We developed a human activity monitoring system in the smart apartment room, which estimates the motion trajectory of a person based on manual power control events. The system can be used, for example, for children living away to monitor the degree of activities of their old parents without directly intruding into their privacy. Moreover, patterns of manual operations of

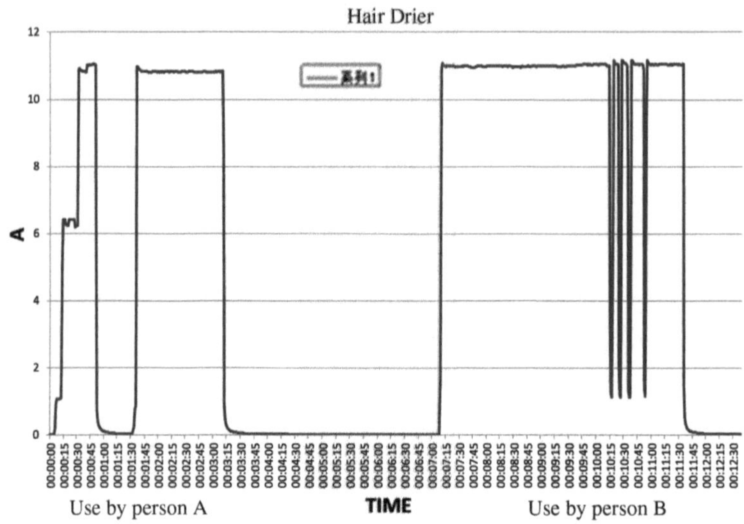

Fig. 7 Usage patterns of a hair drier

appliances vary a lot person by person. Figure 7 shows power consumption patterns of a hair drier by different persons. By learning such appliance usage patterns, it is possible to estimate who is using an appliance.

The demo video showing fundamental functions of the smart tap and the real-time power consumption monitoring system with the appliance recognition and the malfunction detection is available at http://www.youtube.com/watch?v=QRQ72xtzDHE.

4.2 *Energy on Demand Protocol for Intelligent Power Management*

While raising awareness about energy saving by visualization systems like the one described above does lead to a reduction in wasted energy, the effectiveness of such manual power control methods is nevertheless limited and moreover hard to maintain for a long time; people are busy and prone to get lazy. Accordingly, the second step of the *i*-Energy concept explores an automatic power control mechanism.

The *i*-Energy concept is best characterized by a novel automatic power control method named *Energy on Demand* (EoD) that supplies electric power based on power demand messages issued from appliances [3]. Its basic protocol is as follows (Fig. 8):

1. When an appliance is switched on or changes its operation modes, a power demand message describing the required power and its priority, which we call *Quality of Energy* (QoEn), is issued from the appliance to the power manager.

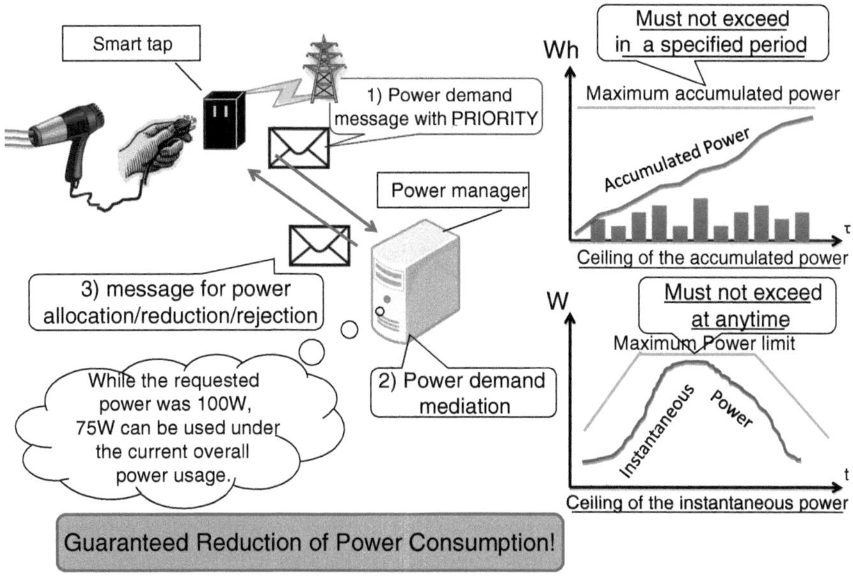

Fig. 8 Scheme of the energy on demand

Note that for simple appliances such as heaters and fans, smart taps attached to them detect switching events and issue messages, while so-called smart appliances, which can directly communicate with the power manager via a network, can issue messages by themselves as well as monitor and control consuming power according to requests from the power manager. Note also that the power manager may be installed inside a house or a cloud energy service provider may support its functions.

2. The manager mediates such power demands from appliances taking into account available energy sources and appliance priorities as well as human activity patterns, which have been learned using the sensor network described above. This mediation is conducted based on *the best effort policy*; some demands with low priorities may not be satisfied. For example, garden lamps might not be supplied power or 75 W might only be supplied in response to a 100 W demand.
3. Power is supplied to the appliance only after a power supply permission message is received by the smart tap attached, which in turn controls the power as specified in the permission message.

To attain substantial power savings by the priority-based best effort power supply mechanism described above, ceiling values in both W and Wh for the power demand mediation are specified by a human power manager. The red lines in the right graphs in Fig. 8 illustrate such ceiling value profiles. At the power demand mediation process in the second point of the basic protocol given above, the power manager supplies power only when the ceiling constraints are satisfied. That is, if the acceptance of a new power demand leads to the violation of the

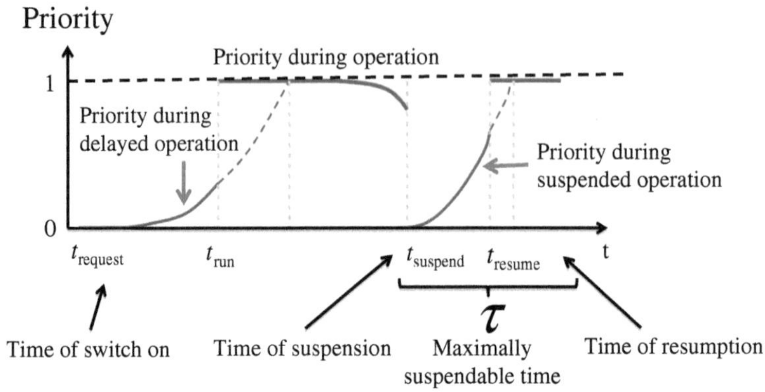

Fig. 9 Dynamic property of the priority of an air conditioner

ceiling constraints, the manager selects the appliance with the least priority and asks it to reduce or cut off the power so that appliances with higher priorities can be supplied power. In other words, appliance operations with low priorities may be interrupted when a new demand with a higher priority is issued. Consequently, the EoD system can guarantee the user-specified reduction rate of power consumption in both W and Wh.

As described above, the power saving rate is specified by a user and its attainment is 100 % guaranteed by the EoD system. So the power saving rate is not a goal to achieve or a performance evaluation measure of the EoD system. Instead, the performance of the EoD system should be measured by how well the user's quality of life (QoL) is maintained. To this end, the priority specification of each appliance is crucial.

In the EoD system, the priority changes dynamically depending on appliance characteristics and human activity patterns in individual households, offices, and factories. For example, Fig. 9 illustrates the dynamic priority profile of an air conditioner. Even if switched ON, its operation can be delayed (time-shiftable characteristic). Moreover, the amount of power supply can be changed (power controllability) and the power supply can be suspended during its operation (temporary interruptible characteristic). The priority increases during delay and suspension periods to start/resume operation when the priority becomes high enough. On the other hand, the priority decreases after sufficient operation, because the air temperature is maintained for a while even if an air conditioner stops. This dynamic priority modification mechanism allows the EoD system to make full use of appliance utility even under the power ceiling control. Currently, we are studying modeling and learning methods of human activity patterns from power consumption data measured by the smart tap network, which will enable the EoD system to adjust the appliance priority based on human activities.

Figures 10 and 11 show the experimental results of one-day real-life activities by a university student at the smart apartment room, where the EoD system with

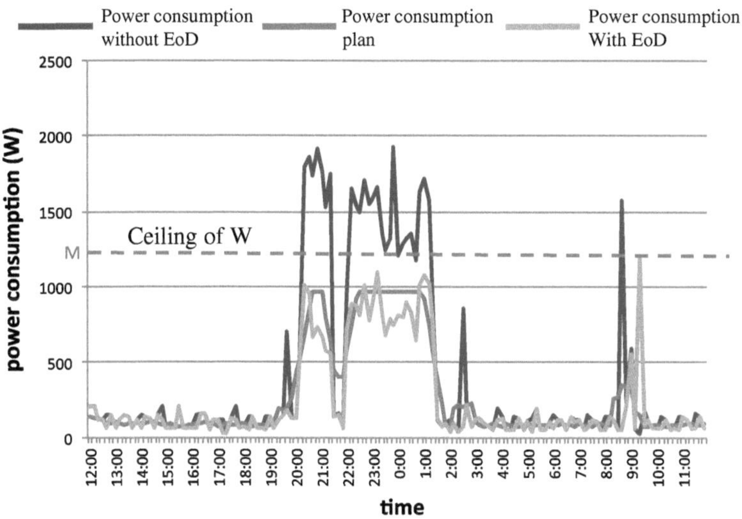

Fig. 10 Instantaneous power consumption profiles of one-day real-life experiment at the smart apartment room (see text)

smart taps and smart appliances is implemented. Figure 10 illustrates the instantaneous power consumption profile and Fig. 11 the accumulated one. The blue lines in the figures indicate the power consumption data without the EoD system, the red the power consumption plans to comply with the ceilings (30 % reduction of the one-day accumulated power and 1,200 W as the maximum instantaneous power), and the green the actual power consumption data with the EoD system. Note that while the constant value, i.e., 1,200 W, was specified as the instantaneous power consumption ceiling in Fig. 10, an arbitrary shaped ceiling profile can be set by a user taking into account daily activities and dynamic pricing data. We conducted a series of weekly real-life experiments at the smart apartment room to evaluate QoL by setting the ceiling of W as 1500 W and the accumulated power reduction rate a day as 10, 30 and 50 %, respectively. The subjective evaluation proved that QoL was not damaged significantly even with 50 % power saving. Currently, we are developing a quantitative QoL evaluation measure, whose validity will be tested with a variety of lifestyles as well as living environments.

In most smart grid projects, demand response and/or dynamic pricing mechanisms have been introduced to reduce the peak power demand. As is well known, however, the actual power reduction is uncertain and fluctuates from time to time. Moreover, the power reduction cannot be attained in real time. With the EoD system, on the other hand, the instantaneous power consumption can be automatically reduced in real time by setting the ceiling value for W. That is, when utility companies are allowed to set and modify ceiling values based on contracts with consumers, EoD systems work as real-time automatic demand response systems.

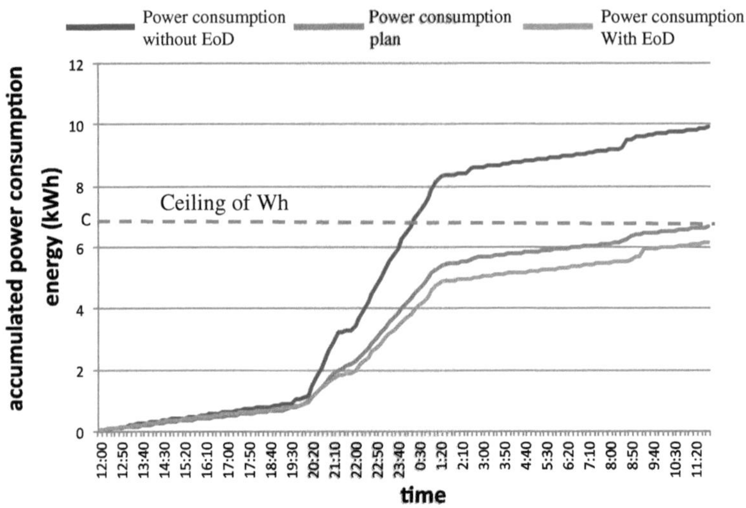

Fig. 11 Accumulated power consumption profiles of one-day real-life experiment at the smart apartment room (see text)

The demo video of the EoD system at the smart apartment room is available at http://youtube/rTgxpD7mAwU. We believe that the EoD can be a driving force to introduce revolutionary changes in our energy infrastructures just as the packet data transmission and the IP protocol have done in communication infrastructures.

4.3 EoD-based Battery Design and Management and Power Flow Coloring

The third step of the *i*-Energy concept realization is the management of multiple power sources: power generators and storage batteries. Toward this end, we first developed the EoD-based battery design and management system [4]. The system consists of two processes: One is the battery design and management plan generation and the other the real-time battery management (i.e., charge/discharge control).

4.3.1 Battery Design and Management Plan Generation

Figure 12 illustrates the basic idea of our EoD-based battery design and management plan generation. Given an expected power consumption profile without the EoD system ($D(t)$, the dotted line in the upper graph of Fig. 12) as well as the ceiling specifications in W and Wh, the EoD system first makes a power usage plan satisfying the ceilings, which is denoted as $PD^{PLAN}(t)$ and illustrated by the solid line in the upper graph of Fig. 12 (see also the blue and red lines in Fig. 10).

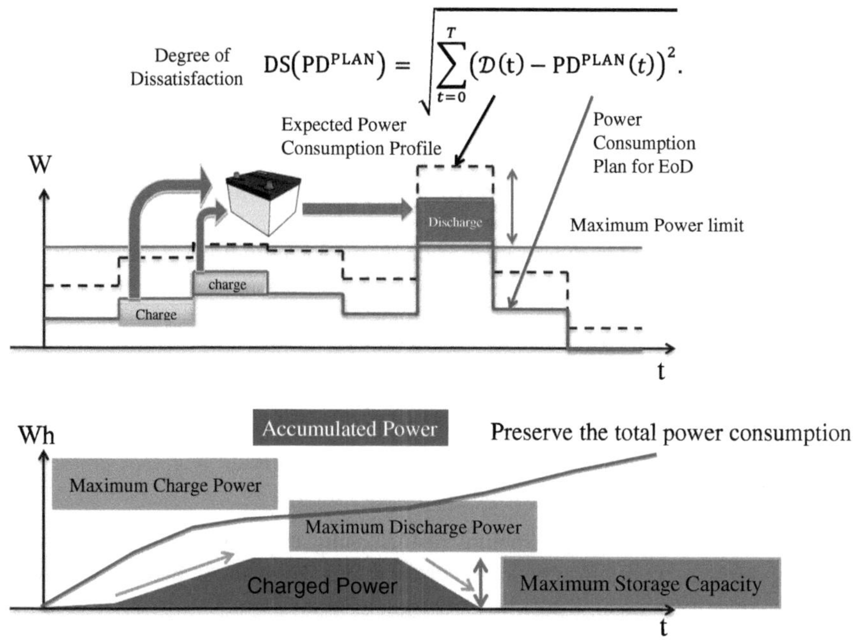

Fig. 12 Optimal charge/discharge planning and battery design to minimize the degree of dissatisfaction

The gap between these profiles, $DS(t) = D(t) - PD^{PLAN}(t)$, implies the degree of user's dissatisfaction; while a user wants to use appliances, some of such demands are not satisfied due to the power ceilings.

Our aim of employing a storage battery is to reduce this degree of dissatisfaction. That is, the battery design and management plan generation process conducts the minimization of the totally accumulated degree of dissatisfaction by using a battery, which gives us a new power usage plan $PLAN^*$ with much less user dissatisfaction while complying with the ceilings, the optimal battery characteristics, and its optimal charge/discharge plan:

$$PLAN^* = \arg\min_{PLAN} DS(PLAN) = \sqrt{\sum_{t=0}^{T} (D(t) - PD^{PLAN}(t))^2}.$$

The basic idea to conduct the optimization to derive $PLAN^*$ is to charge when $DS(t)$ is small and discharge when $DS(t)$ is large while keeping the balance between charged and discharged powers as well as considering charge/discharge losses. Note that constraints on the battery capacity or charging/discharging capabilities need not be specified in this optimization process; they will be derived from the optimization process as described below. Note also that the ceiling of Wh, which was set for the EoD system, is preserved. That is, while the total consumed power is reduced to comply with the pre-specified ceiling of Wh, the degree of dissatisfaction can be reduced with the battery management. The bottom graph in Fig. 12 illustrates the profiles of the accumulated power to be supplied (consumed) from the utility line and the accumulated charged power in the battery.

This optimization process gives us

1. the optimal power usage plan, $PLAN^*$, with much less user dissatisfaction than that by the original EoD system and
2. the optimal charge/discharge plan of the battery (the bottom graph in Fig. 12), which then gives us the following design specifications of the battery:
   (a) the required battery capacity = the peak value of the accumulated charged power in the battery,
   (b) the required charging capability = the maximum positive gradient of the accumulated charged power in the battery, and
   (c) the required discharging capability = the maximum negative gradient of the accumulated charged power in the battery.

Figure 13 shows the experimental results using one-day real-life data at the smart apartment room, which demonstrate that a very small battery with only 420 Wh capacity is enough for the EoD system to attain a 30 % reduction in Wh and 1,000 W ceiling. Note that the total power consumption a day without the EoD system in this experiment is about 12 KWh, which is the standard in Japanese households. Figure 14 illustrates how $DS(PLAN^*)$ changes by reducing the ceiling of W, which proved that the degree of dissatisfaction stays constant even if the ceiling is reduced by more than 50 % from that without the EoD system. It should be noted that this significant peak-cut gives large energy cost savings for those consumers whose energy pricing plans include peak demand charging (W) as well as accumulated power consumption charging (Wh).

4.3.2 Real-Time Battery Management

To realize the real-time battery management, i.e., the dynamic charging and discharging control, the EoD system described above, which only mediates power consumption demands from appliances, should be augmented so that it can mediate power supplies from multiple power sources. We first introduced the *load factor profile* to characterize the availability of a power source (Fig. 15):

(a) The profile takes a monotonically increasing curve ranging from 0 to 1.
(b) The horizontal location of the profile is specified and dynamically shifted left and right depending on the planned power supply at t of the power source, which is depicted by the vertical blue dotted line in Fig. 15. In the case of a battery, when its planned power supply is charging, its profile is shifted left into the negative power supply area for the battery to operate like a power-consuming appliance, while in discharging, the profile is shifted right in the

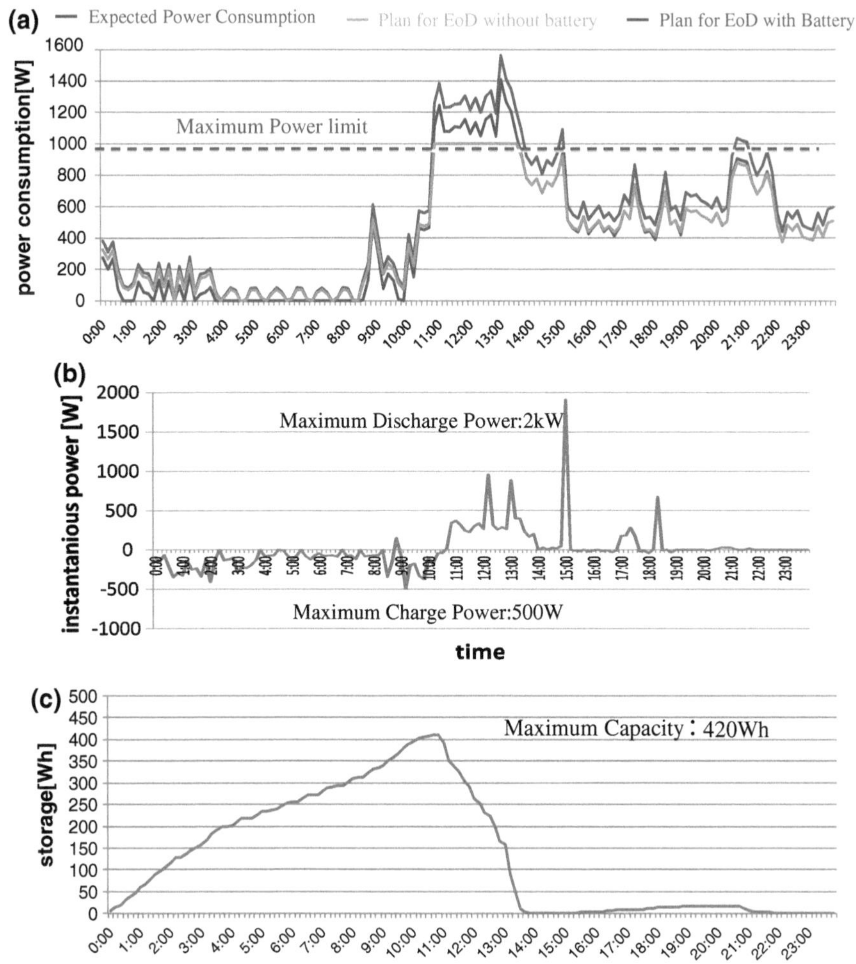

Fig. 13 Experimental results of the battery design and management using one-day real-life data at the smart apartment **a** power usage plans **b** charge (negative power usage) /discharge (positive power usage) plan for the battery **c** accumulated power profile in the battery

positive power supply area for the battery to work as an ordinary power supply.

(c) The gradient of the profile is changed depending on the available accumulated power to be consumed by the end of the planed period, e.g., a day or a week. That is, if more power than planned has been consumed by t, the gradient gets steeper, while if less power has been consumed, the gradient gets gentler.

We developed the real-time power mediation algorithm among both power sources and appliances using appliance priority values and power source load factors:

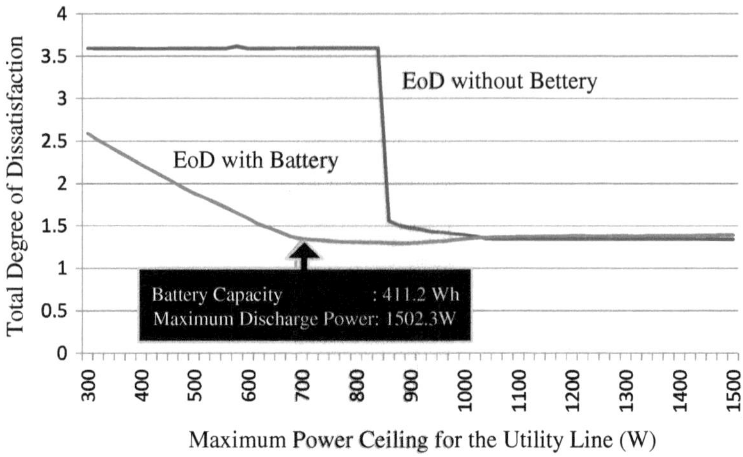

Fig. 14 Relation between the degree of dissatisfaction and the maximum power ceiling

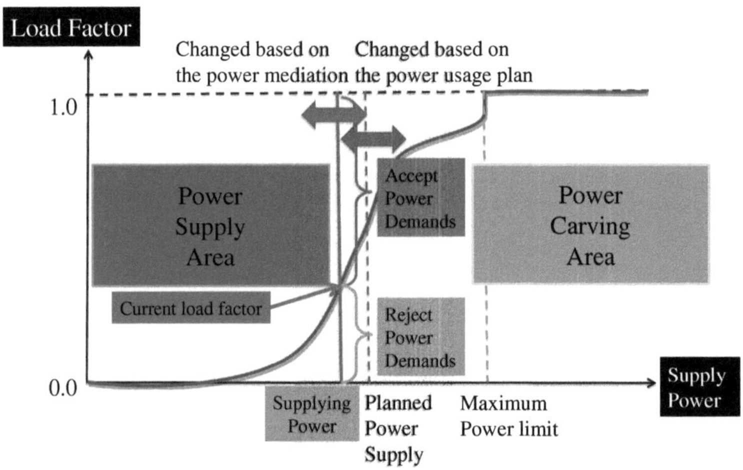

Fig. 15 Load factor profile (see text)

1. Given the currently supplying power denoted by the vertical red line in Fig. 15, the intersecting point with the load factor profile specifies the current load factor of the power source.
2. Intuitively, a power demand request from an appliance whose priority value is larger than the current load factor will be accepted for the power source to supply power, while that with a smaller priority value will be rejected.
3. When multiple power sources can supply power in response to a new power demand request from an appliance, the power source with the least current load factor is selected to supply power to that appliance.

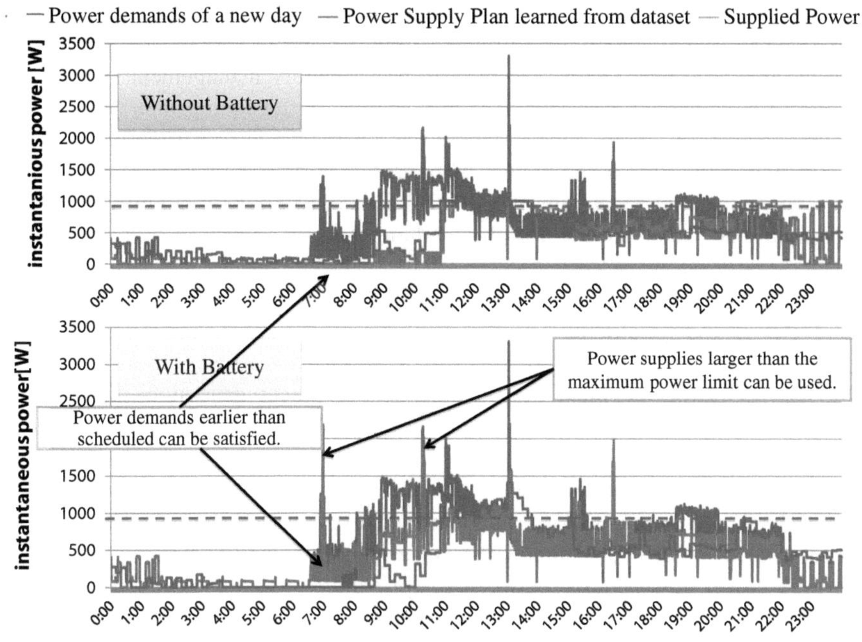

Fig. 16 Experimental results of the EoD systems with and without battery (see text)

As for technical details of the real-time battery management system implemented based on this power mediation algorithm, see [4].

Figure 16 shows the experimental results using one-day real-life data at the smart apartment room, where the power supply plans with and without a battery (the red lines in Fig. 16) were generated from the same daily power consumption dataset at that room and a power demand profile of a new different day (the blue lines in Fig. 16 representing the same data) was used to evaluate the performance. We can observe the effectiveness of the battery management system:

(a) In the power supply plan for the EoD system with a battery, more power than the ceiling, 1,000 W, can be supplied to reduce the degree of dissatisfaction.
(b) The green lines in Fig. 16 illustrate the real-time power supply profiles through the power mediations by the EoD systems with and without a battery. The real-time load factor- and priority-based battery management system can dynamically mediate power sources as well as power consumption demands to comply with the power demand (the blue lines in Fig. 16) as faithfully as possible reducing the degree of dissatisfaction:

    1. With a battery, the system can supply enough power even if power demands are issued earlier than planned (e.g., get up earlier than usual).
    2. With a battery, power demands larger than the plan as well as the ceiling of W can be satisfied (e.g., multiple appliances happen to be used simultaneously).

In short, even with a small-size battery, the augmented EoD system can improve QoL while complying with the ceiling constraints for power saving. Note that the real-time load factor- and priority-based battery management system can be used for managing a variety of power sources by designing their load factor profiles based on their power generation characteristics. Currently, we are implementing the EoD-based battery management system in an independent smart house to evaluate its effectiveness in the real lives of a variety of families as well as designing load factor profiles for photovoltaics and fuel cells.

4.3.3 Power Flow Coloring

Our idea to further augment the management of power generators and storage batteries described above is illustrated in Fig. 17, where what we call *power routers* are attached to individual power sources in addition to smart taps attached to appliances. Both types of power control devices are connected through wired and/or wireless networks to form a cooperative distributed power network. A novel power control mechanism named *the power flow coloring* enables us to design versatile power flow patterns between power sources and appliances. That is, we can "color" power flows depending on their sources and apply different controls based on the colors. For example, watch a TV with the utility supplying power, use an air conditioner with photovoltaic power, make coffee using power stored in a battery, and mix 60 % photovoltaic power and 40 % battery power to operate a heater. As the ubiquitous society described in Sect. 2 has done, this ID-ing of electric power will change our lifestyles as well as open exploration into new energy services and businesses which will drastically reduce energy costs and CO_2 emission in our society.

While power flow coloring is physically impossible, it can be realized by making full use of the concept of virtualization, which has been employed widely in computer systems and the Internet to enrich computational functionalities and communication services even with limited physical resources. Currently, we are developing power routers which can control powers cooperatively with each other as well as with smart taps. We believe that with power routers, the real-time load factor-based power source mediation method developed for the EoD-based battery management can be augmented to realize the cooperative control of multiple power sources. To realize the power flow coloring as illustrated in Fig. 17, a power demand message from an appliance should include the specification of the amounts of power to be supplied from individual power sources as well as the power demand priority. The power manager mediates such demands to determine how much power should be supplied from which power sources and issues power control messages to power routers and smart taps. The demo video of an early version of the power flow coloring with smart taps alone is available at http://www.youtube.com/watch?v=v41NZOmimQc.

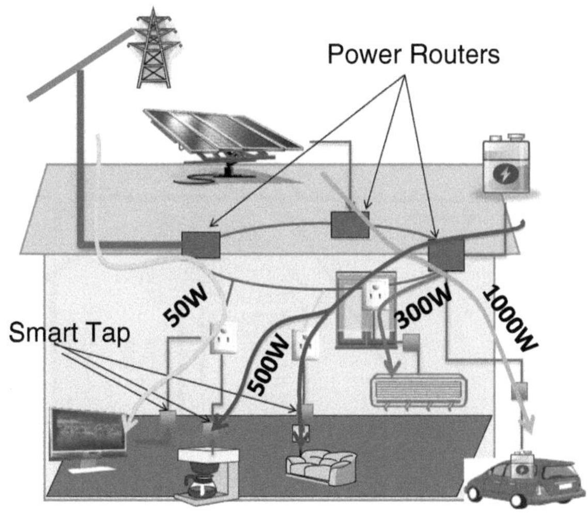

Fig. 17 Power flow coloring

4.4 Smart Community for Bidirectional Energy Trading Market

While the CO_2 reduction effect by a single household might not be that significant, greater reductions are possible if the in-house EoD system described so far is expanded to a local community. This is the fourth step of the *i*-Energy concept realization. When a group of in-house EoD systems are linked with each other to form a *smart community* system, a new energy trading market will be created, where colored powers including positive powers from different power sources, negative powers (i.e., power saving), and available battery capacities (i.e., storable powers for later use) will be exchanged bidirectionally among individual households (Fig. 18). Note that whereas such bidirectional power exchanges over utility power lines are not allowed in many countries like Japan, we can implement them without any problems inside condominium buildings and local areas with private power lines.

A smart community is not a just physical power network that delivers and receives energy, but also a cyber economic network that allows participants to trade energy, thereby making it possible to offer each household large incentives to save power and cut CO_2 emissions. For example, a household may largely lower the energy ceiling setup of its in-house EoD system, even if that may introduce the degradation of QoL (e.g., slightly colder in winter or slightly hotter in summer) to get the economic benefit of selling leftover energy and/or megawatts (i.e., guaranteed power reduction can be regarded as an item with economic value to be traded). Such economic incentives will make large energy savings and CO_2 reduction possible to a technically difficult level. It is exactly the goal of our *i*-Energy project, in that it will create a new social infrastructure and a new way of living. Currently, we are developing a power trading system among in-house EoD

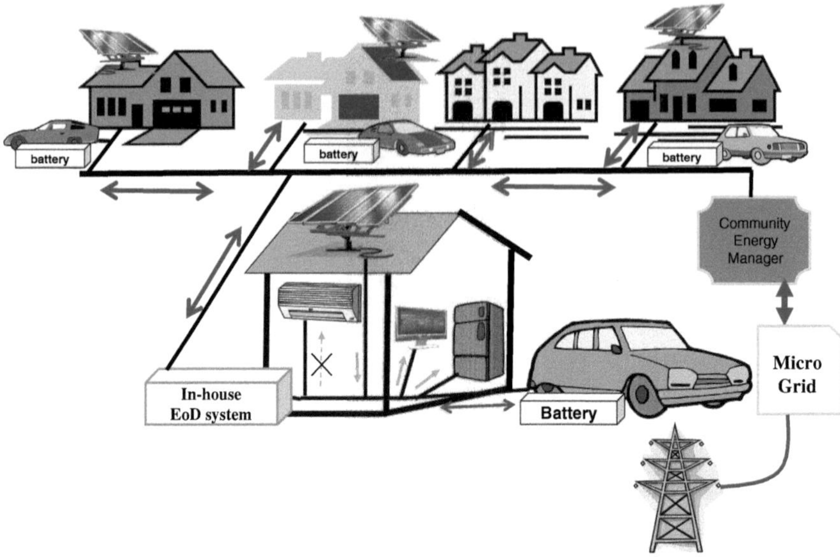

Fig. 18 Smart community for bidirectional energy trading

systems, where mechanisms to guarantee the stability of the physical power network as well as to allow us to implement versatile energy services are being studied.

5 Concluding Remarks

As noted at the beginning of this chapter, while cyber-physical integration plays a crucial role to design new social infrastructures in the twenty-first century, a unified theory has to be created to seamlessly bridge computation and information models and physical models. Moreover, to change social infrastructures dramatically, systematic efforts in business as well as in R&D must be made to bring together activities in many fields. To implement the *i*-Energy concept, for example, collaboration among a variety of industrial fields such as home appliances, storage batteries, electric cars, housing, ICT systems, and utilities should be established. To this end, we have set up the *i*-Energy Working Group (http://www.i-energy.jp) which coordinates activities among industry, academia, and government. The smart apartment room and smart house systems described in this chapter have been developed by collaboration among Working Group members, and we welcome any parties who can share their interest with us.

References

1. Kawashima H, Matsuyama T (2005) Multiphase learning for an interval-based hybrid dynamical system. IEIC Trans. Fundam E88-A(11):3022–3035
2. Kato T, Cho H, Lee D, Toyomura T, Yamazaki T (2009) Appliance recognition from electric current signals for information-energy integrated network in home environments. Int J Assistive Rob Syst 10(4):51–60
3. Kato T, Yuasa K, Matsuyama T (2011) Energy on demand: efficient and versatile energy control system for home energy management. Second IEEE international conference on smart grid communications, pp 410–415
4. Kato T, Tamura K, Matsuyama T (2012) Adaptive storage battery management based on the energy on demand protocol. Third IEEE international conference on smart grid communications, pp 43–48

Switching Perspectives: Creating New Business Models for a Changing World of Energy

Michael Valocchi, John Juliano and Allan Schurr

Abstract Traditional electric utility business models are becoming rapidly outdated. Policy changes, new energy technologies and more demanding consumers are driving the need to move away from purely operation-oriented approaches to more consumer-driven models. With these new business imperatives and a resulting need to focus on industry-level business model innovation, we explore how new models could evolve, and their benefits to consumers, utilities and other stakeholders. We also discuss the challenges ahead and the capabilities required for realizing this transformation.

1 Introduction

A century ago, the first great business model innovation in the electric power industry was set in motion with the move from small local plants delivering power over short distances to central generating plants delivering power great distances over high-voltage wires. This innovation was followed by a long period dominated by a "grow-and-build" philosophy that drove the development of near-universal access to electric power in much of the world through the mid-twentieth century. This philosophy reached its practical limit during the latter part of the century, due to factors such as plateaued economies of scale and practical optimum sizes of generating units,—but since then, there has been little evolution of business models from that of the "grow-and-build" years.

M. Valocchi
Philadelphia, PA, USA

J. Juliano (✉)
Washington, DC, USA
e-mail: juliano@us.ibm.com

A. Schurr
Denver, CO, USA

D. Mah et al. (eds.), *Smart Grid Applications and Developments*,
Green Energy and Technology, DOI: 10.1007/978-1-4471-6281-0_9,

Long-standing electric utility business models are rapidly becoming outdated in light of new technologies, policy changes and more demanding consumers. In the first part of the twenty first century, sustainable growth and efficiency have become imperatives newly demanded of the industry. Roles along the value chain are shifting, with traditional buyers gaining a foothold as value providers. These dramatic shifts all involve technological changes—changes that most energy providers understand. However, these shifts also require leaps into business model transformation that are new to most. To succeed in this new environment, industry model innovators will redefine traditional business models, as well as the infrastructure, rules and standards to facilitate the transition from traditional energy generation and delivery to more complex business models which incorporate emerging products and services enabled by new technologies. Important decisions on how best to make these moves—and the resulting rise to prominence of companies that successfully do so—will occur over the next decade. Understanding the evolving industry dynamics and how they drive transitions to new business models are clearly important from the business perspective. However, policy-makers who share with energy providers the common goal of transitioning to smarter grids must also understand these to and shape an environment that supports such development.

In this chapter, we first will present an overview of the ways in which the industry value chain is shifting in response to policy-, technology- and consumer-driven forces. We then will explore the new business model—industry model innovation (IMI) and focus on its structure, opportunities that IMI is poised to unlock for electric providers and strategies to realize the transition to this new business model in preparation for smart grids. This chapter will be concluded with policy recommendations for an efficient transition toward the new business model for smart grids.

2 Business Model Transitions and Drivers

With today's capital- and carbon-constrained environment, the imperatives of the industry have evolved from those relevant to "grow-and-build" economies, passive end users and one-way delivery of power toward those rooted in sustainable growth and efficiency, as illustrated in Fig. 1. Table 1 highlights the key differences between these two imperatives.

The main cause of such transition is the relentless pressure the industry faces to reassess its business models to accommodate transformations occurring in several key areas:

- *Government policy shifts*: Efficiency, conservation and renewable generation are receiving tremendous attention from governments attempting to meet goals related to climate change, energy security, and economic and job growth. At the same time, most industry revenue models are still based on a careful balance of the fixed nature of capital expenditures and variable cost recovery.

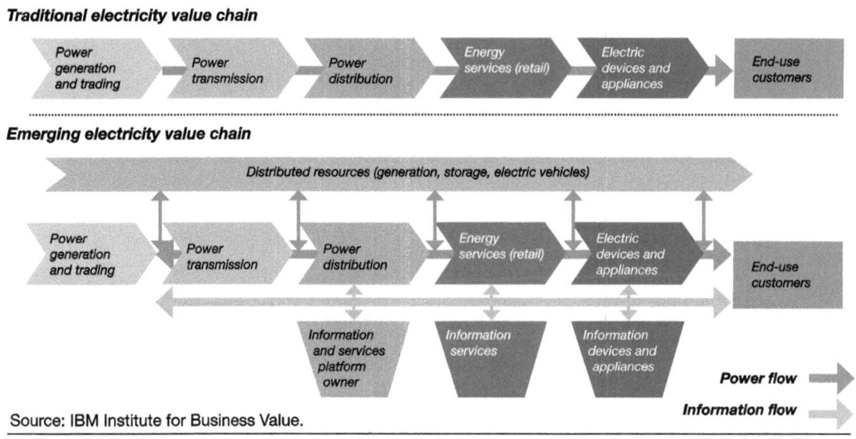

Fig. 1 A comparison of the traditional and emerging electricity value chains

Table 1 Difference between "grow-and-build" and "sustainable growth and efficiency" imperatives

	"Grow-and-build"	Sustainable growth and efficiency
Cost of Inputs	Low	Volatile
Infrastructure growth	Relatively unconstrained	Constrained by environment and public policy
Impacts of technology improvements on cost efficiency	Reduce generation cost	Little impact on generation costs
Generation basis	Large, centrally located units	Small-to-mid-size distributed units with storage
Perspective on resource acquisition	Assumed inexhaustible and benign resources	Assumed finite and harmful resources

The more successful these policies are in slowing growth in overall consumption from centrally generated sources, the stronger the need will be for new pricing models that can balance electric power companies' desires to support public policy objectives with revenue requirements to maintain service and reliability levels.

- *Changing consumer demands*[1]: As demonstrated in our previous reports "Plugging in the consumer: Innovating utility business models for the future," "Lighting the way: Understanding the smart energy consumer" and "Knowledge is Power: Driving smarter energy usage through consumer education," customers are now demanding more from their providers than merely reliable power at reasonable rates. Our global utility consumer surveys show consumers want more

[1] "Consumers" is used to refer specifically to residential and small commercial customers; "customers" is a more general term including large commercial and industrial users.

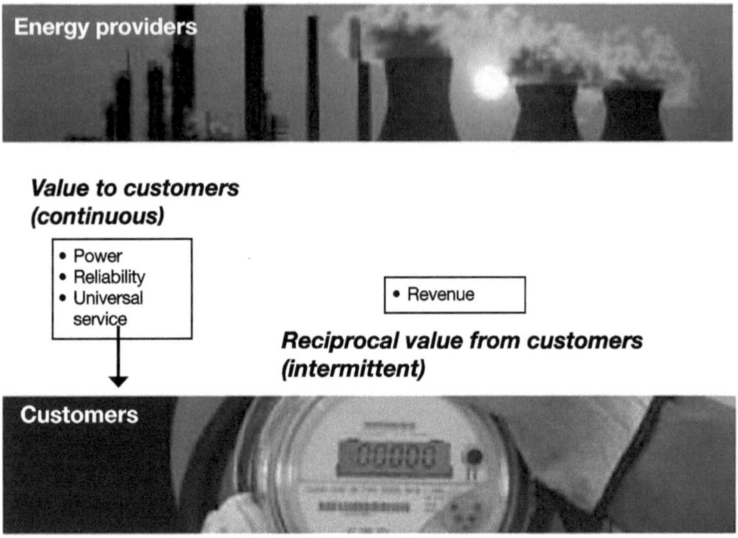

Sources: Jansen, Wendy, Wilchard Steenbakkers and Häns Jagers. *New Business Models for the Knowledge Economy*. Gower Publishing. 2007; IBM Institute for Business Value analysis.

Fig. 2 Traditional industry value model

control over their expenditures and environmental impact and more information about their energy usage—both in content and frequency [10–12].

- *The emergence of new technologies*: The introduction of smart grid and other distributed energy technologies will add complexity to the network, moving power and information in multiple directions and enabling a host of new participants and business models. Distributed energy resources such as customer-owned renewable generation, plug-in electric vehicles and energy storage will extend the value chain to include assets operated closer to the end user. The end users themselves, who may be capable of providing some combination of demand response, power or energy storage to the system, will also be an active and empowered integral part of the new value chain (see Fig. 1). The new participants and business models derived from this new value chain will provide strong competition for existing revenue streams.

This recharacterization of the industry value chain will dramatically reshape the value proposition among energy, service and product providers, as well as customers of these enterprises and the *value model* of the industry as a whole (Fig. 2). A value model is the combination of value provided to customers and the *reciprocal value* received from customers in return [8]. In the case of the electric power industry, the traditional value model involves consumers receiving reliable and universal power at reasonable rates, for which they offer providers reciprocal value in the form of intermittent (usually monthly) revenue. In response to the

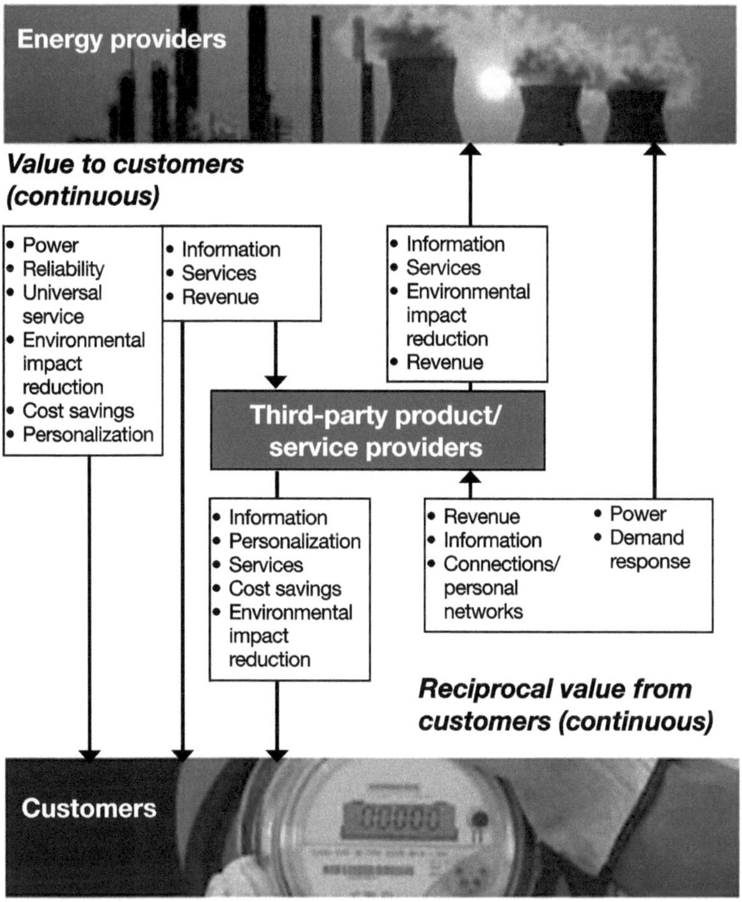

Sources: Jansen, Wendy, Wilchard Steenbakkers and Häns Jagers. *New Business Models for the Knowledge Economy*. Gower Publishing. 2007; IBM Institute for Business Value analysis.

Fig. 3 Emerging industry value model

changes in policy, customer demand and technologies, the emerging value model is more complex because customers have more to offer in return to power providers and other participants than just payment for energy consumed (see Fig. 3).

Some of these new elements of reciprocal value are primarily operational in nature; demand response, load profile flexibility, and distributed power and storage (where customers have these on their premises) allow for optimization of system performance and asset utilization. Others, such as information on energy consumption patterns, other consumer demographic and behavioral information, and access to personal connections/networks for social content delivery are the foundation for new revenue sources for companies able to effectively leverage the information.

Not only are there many more types of reciprocal value, but also the very nature of the value has changed from an intermittent source of reciprocal value to a *continuous flow*. As the number and frequency of reciprocal value exchanges grow, the complexity of the ecosystem increases and the total amount of value in the system available for capture by ecosystem participants increases dramatically.

The flow and volume of information itself, along with new services it enables, are strong contributors to this continuous flow of new value. At present, there is little financial or operational value to the data generated by consumers (essentially total usage on a monthly basis) because it is too limited in scope and frequency of delivery to be of value to parties other than the electric provider's own billing and operations departments. However, the quantity, frequency and quality of data generated by consumers—and its usefulness to energy providers and third parties alike—are set to grow exponentially as smart grid infrastructure is deployed. Devices and software that capture, analyze and present this information to consumers and energy providers are already proliferating, and services that make use of this data are rapidly emerging.

3 Methodology and Framework

Using an extensive literature review, as well as our previous industry surveys and consulting experiences, we evaluated the decisions facing electric power companies as they address the business model-related challenges and opportunities before them. As the basis for this analysis, we chose a framework based on IBM Institute for Business Value research summarized in the report, "Paths to success: Three ways to innovate your business model" [6]. This report presented an analysis of 35 best practices, in which fifteen of them were from the Business Week list of leading business model innovators, including Apple, IKEA, Southwest Airlines and others [1]. The remaining 20 cases are selected based on a company's reputation for leadership in business model innovation according to analyst reports, interviews with experts in diverse industries and a broad literature review. The cases represent nine industries from different regions and with diverse types of business model innovation. Each of the cases was assigned with values of high, medium or low indicating the level to which business model innovation of specific types (industry, enterprise, and revenue) were leveraged. By taking into account financial performance and various factors, such as industry types, period of innovation, age of company and size of company by number of employees, revenues and assets, the keys to success in business model innovation were identified from these cases.

As a result of this work, we identified three main types of business model innovation strategies:

- *Industry model innovation*: Innovating the industry value chain by moving into new industries, redefining existing industries to serve new markets or creating entirely new industries
- *Enterprise model innovation*: Innovating around the structure of the enterprise and the role it plays in new or existing value chains, with focus on those areas of the business where it has an advantage and delivers value
- *Revenue model innovation*: Innovating how revenue is generated through offering reconfiguration (product/service/value mix) and pricing models [6].

4 Industry Model Innovation: Elements, Opportunities and Challenges

In our earlier publications "Plugging in the consumer," "Lighting the way" and "Knowledge is Power," we envisioned a future for energy providers driven by technology evolution and increasing consumer control. Analyzing the impact of different levels of progression in these two areas suggests four states through which the industry will migrate [10–12]:

- *Passive Persistence*: Traditional utility market structures still dominate, and consumers either accept or prefer the historical supplier-user relationship.
- *Operations Transformation*: Some combination of network and communications technology evolves to enable shared responsibility, but consumers either cannot or elect not to exert much control.
- *Constrained Choice*: Consumers take decisive steps toward more control but are limited to certain levers (technologies, usage decisions or choices in providers) by regulatory and/or technological constraints.
- *Participatory Network*: A wide variety of network and communications technologies enables shared responsibilities and benefits [10, 11].

Because of the high likelihood of increasing demand for control and information by customers and continual technological improvement and deployment, these two reports emphasized our belief that the end state for the industry is likely to be a Participatory Network [10, 11].

In the last few years, the pace of progress toward this new model has increased. Government mandates to upgrade and incentives to invest in the existing twentieth-century infrastructure have helped push aside some of the most critical barriers for moving toward a Participatory Network—particularly in places like the United States and China where direct government investment is being made. In addition, national and global efforts to standardize technological and communications specifications by organizations such as the International Electrotechnical Commission (IEC) and the U.S National Institute for Standards and Technology (NIST) help remove another barrier to progress. These developments—among others—have strengthened our conviction that some form of a Participatory Network is a logical destination in the next decade. The most likely path is through IMI that results in extraordinary change to the *platforms* on which electric providers operate.

The term *platform*, as used here, refers to a common architecture (essentially, a design for products, services and infrastructure facilitating users' interactions) and set of rules (protocols, rights and pricing terms) that provide a standard foundation governing transactions among two or more parties [2]. In general, platforms provide a means for providers and buyers of products and services to interact and create value that could not be created otherwise. The platform lowers the costs of providing services by offering some level of standardization for transactions and reducing duplication.

In this sense, the electricity network was one of the earliest technology platforms. It provided a means for power generators to move their output to buyers, a means for buyers to accept delivery of the output, and a standardized technological specification (e.g., the 120 V/60 Hz and 230 V/50 Hz standards for electric power in the Americas and Europe, respectively) around which thousands of applications (for heating, cooling, lighting, mechanical power and so on) would be built over the years.

Many types of platforms exist in consumer and business information technology (IT), exemplified by the broader Internet platform and today's popular social networking sites. The use of platforms is sporadically seen in the telecommunications sector (e.g., the NTT DOCOMO i-mode platform) as well [4]. In each case, there are diverse participants and a common set of business processes that enable competition and new value creation.

4.1 Single-Sided Versus Multisided Platforms

Many platforms are *single-sided platforms*, with a seller at one end and a buyer at the other and, often, intermediaries (distributors) between them that transfer the product from buyer to seller without changing it substantively [4]. The electric power network has historically operated as a single-sided platform. Until the advent of wholesale generators, the business operated as the simplest possible form of a platform—the manufacturer (generating utilities), by virtue of owning the entire value chain from the point of input of fuel to the point of entry into the user's premises, sold directly to the customer with no intermediaries; in fact, some utilities also controlled the fuel production itself. The emergence of independent generators and pure energy retailers moved the power transmission and distribution network closer to a position where it did act in an intermediary fashion, transporting power from wholesalers for purchase and use by end users.

As new value is generated in the network through expansion of the value and reciprocal value exchanged, industry model innovators will develop new businesses that more closely resemble *multisided platforms*. In a multisided platform, there may be multiple types of buyers and/or sellers—in fact, a single party can be both a buyer and a seller.

A shopping mall is an example of a multisided platform: manufacturers, retailers and shoppers all benefit from having a single location where they can meet and transact business. Malls provide common facilities, like restrooms and parking, which help lower costs to stores that otherwise would have to individually provide them. Since these economies help reduce costs to retailers, prices can be lower, benefiting shoppers [4]. A wider variety of stores and services brings more shoppers; more shoppers bring higher sales volumes for manufacturers and lower costs for retailers (and, in theory, also lower prices for shoppers). Thus, some element of network economy is bundled into the shopping mall value proposition. The platform owner (the mall operator) extracts some of this value in the form of rents to store owners and, in some cases, service fees to shoppers. (There are also organizations not directly involved in the mall transactions—credit card issuers, for example—that benefits and take revenue from the transactions.) But without all of the parties being involved, none would get any of the benefits.

Other examples of multisided platforms include newspapers (with readers serving as one side and advertisers another) and health maintenance organizations (with patients being one side and doctors and pharmaceutical companies serving as other sides). Yet another example is video games (with players being one side and developers, publishers, content providers, licensors, tools and middleware providers making up the other sides).

In coming years, a smart grid with energy and information flowing in multiple directions will provide support for interactions among all ecosystem participants, facilitating the development of electric power industry multisided platforms. These platforms will link energy suppliers, service providers, device manufacturers, application developers and end users (residential/industrial/commercial). Each group of participants needs access to a platform to reach the other groups, but a platform does not substitute itself for any particular participant.

Multisided businesses provide benefits to the interacting groups—while profiting from the transactions—by increasing and capturing *indirect network externalities* (INEs). Figure 4 shows how a multistep process stimulates these INEs in a two-sided market. In the first step, growth in the number of potential customers on side one for complementary products and services on side two occurs. This leads to an increase in the quantity and diversity of complements made available by side two. Next, because side one users are favorably inclined to a wider variety of products and services on the other side, they are more likely to join the platform. This makes it even more attractive for side two to develop new complements, and the cycle sustains itself [3].

This can be (and has been) a successful and profitable way to innovate industry models, allowing for additional value creation and profits throughout the value chain. However, as the builders of broadband infrastructure in the United States learned in the late 1990s and throughout this past decade, the entities that construct and maintain multisided platforms are not necessarily the ones that will reap all of the profits generated by such a model. For example, companies such as Amazon,

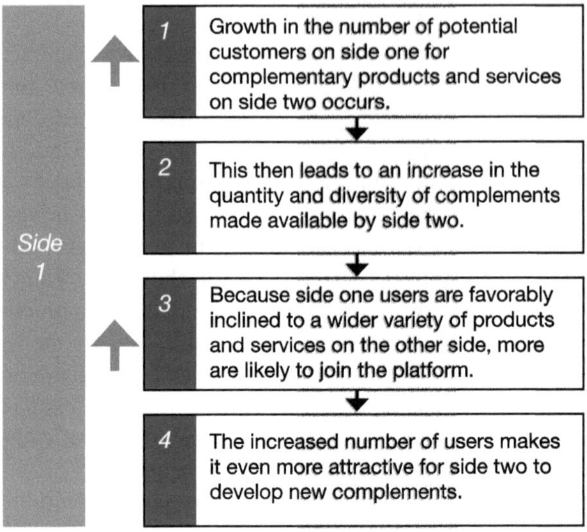

Sources: Eisenmann, Thomas and Andrei Hagiu. "Staging Two-Sided Platforms." Harvard Business Publishing. 2007; IBM Institute for Business Value analysis.

Fig. 4 Indirect network externalities in a two-sided market

Skype and YouTube benefited strongly from others' investments in broadband network infrastructure without taking part in the capital outlay for that infrastructure.

Because of their enormous investments in smart grid and other improvements, today's electric companies will, in a somewhat analogous fashion, be responsible for putting in place most of the infrastructure required for new industry participants to emerge. At the same time, it is likely that new electricity-related business models that leverage the smart grid infrastructure will be launched by entities that did not make direct investments in it. While it is healthy for the industry as a whole to encourage innovation in new products and services, incumbent electricity companies must be well prepared for this. Mapping out business models that take advantage of the new network-enabled capabilities will allow electric companies to reap as much of the ecosystems' new value as their ambitions permit.

4.2 Platform Development in the Electric Power Industry

Up until now, the electric power industry has not had much reason to create multisided platforms because product delivery has been a purely physical process; both energy and information flow have been unidirectional; and the typical end consumer had little desire to communicate with providers other than through

Ecosystem function	Participating sides	Platform providers
Carbon capture and storage (CCS)	Generators, carbon product users	CCS plant operators
Carbon disclosure reporting	Governments, NGOs, consumers, utilities	Third-party reporting organizations
Demand response	Consumers and businesses, distribution companies/utilities	Demand response firms
Electric vehicle charging	Consumers, power retailers, automakers	Public space providers (malls, parking garages, etc.)
Electricity comparison shopping	Consumers, power retailers, advertisers	Portal providers
Electricity transport	Power retailers, energy users, distributed generators	Transmission/distribution companies
Energy aggregator/marketer	Consumers, power retailers	Energy aggregators
Energy broker	Power retailers, energy users, distributed generators, generating companies	Energy brokers/traders
Energy management	Consumers and businesses, energy management service providers, application and content providers	Device/system makers or portal providers
Energy storage	Distributed generators, energy users	Energy storage operators
Information aggregator (device based)	Consumers and businesses, providers of energy products and services, application and content providers	Device/system makers
Information aggregator (portal based)	Consumers and businesses, providers of energy products and services, application and content providers	Portal providers
Renewable/carbon credit aggregation/trading	Renewable generation owners, coal/gas/oil generation owners, power retailers, governments	Third-party market makers

Source: IBM analysis.

Fig. 5 Examples of potential multisided platforms in electricity

service provision, billing and problem resolution. All of this is changing. We expect a number of platforms will develop within the electricity ecosystem in the near future (see Fig. 5).

One-way value, information and money could be exchanged on these platforms is via an *energy marketing portal*, on which customers can shop for the best deals on power or for power that meets specific personal requirements (see Fig. 6). The platform owner creates value by providing the end user with access to various applications (for energy shopping, energy management, etc.) in return for passive usage and preference data, which the customer has approved for use for these purposes. This is delivered back to the platform owner through the applications for aggregation and presentation for the other side of the platform, the energy retailer.

The retailer has access, through the platform provider, to a suite of applications to gain access to and evaluate the customer data. This information is valuable to the energy retailers, and they are willing to pay the platform owner for access to it to build marketing programs for products and services aimed at likely customers. Informed by the platform owner and the retailer side applications, retailers communicate their best offers to the buyers seeking deals or new programs. Ultimately, the retailer gets return for their investment in the platform—paying the platform owner for access and information—in the form of increased revenues from consumers who value the programs and services they offer.

A slightly more complex example involves an *information aggregator* (see Fig. 7). An information aggregator builds a relationship with end users by selling them (possibly at a subsidized price) energy usage display/management devices that are preloaded with useful applications, all of which are purchased from third-party developers. They thus serve as the link between device manufacturers and end users and between application developers and end users.

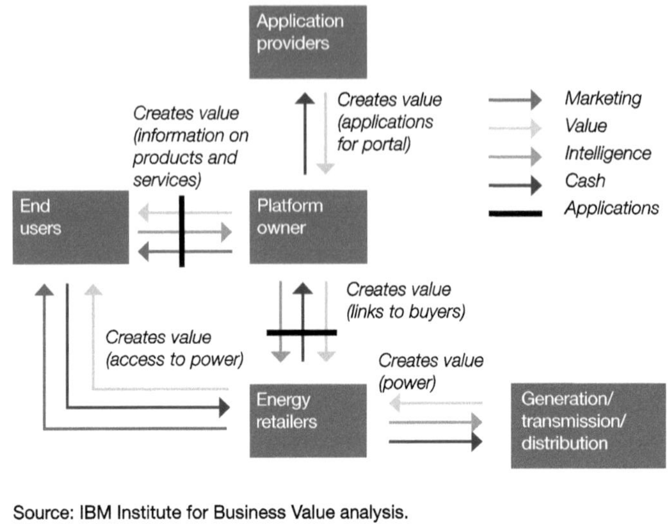

Fig. 6 Platform example: energy marketing portal

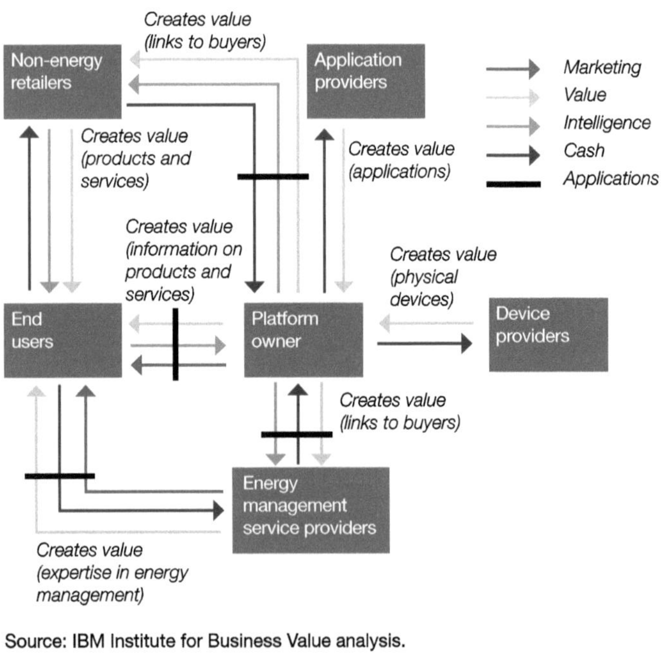

Fig. 7 Platform example: information aggregator

With appropriate permissions from consumers, the platform owner can also collect information about the end users' energy usage patterns, build profiles and market those profiles to energy and nonenergy (e.g., appliance) retailers. As with the energy marketing portal, the retailers are willing to pay for this information because of the benefits they accrue from it. The end users' profiles also include information on demand response they are willing to provide; this can be exchanged with the energy retailers for payment as the need for such response arises. Thus, cash can flow in both directions between retailers and end users, with the transactions in both directions facilitated by the platform owner.

Note that in these two examples, the platform owner is a company purely focused on the operation of the platform and the collection and exchange of data. Except for the end user, any one of the parties in the ecosystem can also serve as the platform owner—visually, this can be seen as "collapsing" the value exchange in the diagram for that party into the platform owner role in the center. For example, a device manufacturer could set up a multisided platform and take on responsibility as platform owner—including all interactions with application providers, end users and energy retailers.

4.3 The Platform Staging Challenge

A critical challenge can be encountered early in the development of new platforms when prospective users on each side are reluctant to actively participate until there are sufficient users on the other side. Often, the platform provider must "stage" the platform in advance, either heavily subsidizing one side to get it on board in sufficient numbers to attract the other side or by bringing on attractive or highly visible transaction partners to affiliate exclusively with the platform. When it works, profits can be enormous. When it does not, failures can be magnified by this investment. Also, in most instances where a platform can profitably exist, the combination of strong network effects and high barriers to entry means there is room for only a few platform owners (and, in some cases, only one).

Incumbent energy providers have an advantage in this market staging, as they already have a relationship with a critical mass of customers. Those customers could serve as one side of the platform in sufficient numbers to attract attention from application, service and device providers. This strategy is both less costly to develop and entails less risk than the approach a company starting from scratch must employ. Success is dependent on potential participants in the new network who are already customers of the platform owner on one side. The value of the products and services they already purchase is not dependent on the presence of a second side—this is certainly the case with electric power [3].

In this scenario, the platform owner facilitates transactions between the existing customer base and at least one new side of the platform, adding new functions, services or capabilities to its offering to the former to encourage transactions between the two. Google, for example, initially launched as a vendor of Web search services (Google.com and others via license). In its first 2 years, Google's only source of revenue was from search engine license fees. However, after amassing a critical base of end users, it was in a position to offer paid-listing advertisements to these customers and transform into one of the most profitable multisided platforms that has emerged [3].

5 Capabilities Required for a Successful Transition

Companies that envision being platform owners will need to have key competencies in marketing, sales and customer relationship management. That fact, combined with the ready-made set of platform participants already present in the form of existing electricity customers, puts retail electric providers (or integrated utilities' retail operations) in a good position to take on the challenge of platform ownership. However, there are other requirements that are not necessarily in-house skills.

Software-based platform owners, like information aggregators and energy shopping portal operators, will need to master systems architecture and application interface development and support—or find a partner that can offer these services seamlessly to platform users. This assumes that the platform owner will cede the job of developing applications themselves to third parties interacting with the platform. If the platform owner instead plans to internalize application development as well, that adds another level of IT complexity (application development and support and IT infrastructure development and support). In either case, the company will need to ensure that its approach gives it a strong enough set of capabilities in these areas to successfully compete against rival platforms.

Service-based platform owners—such as energy management specialists—will have other challenges. These firms will have to transition at least that part of the organization to function more as a professional services firm than a traditional energy supply and delivery company, with requisite skills in solution creation and maintenance, knowledge and intellectual property management, research and development, and contract management. This will require a major cultural shift for a traditional utility, as the focus of management will be human and intellectual capital rather than physical assets and processes.

Companies willing to tackle IMI and sit at the nexus of new complex relationships among business partners and customers will be well positioned to create and capture new demand for emerging products and services. Strong growth in revenues and profits—albeit accompanied by some risks—is achievable in multisided business models because of the embedded network economies of scale (i.e., margins increase with network size). While several types of activities can serve as the basis for a multisided platform (as exemplified in Fig. 5), there are some common questions that potential industry model innovators need to address before making major investments.

How many platforms can effectively serve a single purpose? Where strong network economies of scale are in place and the cost of participating in multiple platforms is high for at least one participant, the likelihood is higher that markets will be served by a single platform. If this appears to be the case, a strategic decision about whether to fight for sole platform ownership or to pool resources in a platform shared with others must be addressed early and in depth. While "winner-take-all" economics make sole ownership of a platform attractive in theory, the reality is that only a company with extensive resources (especially for marketing) and strong existing relationships with a large number of potential users on at least one side will be positioned to succeed [2].

What are the incentives and costs for platform participants? Appropriate pricing, support and incentives for participants are critical success elements for any platform. A key question is whether any particular side requires subsidization and, if so, which one. Based on the history of past platforms from a variety of industries, the most price and/or quality sensitive participants and those with high visibility or attractiveness appear to be the strongest candidates for subsidization [4]. A firm able to leverage existing relationships as a vendor to one side may not have to provide strong subsidies. As for pricing, network economies of scale mean that underpricing platform participation will lead to suboptimal platform profitability; overpricing participation for any one participant will choke off growth and leave room for competitors to gain a stronger foothold in the marketplace.

What is the critical mass needed for success? Companies thinking of transitioning to a multisided platform model must be confident that the business model embraced is easily scalable [2]. Potential platform owners hoping to successfully leverage a vendor relationship to get a leg up on competitors should critically examine whether their current customer base is sizable enough to ensure this advantage can provide a meaningful head start and whether other territories can be easily integrated into the same side of the platform.

When should the move toward a platform-centered business model be made? In many platform battles (especially the ones that began in the dot-com era), gaining first-mover advantages was viewed as the most critical element of a business strategy. However, history has shown that later movers may actually benefit from standing back from the first wave. Google was neither a first mover in Web search nor paid-listing models, but it was able to leverage lessons learned from earlier proponents of each to improve on their efforts [2].

When and how should the move toward a platform-centered business model be communicated? Appropriately timing and managing the announcement of a business model shift is critical. Poorly handled or delayed announcements of major changes run the risk of surprising and angering investors, regulators and employees. This is particularly true for companies that have been operating a certain way for a very long period of time, as is the case with many traditional vertically integrated utilities. Conversely, communicating intentions too early may elicit strong competitive responses or lead to unrealistic expectations about future prospects for the business (and possible volatile stock price behavior for publicly traded companies) [3].

What cultural changes are required to successfully transition to a platform-centered business model? Leaders of electric power companies already understand the need for new workforce skills as the transition to a digital, information-driven industry environment takes place. However, companies transitioning to a multisided model should also prepare for cultural changes [3]. As discussed in the previous section, this includes some shifts in focus from physical assets and processes to human and intellectual capital. Additionally, companies might have to rethink their approach to customers, as explained by Amazon CEO Jeff Bezos:

> One of the things we had to learn through zShops [which host small merchants] and auctions was that we needed to think of ourselves as serving two different sets of customers. We pride ourselves on being customer-centric, but for years, 'customers' meant 'buyers.' As we began to operate auctions and zShops, we realized that these third-party sellers were equally important customers. And, it took a little while for the organization to learn what their needs were and how we could best meet them [3].

6 New Policy Approaches to Stimulate Platform Development[5]

While energy providers can and should move forward as answers to the questions above are developed, there are unresolved policy and regulatory issues that often hinder grid modernization, consumer choice and industry business model innovation. While the policy framework for the industry differs in substance and strength across the globe (and even across regions within a country, as in the United States of America), two areas stand out in their relation to acceleration of business model renewal in the industry.

How to reform the cost of service regulatory model: The current model favors investment in traditional grid assets like power plants, substations and "stringing more copper." These kinds of investments are still needed, but are likely to be insufficient to meet the system challenges of the future. This type of model has also over the years largely excluded energy providers from pursuing investments in unregulated businesses, which is clearly counter to the sort of growth-focused platform development outlined earlier in this chapter.

A strong alternative model that is more conducive to the future directions of the industry can be found in performance-based regulation, in which the regulator sets out specific metrics for energy providers and judges the prudence of investments based on success in achieving these metrics. A revenue premium is established to motivate overachievement toward these goals, while penalties are assigned for underperformance. The most effective performance-based regimes reform traditional regulatory models with an eye toward fostering disruptive change in business models by incorporating the following features:

- A realistic investment cycle, such as the eight-year price control period in the United Kingdom's RIIO performance-based regulation scheme [9]
- A clearly understood and measurable system of outputs for the utility for planning and investment, including but not limited to customer satisfaction, availability, reliability, environmental impact
- Strong financial incentives that drive both profitability and utility performance, such as annual revenue adjustments linked directly to outputs and basis point penalties for not meeting minimum regulatory outcomes; flexibility to provide customer choice for new regulated and unregulated products and services, using short-run marginal cost-based transfer pricing, is an important financial-related enablement mechanism

Balancing data privacy protection and market innovation: Appropriately designed industry practices and regulatory guidance can help the energy ecosystem protect sensitive information. On the other hand, enacting rules that make data access too difficult or too costly will effectively limit consumer choice and inhibit progress in delivering efficiency and conservation benefits, consumer choice, and desirable new products and services. Strongly expressed consumer concerns in recent smart meter rollouts [11, 12] (which our consumer surveys foresaw) make attention to privacy imperative, but an approach that also balances the needs of a new industry ecosystem can be achieved by:

- Making regulations flexible enough to accommodate fast-changing technologies and consumer preferences
- Making it easy for consumers to express their preferences
- Considering standard best practices from other industries regarding consumer privacy
- Considering a "Privacy by Design" framework at the operational level [7]

While some form of regulation is likely to persist in many parts of the world, failure by these regulators and policy leaders to allow energy providers the flexibility to develop unregulated businesses be active participants in the new platforms that will develop will put the regulated entities themselves in peril of being made obsolete by the new entrants gaining market share through these ecosystems.

7 Conclusion

The business models that brought the electric utility industry success in the middle of the twentieth century are overdue for revisiting. Much of the basis for their foundation—one-way flow of power and information, declining costs associated with increased usage, undifferentiated and passive consumers, unlimited access to inexpensive carbon fuels for generation and regulatory protection from threats to the core businesses—has already shifted or will do so in the next decade.

For those well positioned to be industry model innovators, the "grow-and-build" years are, in a sense, back—but with a different emphasis. What is being "built" are sophisticated new business platforms to support information exchange, consumer participation and new services. As with the generating units of the post-World War II era, these platforms can increase in profitability as usage increases. "Use more—we will keep building" will return as a marketing message to consumers—but this time around, the emphasis will be on information and services rather than energy.

Acknowledgments This chapter is an updated republication of the paper "Switching perspectives: Creating new business models for a changing world of energy" by Michael Valocchi, Allan Schurr and John Juliano, IBM. The copyright permission for reusing the paper has been granted by IBM.

References

1. Businessweek (2006) The World's Most Innovative Companies. Bloomberg Businessweek Magazine. Retrieved 3 Oct 2013 from http://www.businessweek.com/stories/2006-04-23/the-worlds-most-innovative-companies
2. Eisenmann T, Parker G, Alstyne MWV (2006) Strategies for two-sided markets. Harvard Business Review. Retrieved 6 Oct 2013 from http://hbr.org/2006/10/strategies-for-two-sided-markets/ar/1
3. Eisenmann TR, Hagiu A (2007) Staging two-sided platforms. Harvard Business School Background Note 808-004. Retrieved 6 Oct 2013 from http://www.hbs.edu/faculty/Pages/item.aspx?num=34746
4. Evans DS, Hagiu A, Schmalensee R (2006) Invisible engines: how software platforms drive innovation and transform industries. Massachusetts Institute of Technology, London
5. Futch M (2013) Power shift: smart grid transforms electric power for the 21st century. Georgetown J Int Aff 14(1)
6. Giesen E, Berman SJ, Bell R, Blitz A (2007) Paths to success: three ways to innovate your business model. IBM Global Business Services, U.S.A. Retrieved 5 Oct 2013 from http://www.businessmodelcommunity.com/fs/Root/8v2kj-Paths_to_Success_paper.pdf
7. IPC (2011) Privacy by design: from poicy to practice. Information and Privacy Commissioner, Ontario, Canada, Ontario, Canada. Retrieved 12 Nov 2013 from http://www.ipc.on.ca/images/Resources/pbd-policy-practice.pdf
8. Jansen W, Steenbakkers W, Jägers H (2007) New business models for the knowledge economy. Gower Publishing Ltd., Hampshire
9. Ofgem (2013) Network regulation—the RIIO model. Retrieved 12 Nov 2013 from https://www.ofgem.gov.uk/network-regulation-%E2%80%93-riio-model
10. Valocchi M, Juliano J, Schurr A (2009) Lighting the way: Understanding the smart energy consumer. IBM Institute for Business Value, New York. Retrieved 6 Oct 2013 from http://public.dhe.ibm.com/common/ssi/ecm/en/gbe03187usen/GBE03187USEN.PDF
11. Valocchi M, Schurr A, Juliano J, Nelson E (2007) Plugging in the consumer: Innovating utility business models for the future. IBM Institute for Business Value, New York. Retrieved 6 Oct 2013 from http://www-05.ibm.com/de/energy/pdf/plugging-in-the-consumer.pdf
12. Valocchi M, Juliano J, Schurr A (2012). Knowledge is power: Driving smarter energy usage through consumer education. IBM Institute for Business Value, New York. Retrieved 6 Oct 2013 from http://public.dhe.ibm.com/common/ssi/ecm/en/gbe03475usen/GBE03475USEN.PDF

Part IV
International Case Studies

Smart Transmission Grids Vision for Europe: Towards a Realistic Research Agenda

Luigi Vanfretti, Dirk Van Hertem and Jan Ove Gjerde

Abstract Smart grids have attracted significant attention lately, and one can even speak of hype. However, much of the attention is paid to the distribution side and consumer interaction. Nevertheless, also at the transmission-level important improvements can be achieved through farsighted and careful intelligent grid design and implementation. This chapter describes and proposes a realistic research agenda in which smart transmission grid (STG) research may operate, with focus on operational planning and operations of the pan-European electricity grid. Firstly, a research outlook seen from current European policy is laid out to redefine the most consequential research directions liked to the needs of transmission systems from real time up to planning. Secondly, operations (real time to hours) of the transmission system are discussed, and monitoring and control technologies that can be achieved through the application of synchronized phasor measurement technology are highlighted as a means toward a STG. Next, challenges related to planning (hours to years) are discussed keeping in mind the need of flexibility, coordination, and new methods for assessing system security. Going beyond the purely academic point of view, this chapter specifically aims to bring a realistic approach toward research for electric power transmission to be able to transform into a STG.

L. Vanfretti (✉)
KTH Royal Institute of Technology, Stockholm, Sweden
e-mail: luigi.vanfretti@ee.kth.se; luigi.vanfretti@statnett.no

L. Vanfretti · J. O. Gjerde
Statnett SF, Oslo, Norway
e-mail: jan.gjerde@statnett.no

D. Van Hertem
Catholic University of Leuven (KUL), Kasteelpark Arenberg 10,
3001 Heverlee, Leuven, Belgium
e-mail: dirk.vanhertem@esat.kuleuven.be

D. Mah et al. (eds.), *Smart Grid Applications and Developments*,
Green Energy and Technology, DOI: 10.1007/978-1-4471-6281-0_10,

1 Introduction

Smart grids have attracted significant attention lately, even so that one can speak of a hype. Smart grids can be seen as energy systems that enable optimal operation of the energy system through improved measurements and better control, and this while taking into account all stakeholders involved. The raising interest in "smart grids" has prompted an ever-increasing wave of discussion regarding a more disruptive introduction of information and communication technologies (ICT) to increase efficiency in electricity delivery and power network management. However, much of the attention is paid to the distribution side and the consumer interaction [6, 26]. In particular in Europe, the STG has remained in the background, often using the argument that "the transmission system is already smart."

Synchronized phasor measurement units (PMU) are a notable exception to this. Together with their supporting ICT infrastructures, PMU data applications form an important of many smart grid implementation plans. Nevertheless, the penetration of this technology in the European power system is lagging the development in the USA [5]. However, at the transmission level, there are also other important improvements that can be achieved through farsighted and intelligent grid design and implementation. These improvements are indispensable to operate future electric power systems. This future system will be challenged by unpredicted uncertainties brought upon by a higher penetration of renewable and variable energy sources, limited investments in transmission assets, and an ever higher demand for a more secure supply of electric energy at the lowest possible cost.

Many of these improvements are not ready to be immediately implemented in the current system. There is a specific need for research on some key aspects of the smart transmission system which are deemed essential for the full development and utilization of the future grid. The demonstration of research in real applications is essential in order to get the results validated and the innovations be used in the field. However, a significant portion of the ongoing research seems to be out of touch with reality. This problem arises from the fact that many theoretical approaches are not benchmarked in a lifelike simulation environment using realistic test systems, approaches that do not sufficiently take into account current installations, communication protocols, practices, and operation. Also the non-technical and organizational (or regulatory) background is not sufficiently taken into account. In a nutshell, "artificial" research environments, while useful for discussing theoretical concepts, limit practical implementations and may be unfit to cater to measurements, data, and organizational aspects arising in the real power network. It becomes climacteric to realize that several limitations and boundaries conditions exist and must be taken into account in order to avoid oversimplified or overly optimistic solutions that are not applicable in reality.

In this chapter, the desired framework in which STG research must operate is described, and some policy incentives are outlined. For this, the focus is on the transmission system itself, and more specifically its operational planning and operations. Section 2 lays out the research outlook as seen from current European policy, hence helping redefine the most consequential research directions. A STG framework from real-time operations up to planning is developed. Section 3 focusses on the operations (real time to hours) of the transmission system. We discuss several considerations that need to be considered so that monitoring and control of smart transmission grids (STGs) can be practically achieved through synchrophasor measurement technology (SMT). In Sect. 4, we discuss a number of challenges that are related to system planning (hours to years). The changes that are needed because of the rapid increase of generation from variable and less controllable energy sources are discussed. Increased flexibility, coordination, and a new way of looking at system security are discussed.

Within this chapter, it is not the authors' intention to give a comprehensive overview of all the outstanding issues and their binding research, but rather to cherry-pick some key challenges and potential pitfalls within smart transmission system research and development.

2 Smart Transmission Research Defined

2.1 A European Perspective on Research Needs

A definition of smart grids is given by the SmartGrids European Technology Platform [48]:

> A Smart Grid is an electricity network that can intelligently integrate the actions of all users connected to it – generators, consumers and those that do both – in order to efficiently deliver sustainable, economic and secure electricity supplies.

The same organization identifies key research areas in its strategic research agenda for 2035 [49], where the main transmission challenges are listed as follows:

New technologies: that improve flexibility and storage and allow long-distance energy transmission. The development of these new technologies will include upgrades in terms of materials, component reliability, and automatic controls.
Improved ICT technologies: to enhance monitoring, control, and modeling of the grid.
Smooth transition path: through investments at the current stage that are not compatible with developments of the future system.
Legal framework and market structures: to enable the different stakeholders to optimize the use of the energy system in the most optimal, "smart" manner including correct allocation of costs and benefits among stakeholders.
Socioeconomic incentives: to allow the necessary infrastructure investments those are required to be performed.

The European Commission communicated in its "Blueprint for an integrated European energy network" [16] that it is necessary to have "rapid investments" in order to ensure "(1) a competitive retail market, (2) a well-functioning energy

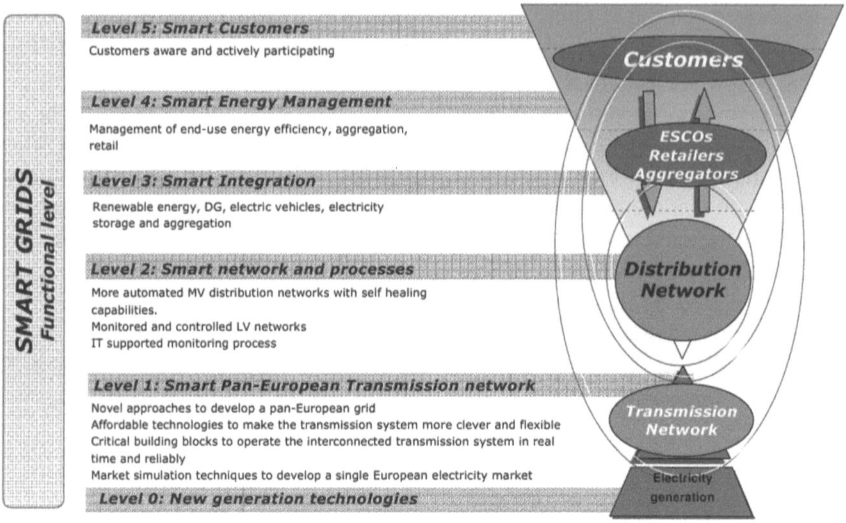

Fig. 1 Functional levels of the smart grid according to European Electricity Grid Initiative [17]

services market which gives real choices for energy savings and efficiency, (3) the integration of renewable and distributed generation, and (4) to accommodate new types of demand, such as from electric vehicles." Next to these technical incentives, there is also a need for an update of the legal framework and to adapt legislations taking into account smart grids and smart meters [3, 51]. A higher transparency within the smart grids and an information platform is needed [16]. Much of this vision is also shared by the European Regulators Group of Electricity and Gas (ERGEG) position paper on smart grids [8, 47]. The regulating authorities are the responsible to provide the correct framework in which the different stakeholders can develop and use the smart grid. Performance indicators are seen as an important aspect in order to the development of a correct framework. Furthermore, a further harmonization of regulation and a regulatory framework which is consistent on a longer-time basis is an absolute requirement for the successful development of efficient smart (transmission) grids.

The European Electric Grids Initiative (EEGI) has indicated in its roadmap for RD&D 2010–2018 [17] a number challenges for the upcoming years. The document identifies 3 main action areas for the development of the future grid: (1) the integration of new generation and consumption models, (2) a coordinated planning and operation of the pan-European grid, (3) new market models to maximize European welfare. The smart grids field is subdivided in different functional levels (see Fig. 1). Within the STG level, the EEGI indicates four main research domains: the pan-European grid architecture, power technologies, network management and control, and market rules. For each of these four areas, specific

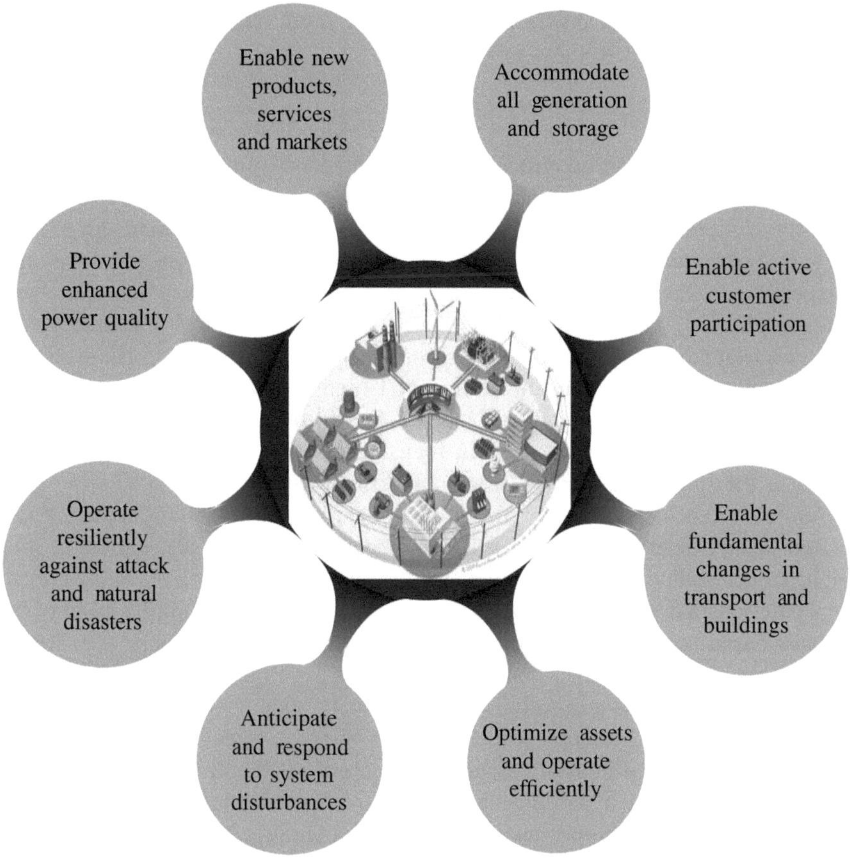

Fig. 2 The different characteristics associated with smart grids. Smart grids can address a number of issues, usually multiple ones at the same time. Based on Department of Energy (DOE) [10]

RD&D activities are proposed. Observe in Fig. 1 the smart transmission network for the increased coordination between system operators (both TSO and DSO).

Numerous other definitions exist, which can be technology oriented, more functional or based on the benefits of the smart grid. It is important to note that in general, smart grids also have a specific regional connotation. While in Europe smart grids are often praised for their ability to provide a transition to a more sustainable energy system with more renewable energy sources and lower energy prices, the emphasis in the USA is more placed on reliability and system automation, and in countries such as India and China, the smart grid is seen as an enabler to manage the rapid growth of new generation, often located far from the load centers (Fig. 2).

2.2 A Transmission System Perspective and Research Framework

From the discussion above, it is relevant to ask what is the impact in maintaining the "status quo" in the activities that a transmission system operator? That is, maintaining the assumption that "the transmission system is already smart." The recently completed ENTSO-E Research & Development Roadmap 2013–2022 [19] and Implementation Plan [18] make it clear that there needs to be a long-term perspective to face the challenges that the European system will meet while adopting the vision of the R&D plan relating to security of supply, adequacy, and energy sustainability. These challenges are coupled with the European climate energy objectives defined in the European Union's "20-20-20" targets and the European Commission's Roadmap 2050. This section provides a framework for identifying key areas that can help in focusing research activities to support the vision of the ENTSO-E R&D Roadmap and Implementation Plan.

As illustrated in Fig. 3, the current grid operation and planning approach and related technologies may not be suited to meet long-term goals, although, they might suffice today's needs. Starting from a timescale perspective, today's technical solutions have to aid in maintaining grid stability from the millisecond to the seconds basis, allow for adequate balancing in the minutes to hours time frame, and guarantee reliability in both operational as well as long-term planning. However, technical and non-technical constraints brought about by uncertainties in production and demand, physical and cyber-vulnerabilities, as well as market forces, regulation, and legislation and can pose unforeseen difficulties to both the philosophy and technical solutions available for a transmission system operator to guarantee security of supply and reliability facilitating a well-functioning electricity market.

Current solutions in the form of methodologies, software tools, and different technologies cannot take into consideration all of the new constraints posed above without substantial harmonization. A negative impact therefore will translate in reduced grid stability, security of supply, inadequacy in power supply and insufficient grid developments to meet with new constraints. Such negative impacts are mapped together with their corresponding timescales in Fig. 4, as it can be observed, these negative impacts are product of today's solutions not being able to meet different constraints.

For a transmission system to be able to meet the constraints listed above, as illustrated in Fig. 3, smart grid solutions need to be developed. This would imply the transition into smart operation and smart operation philosophies and technologies which can enable the adoption of new methodologies for analysis, modular and extensible software for design and optimization, and high-voltage and high-power technologies that can help meeting these new constraints. Such solutions and technologies are identified in Fig. 5 for each timescale, together with a mapping of different technical and non-technical constraints under which they will operate. Starting from the fastest timescale, we identify what are the different

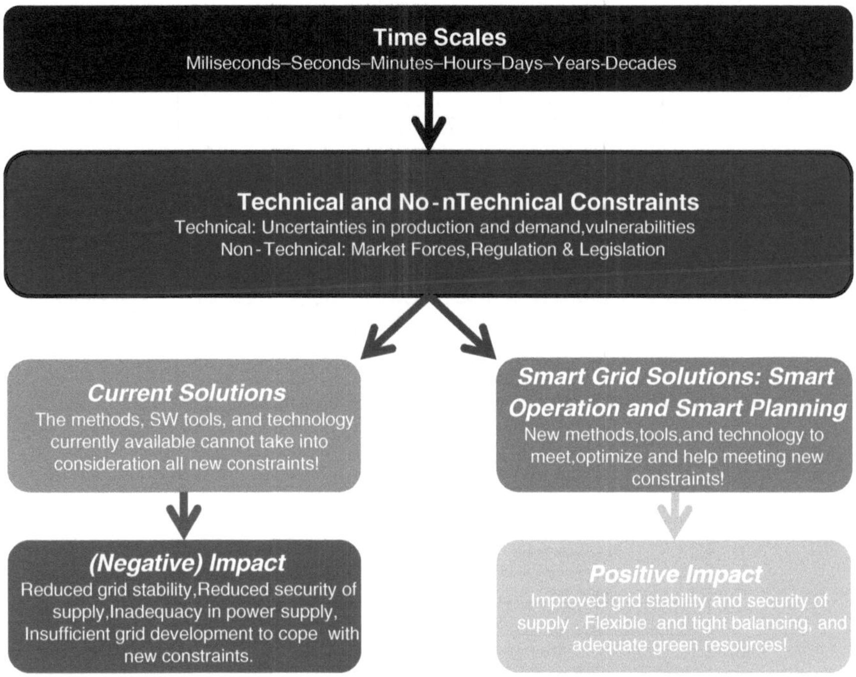

Fig. 3 Need for smart transmission grid solutions: from real-time operations to planning

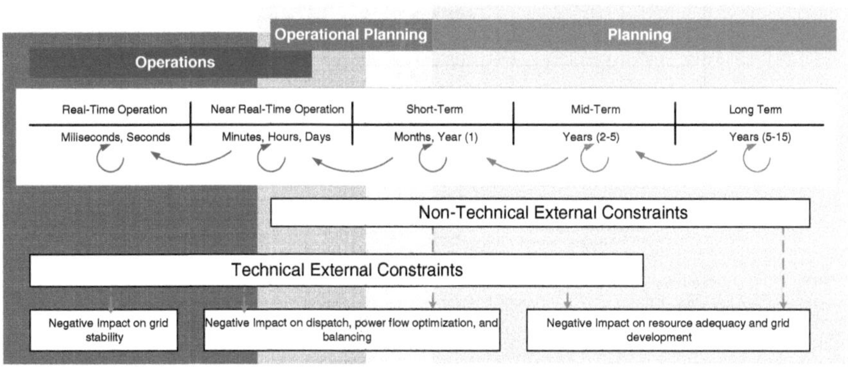

Fig. 4 Technical and non-technical constraints impacts on power system operation and planning

"actions" that a TSO must be able to take; this actions in turn will need not only power transmission technologies, but also software environments allowing for the implementation of new methodologies providing a possible change in paradigm in the operation and planning of the grid.

The result of the adoption of these new solutions need to provide improve grid stability and security of supply, flexible and tight balancing, and adequate usage of

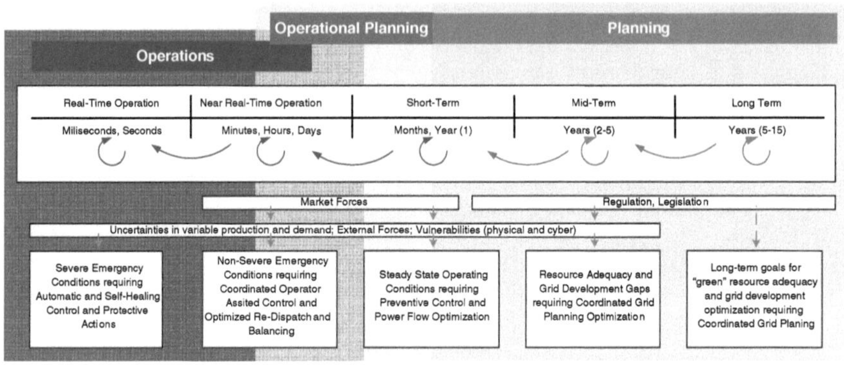

Fig. 5 Smart operation and smart planning solutions for transmission networks

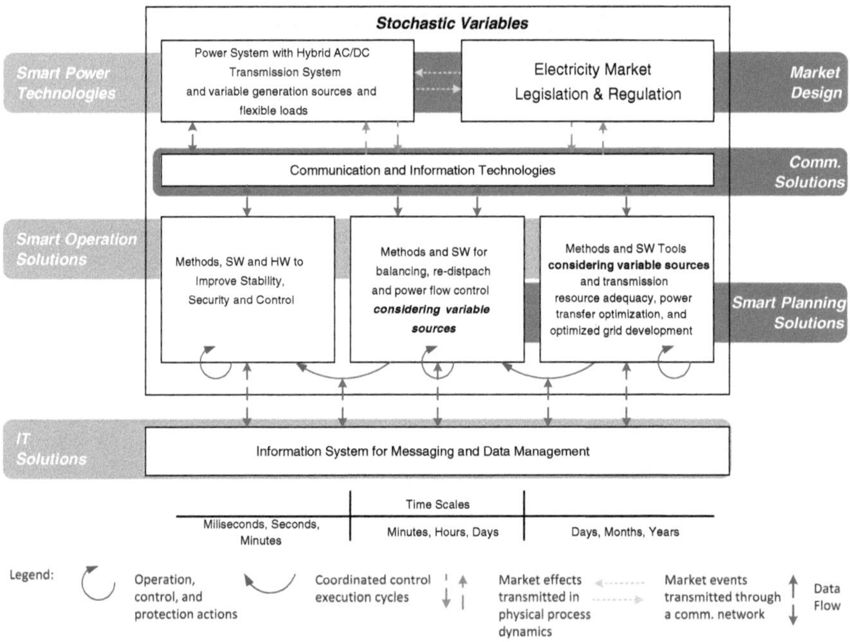

Fig. 6 Smart operation and smart planning research and implementation framework

energy sources. This will in turn facilitate the functioning of the integrated European electricity market and, furthermore, help in attaining the ambitious EU 20-20-20 goals.

A framework that allows the interaction of different solutions is conceptually illustrated in Fig. 6. One could envisage such framework as a massive distributed computer system, interacting between different software (SW) and hardware (HW) solutions which allow the implementation of technologies, methods, and tools that address particular technical and non-technical constraints at every timescale. Such software system is of course futuristic and perhaps not possible to achieve, but the architecture in Fig. 6 provides for a framework that helps identifying what is needed and how each of the solutions will interact to support different roles at the transmission system level. Hence, this framework can be used for identifying research areas that need to be addressed toward the development of STGs.

There are key benefits for a transmission system operator to be derived from adopting such a research framework, it will allow TSOs to identify and support work toward the integration of different grid technologies and users, e.g. renewable energy sources and new demand (electric vehicles). Smart operation and smart planning solutions will help in exploiting these technologies by the following:

- Achieving green goals by enabling the capability of the grid for renewable energy sources integration
- Ensure market competitiveness by making possible affordable electricity pricing with high-quality standards
- Increase grid stability while maintaining flexibility in system operation
- Guarantee security of supply under increase reliability compatible with societal needs

In the next sections, we review research challenges on some of the identified areas in Fig. 6, focusing primarily in the necessary technological building blocks and applications that can aid in improving grid stability, security, and control and that facilitate better resource adequacy while being compatible with regulatory constraints and organizational philosophies.

3 Smart Operation: Enhanced Monitoring and Control of Transmission Grids

Smart operation and smart planning solutions such as those envisioned in Fig. 6 to cope with the technical and non-technical constraints shown in Fig. 5 will require improved awareness of the current system state, and the possibility to act using that additional information. One of the cornerstones of that heightened ability lies in the exploitation of synchronized phasor measurement technology and a supporting IT and communications infrastructure. In this section, we highlight the state of the art, roadblocks, and potential showstoppers that smart operation will face and outline a direction for the development of methods and tools that can have actual applicability for enhancing grid stability, security, and control.

3.1 Synchronized Phasor Measurement Technologies as Building Blocks for Smart Operation Tools

3.1.1 State of the Art in Grid Monitoring and Control

The current approach for power system operations at the transmission level is to perform most of the monitoring and control actions within an energy management system (EMS), which makes use of a supervisory control and data acquisition (SCADA) system as illustrated in Fig. 7. The SCADA system supplies non-synchronous time-skewed measurements every few seconds to a state estimator (SE), which are obtained through round-robin polling. The SE uses these measurements along with the topology determined by a topology processor to provide an approximation of the "state" of the system in its current operation condition. Note that the "state" of the system consists of an estimate of the voltage magnitude and angle, transformer taps, and other quantities related to the SE model utilized. This estimated state is used for initializing several applications such as contingency analysis, optimal power flow, and other applications such as static security analysis, and for the initialization used in dynamic security assessment. Observe that due to the slow rate of acquisition of the SCADA system and the dependency of these operation tools on the starting point from the SE, applications are executed with a time lag as compared with the current operating condition of the power grid.

There are many solutions available and currently used by transmission system companies and system operators. Although these systems are mature and dependable, it has not been until recently that wide-area features have been added to these systems. Wide-area PMU-based features are not broadly adopted and have a reduced number of available phasor data applications, such as to support conventional state estimators as illustrated in Fig. 7. Observe that in this case, the PMU data are merely used to provide additional measurements into the state estimation process; however, the SE does not take the advantage of the time-synchronized measurements from the PMU and simply treats the data as an additional measurement source as done with conventional remote terminal unit (RTU) measurements.

In addition, the existing systems were not developed to withstand the strain of managing the data volumes from the streaming of synchronized phasor measurements in an efficient manner [45]. Despite these limitations, there are initiatives in North America which have created specialized systems exploiting phasor measurements with the aim of enabling new applications of PMU data and increasing the utilization of synchrophasors in operations [44]. New applications must take full advantage of not only the higher sampling rate from PMUs, but also from their time synchronization features and additional information provided by them. To take full advantage of PMU measurements, adequate software systems to support operation applications must be developed with a philosophy of modularity, scalability, connectivity, interoperability, and redundancy, for which current software systems to support PMU applications are not designed.

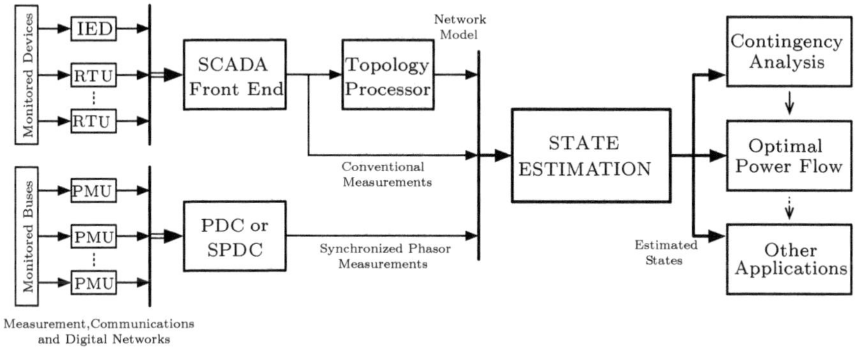

Fig. 7 A conventional SCADA/EMS system supporting operation tools

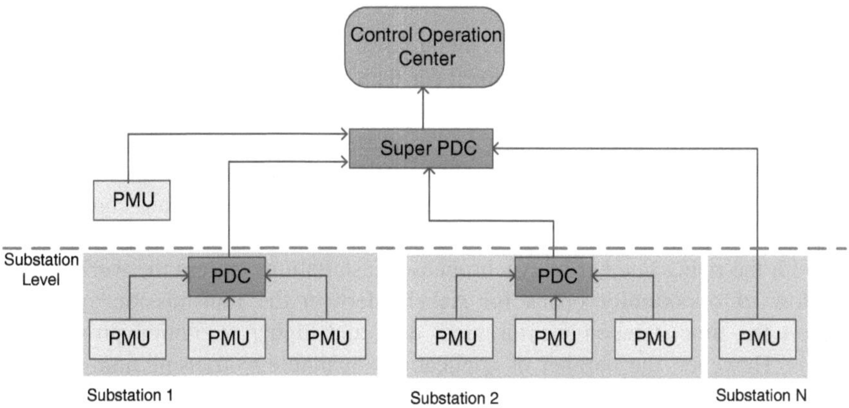

Fig. 8 A typical hierarchical WAMS layout

3.1.2 Synchronized Phasor Measurement Technology

The new enabling technologies of so-called wide-area monitoring systems are PMUs as the measurement device of choice, and their supporting infrastructure which is formed by communication networks and computer systems capable of handling PMU data and other information [usually called phasor data concentrators (PDCs)]. The set of PMUs and their enabling information and communication infrastructures is termed SMT [7].

Figure 8 shows a typical layout of the communication between PMUs and PDCs. The PMUs are the actual measurement units dispersed throughout the power grid, at substations, major interconnection points, and main generator sites. The PDCs then receive the signals from several PMUs and/or PDCs, aligning the measurements and output a stream having the aggregate of these measurements.

PDCs may also include historians archive data for detailed analysis and off-line applications. PMUs can also include other critical state such as breaker position (digital I/O, speed messages, etc.) to record alongside the phasor measurement into the historian or for real-time applications.

PMUs and PDCs can typically produce multiple streams, which among other things are used to feed redundant control centers over redundant communication infrastructure. These streams can also contain subset of available data, such that lower rate streams might be used for some applications while historians and dynamics analysis tools may receive the full rate.

PDCs are typically located in a substation or at a transmission operator center in order to aggregate the traffic from several PMUs within that substation or operation region. These PDCs can forward concentrated output streams of either all or selected measurements to an upper layer. This will continue in hierarchical fashion, and at the end of the aggregation tree sits a PDC often nick-named "SuperPDC" which essentially has all the streams of PMU data available for the operational center analysis.

The challenges faces by this hierarchical model have been realized and NASPINet offers an alternative model for data transfer [38]. The adoption of the NASPINet model in North America has not yet been fully realized, and it is not clear whether this model will be adopted elsewhere in the near future [4]. It is, however, known that each of the TSOs realizes their networking according to similar principles to those of NASPInet, with typically a redundant control operation center and network infrastructure.

With the rising number of synchrophasor installations around the world [46], a window of opportunity opens for stakeholders in the transmission system to exploit the time-stamped measurement data and higher resolution provided by PMUs. However, the number of applications available to transmission operators for exploiting these measurements seems to be insufficient to justify investments in SMT. New applications that support the smart operation of the power system would justify these investments and therefore need to be developed.

3.2 Technological Challenges

3.2.1 Need for Continued Standardization

The IEEE Power System Relaying Committee (PSRC) defined the standard for synchronized phasor measurements in substations in the IEEE Standard for Synchrophasors for Power Systems, i.e. IEEE STD C37.118-2005 [30]. This standard addresses the definition of a synchronized phasor, time synchronization, application of time tags, methods to verify measurement compliance with the standard, and message formats for communication with a phasor data concentrator.

A comprehensive set of tests and calibration methods were conducted on a number of PMUs to assess all aspects of their measurement performance before being deployed in transmission, such as those reported in Moraes et al. [37]. However, in 2009, the IEEE started a joint project with IEC to harmonize real-time communications in the IEEE STD C37.118-2005 with the IEC 61850 communication standard to introduce measurement accuracy under steady-state conditions as well as interference rejection.

As a result, the original IEEE STD C37.118-2005 has been improved and split into two standards, one for measurements (the C37.118.1-2011 [31]) and another one for communication (the C37.118.2-2011 [32]). On the other hand, for the operation of a Phasor Measurement Unit (PMU) to be qualified, the units' performance should comply with the accuracy requirement stated in the C37.118.1-2011. Hence, IEEE PSRC provides guidelines for synchronization, calibration, testing, and installation of PMU in the IEEE PC37.242-2012 [28]. This guide also covers the associated interface requirements for communications testing to connect PMUs to other devices including phasor data concentrator (PDC) as given in the C37.118.2-2011. In addition, the performance and functional requirements of typical PDCs or PDC systems such as synchrophasor data processing, real-time access, and historical data access must be verified to conform to the suggested guide named IEEE PC37.244 [29]. This guide also described PDCs test setups and some user applications.

Despite this recent large effort on standardization, most of the currently available PMUs are not meeting the complete specification of the current standard C37.118.1-2011. This is due to a lack of specific application requirements that have to be met by the instruments. Further, the recent standards on PDCs open now the door for a further discussion on how PDCs should provide standard output to support PMU applications so that they are independent from particular software systems and manufacturers. The standardization work is certainly progressing at a reasonable pace; however, attention must be paid into guaranteeing modularity and interoperability of different software systems that will be supported by PDCs for implementing advanced phasor applications.

3.2.2 Big Data Management

There are several reasons for the lack of SMT-based applications and their limited adoption. These reasons emerge from the two different development approaches currently used: application development using real PMU data and the simulation approach. From a researcher's stand point, obtaining real PMU data from transmission companies involves signing non-disclosure agreements which delays the start of research efforts, and more importantly, they may impose restrictions on the intellectual property of the derived works [41]. Foremost, when developing PMU-based applications, the PMU data itself are not sufficient: knowledge about the transmission system model parameters during the archived data time frame and other data (such as bus-bar level breaker status) are crucial for some applications [33] and may not be easy to obtain or interpret. Despite that the COMTRADE format has been selected for PMU data sharing in North America [1], due to regulations for postmortem forensic analysis [40], these data format may not be the most convenient for application development and straightforward data analysis.[1]

Many applications require large records of phasor measurements (from 1 day to even weeks of archived data [59], and this from multiple PMU). Data availability and correct sharing mechanisms are not only an issue for academic researchers, but may also become important for application developers looking to extract features of data from massive data sets [52]. The industrial development and adoption of these applications can be further delayed if adequate software systems to manage these data sets are not available.

3.2.3 ICT Aspects

Many academics have proposed applications of PMU data based solely on simulations using software commonly used in the power industry (which are mostly positive sequence-based (or phasor) simulations). While this approach is suited for some fundamental research, it might not be appropriate for actual implementation. This is because this approach does not take into account many of the challenges and characteristics of PMUs and the ICT systems. As a result, unreasonable assumptions of what the capabilities of these enabling technologies are made, often through an insufficient of the underlying technology limitations [58], although some of them have been acknowledged [35].

Current approaches for simulating the use of remote data for control purposes are considerably easier than the actual implementation, where appropriate data filtering, transmission to a PDC, processing, and transmission to the (remote) controller are needed. This requires several different stages where practical issues and possible delays occur. Without full consideration of these practical issues, it may not be advisable to install these applications at a control center without going through a thorough testing process. This highlights the need of both more realistic simulation approaches as those used by other communities [13, 42] and the possible uses of co-simulation of cyber-physical networks [50].

Beyond aspects of data transmission, time synchronization over wide geographical regions will continue to pose several challenges. An example of one of such practical issues is illustrated in Fig. 9, where the voltage phasor angle from PMUs installed at three different substations in the Mexican power system is shown. This figure shows the effect of GPS signal loss of in the THP-LBR and LBR-THP voltage angles. The trace of the LBR-GUAT angle shows no issues with the GPS signal. Such issues with PMU measurements, and other similar ones [58], need to be taken into account while researching new time and data transfer architectures that can be adequately employed by industry. This example illustrates

[1] Somehow, the power industry continues to overlook how other fields of research have dealt with massive amounts of data and have developed formats that allow to work and exchange numerical and graphical data efficiently [22].

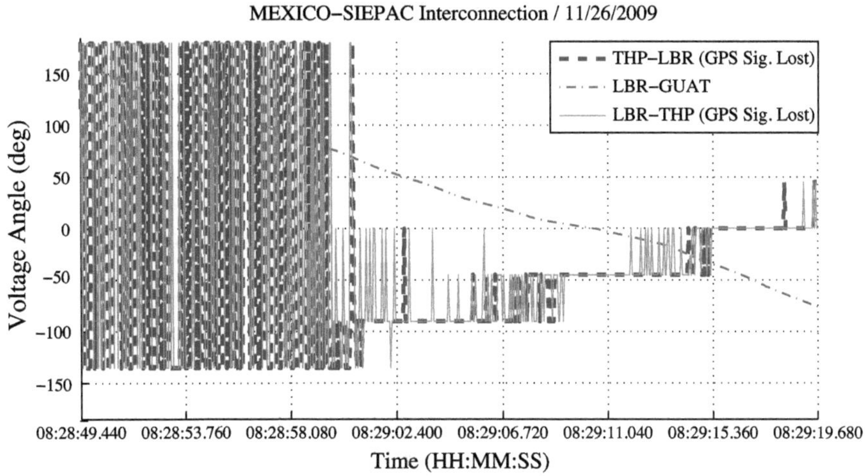

Fig. 9 Effect of GPS signal loss in PMU data

the need for GPS time-independent timing and data transfer, for which technologies developed in other fields such as digital television could be applied [25].

3.2.4 The Requirements Dilemma

With the considerations made above, we realize that there is a dilemma on determining the appropriate ICT design specifications for each particular "application." The dilemma arises because not all the future applications enabling these STGs according to these principles have been developed. In order to be developed correctly, they need "an" ICT infrastructure, which in turn needs application specifications for its design. A new approach for R&D is necessary to flexibly evaluate different ICT paradigms at the same time that the power system operation and control strategies are being developed. Two approaches might be suitable for tackling this dilemma: the proper use of holistic simulation environments [13, 42] and co-simulation [50], and the availability of experimental facilities for testing and validation [57].

3.3 Smart Operation: A Way Forward and Future Grid Monitoring and Control Solutions

3.3.1 Holistic Architectural Analysis

An STG is more than a grid that takes benefit of PMUs and requires ICT for this purpose. At a minimum, a STG should make use of these data in order to exploit all the available "observability" and "controllability" in a power system through closed-loop feedback control and to coordinate system control with protection. As such it can behave as a "self-healing" system, or at least utilizing the system more securely through increased awareness. To this extent, all measurement devices should be capable of producing synchronized and high-resolution time-stamped data that capture the dynamic behavior of the power system and can provide system observability. Controllability can be effectively provided by all those devices that can be in closed-loop control including conventional generation, flexible AC transmission systems (FACTS), high-voltage direct current (HVDC), and tap-changing and phase-shifting transformers.

To accomplish these ambitions, STGs should contain more than the high-resolution measurements provided by PMUs. In Fig. 10, a conceptual diagram of a "centralized model" for a STG is shown. In such STG, synchronized measurements are obtained at transmission substations through time-synchronized measurements not only from PMUs but also from other envisioned highly accurate measurement systems retrieving data from controllable devices and protective device "information sets" (i.e., all available information from within a protective relay). This plethora of data is sent through communication networks, received, and concentrated at a decision and control support system that determines appropriate preventive, corrective, and protective measures. This support system is the cornerstone for enabling STGs using synchrophasor data, and it is here where the newly developed analysis techniques will produce "smarter" decisions allowing the power system to operate more securely, efficiently, and reliably. The decisions determined by this support system will then support operators at control centers to take "smarter operator control actions" or even device "smart-automatic control/protective actions." These actions are translated into feedback signals that are sent through communication networks to exploit the controllability and protection resources of the power system.

Note that although the diagram shown in Fig. 10 is a centralized model, there can be other more decentralized models for STGs. A "decentralized model" of a STG would divide the system into "focal area" systems with different operational functions (some of them might not include a focal area control center for example, implying that only other functions are taken there and thus a lower amount of data with perhaps lower quality of service (QoS) is needed) and a "wide-area" system. The data delivery is done through a publisher–subscriber model, such as Grid-Stat [21], instead of a traditional star communication with round-robin polling model used in traditional EMS/SCADA systems, whose limitations have been acknowledged in Bose [5]. Feasible approaches considered in Bose [2], Gjermundrod et al. [5], and Bakken et al. [21] have great potential and should be further investigated.

However, as mentioned before, the whole architecture of the system faces a dilemma as it will be determined by the requirements from different applications using PMU data, which in turn need the ICT infrastructure to be developed—in other words, how and for what purpose the measurement and other data be used will determine the most cost efficient system architecture. Yet, "an" architecture is needed to obtain data to develop the applications. To find the appropriate

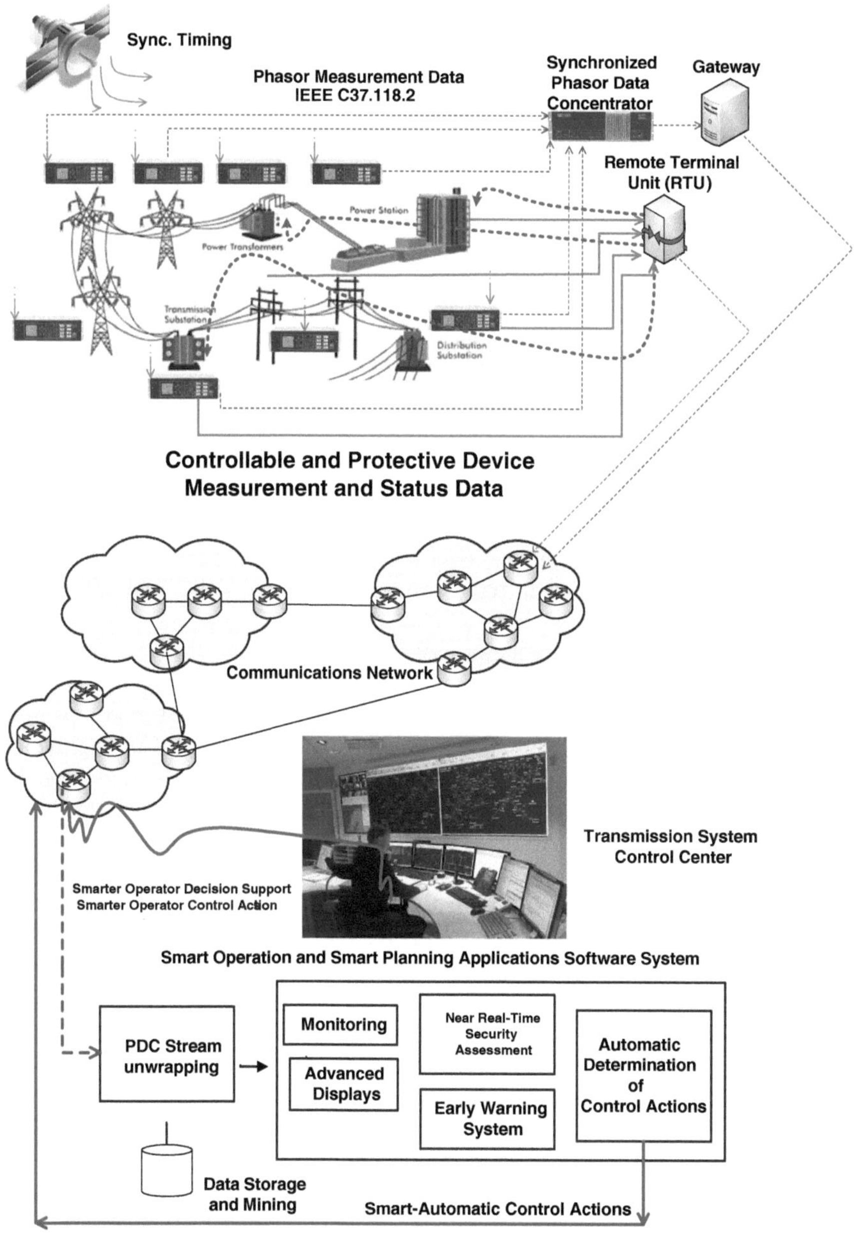

Fig. 10 A centralized model of a smart transmission grid

architecture that fits the needs of future power systems, appropriate experimental platforms for research are needed [54, 57], as well as the use of holistic simulation environments [13, 42] and co-simulation [50].

3.3.2 GPS-Independent Time and Data Transfer

Phasor measurements create real-time traffic, and this traffic needs to be transported over some network infrastructure. The real-time traffic has some timing issues of its own, and this suggests that low loss may be good.

Packet network analysis shows a variety of issues, in which real-time properties such as loss and delay becomes affected by interaction between other traffics. It is also shown that these behaviors have a high degree of unpredictability in them. The microscopic details of protocol interactions are many, but under the assumption that information gets transferred (or can be interpolated), the remaining macroscopic effect for the system is the average delay. The interaction between many other traffic streams can cause this average (over the control-loop bandwidth/time constant) to change over time.

Means to reduce jitter and to reduce loss will increase delay and still do not provide predictable results.

Delay and delay stability are major issues in ensuring control-loop stability and meaningful reaction time to achieve the control goal. Power grid controls, such as damping controls (e.g., PSS and POD), could potentially be tuned to compensate for delay, but large variations of delay over time would require self-tuning, which would add to the system complexity. An alternative approach is to reduce the control-loop bandwidth, which makes it too slow to react to actual problems in the power grid.

The involved signals have real-time properties; this means that low delay and low jitter are required. Loss and high jitter will require additional delays to the signals, and this is clearly not a good property for the overall system behavior.

In order to meet the requirements, an alternative network solution is proposed, where technology developed for the real-time networks of radio and TV broadcast networks can be utilized. Such network has similar requirements on low latency to handle long-distance live broadcast (interviews, sport events, etc.). The network solution offers stable latency requirements to handle the low jitter tolerance of broadcasting and production equipment. It has the low loss requirement typical of live transmission, as there is no time to do re-transmit of information.

There are similarities between the power grid needs and the properties provided by such communication network solutions [14, 25]. Among the similarities lies a high QoS need for the real-time streams, bounds on propagation delay, low jitter, low loss, and high reliability. It distinguishes itself by providing significantly higher real-time properties compared to typical IP SLAs and even MEF 2.0 requirements.

Another aspect is the need for precision timing, which may be available in the communication network solution [25, 39]. The detailed requirement varies from network to network, but lies in general in the region of ±1–4 μs from the reference time, over a 10-hop network. Comparing this to the PMU need of ±26 μs [32], we see that this solution can support the PMU needs.

A detailed analysis and other communication network challenges for synchrophasor-based wide-area applications are presented in [9].

3.3.3 Software Development for Real-Time PMU Applications

Today, researchers need to devote large efforts developing mechanisms that allow them to use PMU measurements, e.g. PMU data extraction for off-line applications and real-time data mediation for online applications. These tasks are especially difficult in the case where PMU equipment and PDCs are provided by different vendors.

Figure 11a shows the main difficulties in today's monolithic vendor-specific PMU application software development environments. As it can be observed, once the infrastructure is put in place with PMUs installed and networked through a communication network, all real-time data arrives to a PDC. At this stage, all of the data are locked into a vendor-specific software system which may or may not provide its users with the necessary tools to implement applications. If these tools are available, they are heavily reliant on the software libraries provided by the vendor and offer limited integration options. On the other hand, historical data that can be used for off-line analysis and applications (e.g., data mining) are also locked into a proprietary time series database system specific to the software system. As a result, the user has few possibilities in implementing new applications without relying on the software system provider. This limits the possibilities of exploration and interfacing with external tools and software systems. Hence, it is realized that once the data arrive at the PDC, and concentration and alignment functions have been carried out, the PDC could be interfaced with standard protocols to a flexible development environment.

In order to develop applications for monitoring and control based on synchrophasor measurements, it is important to have real-time access to the individual quantities (phasor/analog/digital) of each PMU, which are wrapped inside the real-time PDC stream. The IEEE C37.118.2 standard has provided the specification for the creation of "concentrated output streams" from a PDC, with PMU data coming out aligned and concentrated into a single stream routed to another larger PDC (i.e., the SuperPDC). Although this facility was meant to be able to concentrate and align the PMU data from different PDCs, it offers the possibility to decouple the development of applications from the PDC.

The use of concentrated output streams provides a mechanism for standardized real-time data sharing that can be used to interface with alternative software systems from that of the PDC provider. Thus, instead of building monolithic software architectures and systems, it possible to develop a modular approach to software development by exploiting synchrophasor standards for real-time data communication. As such, applications can be deployed in different clients and

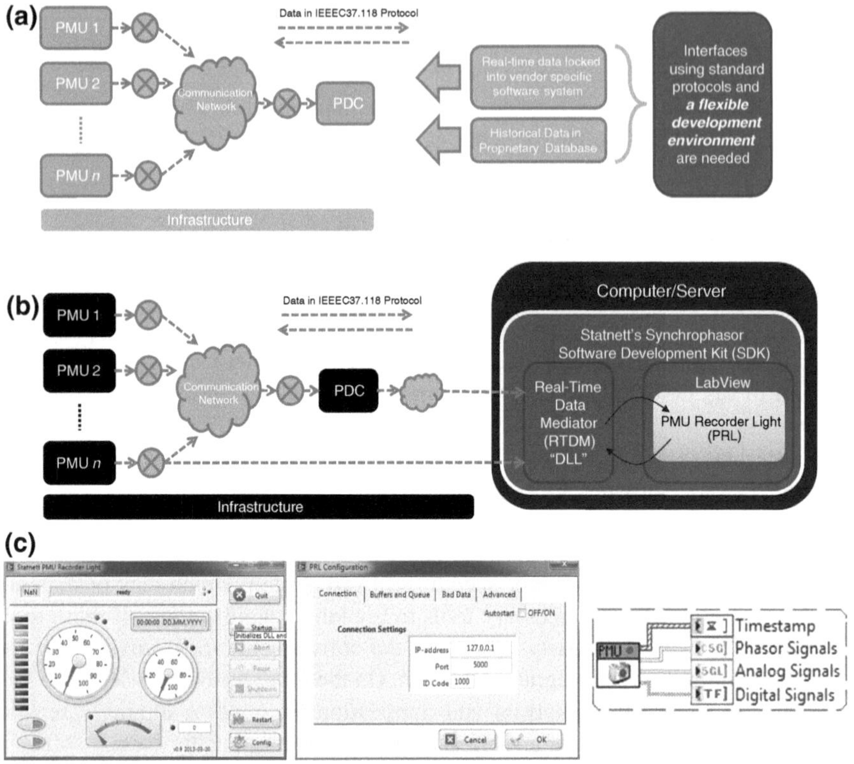

Fig. 11 Statnett's synchrophasor software development kit (*SDK*) for modular, real-time, PMU application development. **a** Limitations in monolithic software environments for developing PMU applications. **b** Statnett's synchrophasor SDK for modular PMU application development. **c** Statnett's SDK PRL configuration and user interface

meeting different application requirements. This is not yet possible with today's software solutions available in the market.

To exploit this possibility, a software development toolkit was developed by Statnett SF (the Norwegian Transmission System Operator) [56].

The main components of the SDK along with the possible interfaces to multiple PDCs or PMUs are illustrated in Fig. 11b. The aim of the SDK facilitates research, fast prototyping, and testing of real-time synchrophasor applications. The SDK enables the usage of high-level programming languages such as LabVIEW, regardless of the equipment used and its manufacturer, thus providing platform independence for research and development. This property enables users to be more focused on developing synchrophasor applications and not on platform-specific implementation issues. The SDK is capable of connecting to an arbitrary number of PMUs or PDCs compliant with the IEEE C37.118.2-2011 protocol.

The SDK provides a real-time data mediator that reads and stores real-time data in a configurable buffer (RTDM in Fig. 11b). The RTDM is built on a client server architecture allowing the connection of multiple clients providing data in the IEEE C37.118.2-2011 protocol, and enabling the access of these data by methods. The RTDM is compatible with several operating systems, and in the Microsoft Windows operating system, the RTDM is compiled in a dynamic link library which is accessed by a client. Further details on the RTDM will be available in a future publication.

The clients can access the data and other information from the PDC using a library of methods. Currently, as shown in Fig. 11b, the content of the buffer is accessible in the LabVIEW environment through different functions in a library named PMU Recorder Light (PRL), which provides a standard LabVIEW function control (VI). These libraries are illustrated in Fig. 11c, showing the GUI for data access, the configuration for connection to output streams, and one of the LabVIEW blocks providing access to real-time data. The LabVIEW platform is selected because it provides easy integration with different hardware equipment as well as intuitive graphical programming language (G language) which supports integration with MATLAB, C++, and other programming languages.

The PRL has two major components:

- Data collector: This component reads the data from the PDC/PMU and stores them in configurable buffers.
- Data extractor: This is a collection of functions (VIs) that allows the user to access the buffers and queues in the PRL. It reads the data from the buffers and provides the user with control over the data streams in a form suitable for further processing (i.e., as a signal data type in LabVIEW).

Such modular approach for accessing real-time streams provides large advantages for prototyping PMU applications and application deployment possibilities. We illustrate this for the case of real-time display of PMU data, other real-time applications developed can be found in Vanfretti et al. [57]. With current monolithic approaches, the display of PMU data is confined to the control center and tied to the PDC system receiving the data. In contrast, the SDK offers multiple deployment possibilities. Figure 12 presents two of them: integrated computer/server environments in Fig. 12a which can be deployed in multiple computers, and the deployment of PMU applications in mobile devices. In the first case, the SDK along with a custom application for visualization is deployed in a single or multiple computers/servers, allowing multiple users to visualize the data.

In the second case, the SDK is deployed together with a custom application offering a publishing mechanism to feed the visualization application in a mobile device. Figure 13 shows the applications running on Apple's iPhone smartphone and the iPad tablet.

Smart Operation Tools: Monitoring and Control Applications

Statnett's Synchrophasor SDK illustrates how the development of measurement-based real-time PMU applications can be deployed in different environments. This

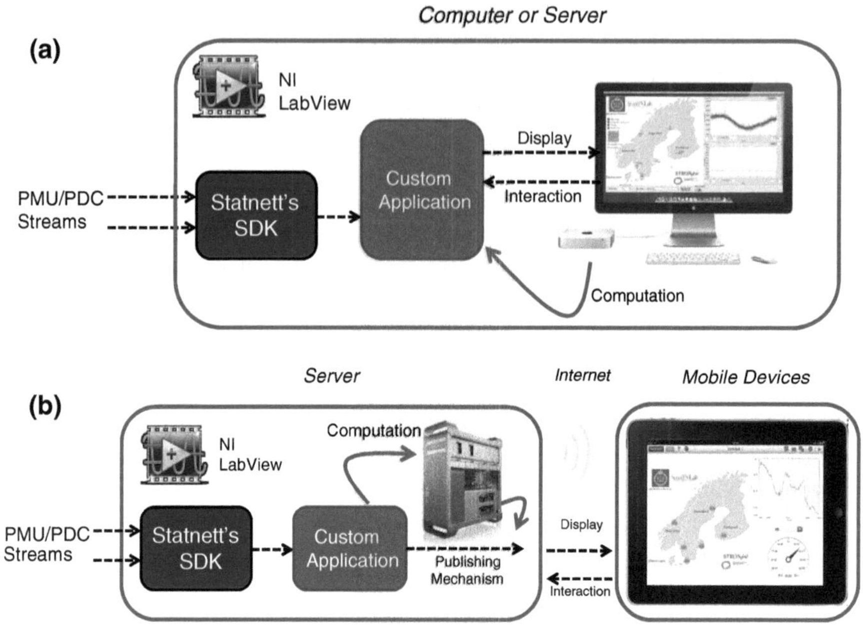

Fig. 12 Deployment of Statnett's SDK in different computer system architectures for powering PMU applications. **a** Development of Statnett's SDK in integrated computer/server environments. **b** Development of Statnett's SDK powering mobile applications

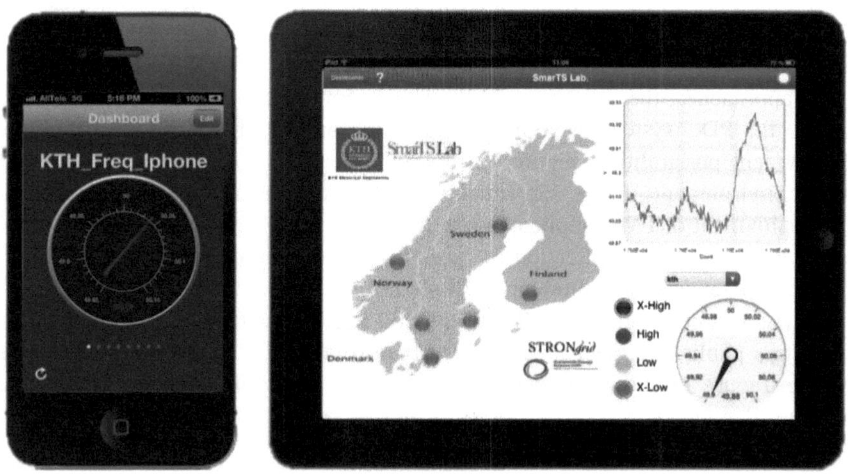

Fig. 13 iPhone and iPad PMU data visualization Apps powered by Statnett's SDK PRL interface

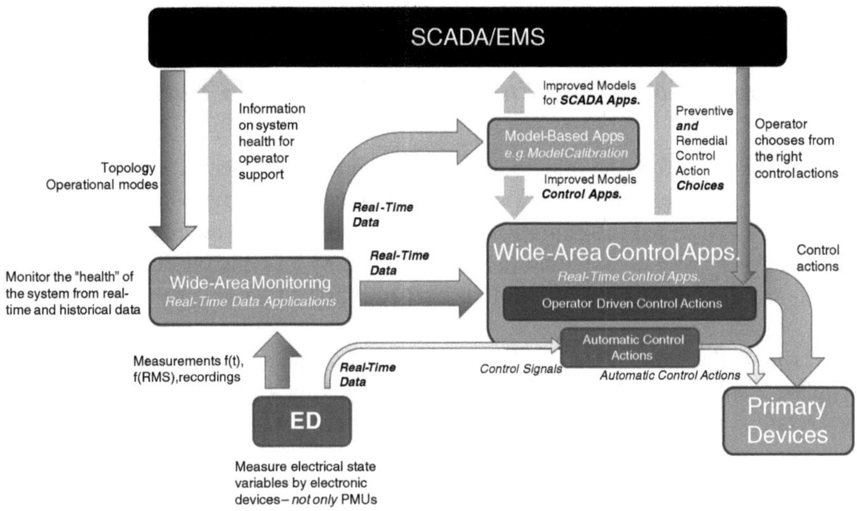

Fig. 14 Holistic system architecture for smart operation applications

flexibility would allow to achieve smart operation applications software systems as envisioned in Fig. 10. This system should be able to holistically exploit both measurement-based as well as model-based applications as illustrated in Fig. 14.

Such architecture should take advantage of tools such as Statnett's SDK for deploying real-time data applications for measurement-based monitoring and stability assessment, near real-time, and real-time control actions through operator-driven control and automatic control loops. Although there are no flexible solutions to integrate model-based applications, the architecture should support model-based stability assessment tools and allow for the interaction between measurement-based and model-based tools. One example is the provision of a model calibration application that uses real-time measurements to calibrate the dynamic power system model which is used in both stability calculations derived from models as well as in real-time control applications.

The emergence of smart operation applications will depend on a software architecture that provides these features and allows for a flexible deployment of the different components needed in disparate software and hardware architectures. To achieve this, such architecture should be modular, extensible and be able to re-use components in different targets for deployment. Statnett's SDK is a step forward in this direction; however, much work is needed to transition from today's monolithic and vendor-specific software systems, to a modular and flexible software system to support power system smart operation.

4 Operational Planning of the Smart Transmission Grid

4.1 Organization of the Energy Supply Chain in Europe

In earlier days, before the mid-1990s, the operation of the electric power system in Europe was rather straightforward, with vertically integrated companies that owned and controlled the entire value chain. As a result, the entire decision-taking process was largely done within in a single company and this for the different time frames. At that time, the security of the energy supply at an acceptable cost was the main objective.

Through the process of liberalization, the European energy system became regulated and unbundled, with separated generation, transmission, distribution, and retail. A regulator sets the rules of the grid operation. In view of the smart grid, the regulator plays a vital player as he forms the main influence on grid rules and tariffs and has a strong voice in the investment process. In a sense, the regulator even defines the different players. Within a smart grid, all stakeholders are able to optimally perform their tasks, taking into account the limitations associated with stakeholder interaction. Interaction between stakeholders happens between similar stakeholders (e.g., between TSOs in different countries) or between different ones such as generators and the TSO. The required interactions (and communication) has significantly increased, for instance related to the use of ancillary services.

When looking at the interconnected European power system and its stakeholders, a complex patchwork of different entities with different tasks, objectives, and non-overlapping geographical areas can be seen. The non-harmonized regulatory framework adds to the complexity and is a serious constraint to the development of a true smart grid at an international level. Ongoing efforts of the EU (through the Third Energy Package [15]) are a first step toward a more harmonized energy policy. The Third Energy Package caused the creation of ENTSO-E, the association of transmission system operators in Europe, and ACER, the agency for the cooperation of energy regulators. The actions of these organizations aim to manage the international interactions between guidelines and to harmonize the framework throughout the European power system. When considering this international power system as the setting for the STG, it is essential to recognize the importance of the complex interaction between the different regulatory frameworks and the roles of the stakeholders.

4.2 System Working up to Its Limits

In the pre-liberalization era, the generators and grid were part of the same company that owned, maintained, and operated the grid. The grid owner had the means to make the necessary investment decisions based on a coordinated planning (generation investments coordinated with necessary grid reinforcements).

These generation investments were mostly large, traditional generating units following the principle of economy of scale. If possible, the generation was planned located close to the load centers.

The system operator could control all aspects of the power system: generation for unit re-dispatch and managing the grid and its elements (line opening, capacitor switching, etc.) to control the system flows and avoid line overloads.

After being unbundled, the TSO still makes the investment decisions and manages the power system. However, he experiences several limitations in this respect. Generation investments are no longer coordinated with grid expansion, or rather, they are performed by independent organizations. The result is the shift of generators closer to the source of energy, e.g. the harbor as a location for coal-fired power plants.

The newly installed generation capacity is often also of a different type than it was before. Generation units from variable power energy sources are more commonplace due to the strive for more renewable energy generation and the emergence of other small-scale generation such as CHP (combined heat and power). These devices are not only not predictable to a large extent, but also uncontrollable by the operator. A more unpredictable power injection pattern will cause a higher uncertainty of the energy flows in the system. The increased market working has also led to a higher volatility of the energy flows. At the same time, there has been a lack of investments in the transmission system. The effects are seen most prominently when looking at the interconnections between zones. Additionally, the permitting process for generation is often considerably shorter than that of transmission, mainly due to projects that are postponed due to opposition from public, ecologists, etc.

The variable power injections in the system cause fluctuating flows on the AC grid. The system operator can still control the grid to manage these flows, but redispatching generation has more difficult and costly. As a result of the increase in variable energy flows, the limited grid investments, and the reduced control options for the system operator, the grid is being operated at a higher uncertainty [43]. This means that it in some cases is operated closer to its limits, with potentially serious consequences for grid security, while on other cases the system might not have been used up to its potential, with a negative influence on the social welfare. In order to manage the system under these circumstances, three main innovations are needed. The grid operations need to become more flexible, increased coordination is required, specifically between different zones, and the manner in which the security is dealt with needs to be redefined.

4.3 Flexible use of the Power System

Simply put, the transmission of electrical power encompasses two fundamental aspects: on the one hand, the balance between generation and load needs to be maintained, while on the other hand, the system has to remain within the security limits. For both aspects, the requirements with respect to flexibility have significantly increased due to higher uncertainties and fewer control means [60].

4.3.1 Flexibility of Generation and Load

Maintaining the balance between generation and load is one of the classical problems within electrical power engineering, specifically because storage of bulk quantities of electrical energy in a cost-effective manner is difficult. However, a number of aspects have changed in the modern power system, both on the generation and the load side.

First of all, a larger proportion of energy is delivered by sources with a fluctuating output. Furthermore, these fluctuations occur at a higher frequency than those originating from classical generators resulting in faster control requirements and extensive balancing services. A second effect is that a significant part of the generation is either not flexible (e.g., most nuclear power plants) or not centrally controllable (most renewable energy generation). As a result, little the remaining "flexible" generation is responsible to take care of all the fluctuations, which can have an influence on the economics of those generators. In some cases, the non-flexible generation might even surpass the load, e.g. at moments of low load and high-renewable infeed. At such moments, there is no or insufficient downward regulating capacity [12]. A third change is the increased use of power electronic converters as an interface between the generation and the grid. These converters normally decouple the generation, and with that also the inertia, from the grid. As a result, the system inertia decreases which causes an inherently different dynamic behavior of the grid and higher-frequency deviations in case of a disturbance.

Also the load perspective has changed considerably. On the one hand, there is renewed attention for demand response or demand-side management. The potential of aggregating smaller loads on the distribution side and to use them to manage the frequency is expected to bring a considerable contribution to the balancing needs. Similar as for the generation, also loads are increasingly connected to the mains supply using power electronic converters. The effect of reduced load inertia is however smaller than that of the generation. Also the ongoing developments of storage devices make that electrical energy storage can become a more prominent option, next to the already existing pumped storage facilities. Whether storage will be used in a distributed manner or centralized, and whether the main application will lie in short-term or longer-term balancing is yet to be seen. In any case, storage can offer services within the larger system, but it is not a separate service in itself.

The main hurdles for the development lie in the correct integration of the different levels of flexibility among stakeholders, and this in the different operational time frames. This includes the integration of such services into the market and providing all stakeholders the right (price based) incentives. Without these incentives, the necessary investments will not happen. Another field of research is making use of the control capability of power electronic converters to provide ancillary services such as inertia to the energy system. Traditional power engineering approaches to dealing with load and generation might change dramatically where one see the change from the original generation which follows load, toward load flattening and eventually the combination of generation, load, and storage following the profile of renewable generation that operates at a very low marginal cost.

4.3.2 Extended Grid Use through the Dynamic Use of Existing Assets

Grid operators in the pre-liberalized energy system had quasi-full control of the system, including generation redispatch (which came at a cost which was socialized among users). This is no longer the case in the unbundled system. As a response to the lack of control means available to the system operator and to increase transmission capacity, new ways of providing flexible grid operations are sought for. In this subsection, the approach to manage the flexibility of the grid during operational planning is performed in the NetFlex demo, part of the EU FP7 project Twenties.[2]

A first option is the installation of power flow controlling devices (PFCs) to manage the power flowing through the grid in a more dynamic manner. These devices can be used to redistribute the flows from the heavily loaded lines to the less congested lines. The PFC can be used to free capacity on the market (D-2) and plan the system in such a manner that it can be operated securely (D-1) and be used as a means to solve problems intraday (either as preventive or corrective action). Results from the NetFlex demo have shown that the use of PFCs effectively reduces the flow on transmission lines with an insufficient margin and therefore reduce the need for redispatch [24]. It was also shown that through adequate scheduling (D-1) additional margin could be created to manage also uncertain generation patterns. It was also shown that the system itself was capable of "absorbing" significantly more wind energy without hitting the security constraints. Figure 15 shows the system capabilities in the month of January 2013 with uncoordinated operation (purple) and coordination of PSTs in the CWE region against the expected (P50) and high wind (P90), sorted from highest to lowest infeed [23, 24].

A second option is to make better use of existing infrastructure. One example of such techniques is the use of dynamic line rating (DLR). This technology takes into account the actual limit of transmission lines, which is strongly dependent on wind speeds, rather than the conservative seasonal limits. Also within the NetFlex Demo of the Twenties project, it was found that the actual rating of the transmission line is 95–99 % of the time, the gain in capacity beyond the seasonal rating is higher than 10 % and in 90–95 % of the time, the gain is higher than 20 %. It was found that using adequate forecasting tools, the predictions of the line capacity can be used to include them in operational planning. On a two-day and

[2] Twenties project: http://www.twenties-project.eu.

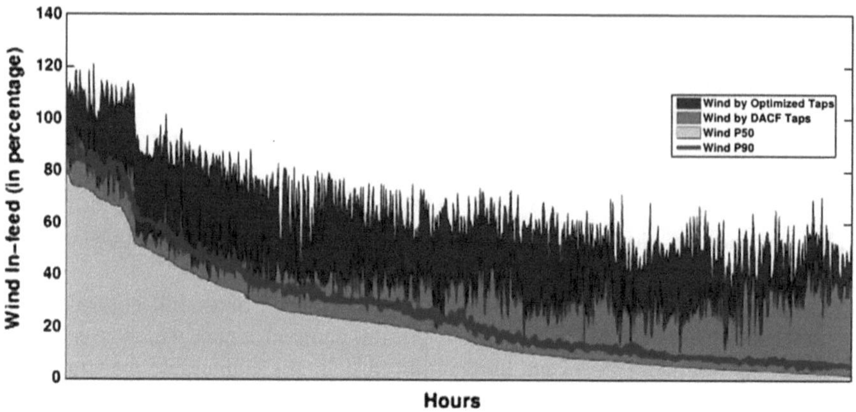

Fig. 15 Increase in system capabilities through the use of power flow controlling devices in the month of January 2013 for the CWE region [24]

day-ahead basis, the gain in capacity over the seasonal rating is higher than 5 % and in over 50 % of the time, the gain is higher than 10 %.

Both the use of PFC and DLR allow the operator to make a better use of the system by working closer to the system thermal limits. However, the remaining security margin becomes lower. Furthermore, by pushing the system to work closer to its limits, other security boundaries might be hit. As such, it is important to improve the monitoring of the system dynamic behavior. Within the Twenties project, a damping prediction monitor was developed, monitoring whether the dynamic limits were not reached while the thermal limits of the system were pushed.

4.4 Coordination in the Power System

As the operation of the power system has become more inter-zonal, with higher and more variable power flows through the system, more coordination is needed. In particular, because zones that are not directly taking part in energy transactions, are also affected by decisions made outside their zone. Currently, TSOs in Europe do coordinate the operation of the power system. There is an exchange in system data and operational information:

- Grid status (important scheduled outages)
- Day-ahead congestion forecasts are made
- Expected available transmission capacities are calculated
- Emergencies with possible effects beyond the local zone are communicated

However, the exchange happens on a very basic level. First of all, not all grid information is exchanged. One uses a "need to know" principle. An example is the load/generation data for the day-ahead congestion forecasts which are not shared independently, but rather as an aggregated quantity. Also information on the exact settings and status of devices (such as protection devices) is not always clearly communicated. One of the hurdles of this system is the exchange of data. Steady-state data are exchanged, but in a format which is generally not the one used for the internal studies where different tools are used, possibly in-house developed, with specific features.

Dynamic grid data are only rarely exchanged, and for this the data format is even more problematic as the different dynamic models used might differ significantly. Currently, no common data format exists, although that there are efforts in developing such a standard model (see CIM [34] and ODM [44]). However, significant improvements are needed in order to make them practical to use when many custom models are needed.

Not only the data sets that are used are different, also the tools and methodologies that are used differ among system operators. A good example is the $N-1$ rule, which is one of the fundamental security rules in the power system and well known to power system engineers. However, when going in detail, it is clear that both the interpretations of "N" and "-1" differ between organizations. Even within one TSO, the concept of $N-1$ usually differs between the grid planning and the grid operation department.

Nevertheless, this system of "need to know" communication works reasonably well, with only a limited amount of grave events occurring due to mis-operation (e.g., Italy blackout, August 2003; Germany-UCTE incident, November 2006). Yet, potential problems remain the following:

- Unidentified loop flows occur
- The uncertainty in the system remains high
- The "limited amount" of grave events mentioned before could have been avoided
- Problems might be solvable on a local level at a high cost, but could have been solved at low expenses elsewhere
- There is no system-wide perspective

The existing coordination in the transmission system is currently sufficient to keep the lights on, but more is needed in order to make optimal use of the available resources. The current framework does not allow an integrated operation of the power system. Considering advanced operating principles without taking the current background of cooperation into account will lead to unrealistic results.

4.4.1 Example: Coordination of PFC

A significant number of PFCs such as phase-shifting transformers (PST) or HVDC lines are installed between different zones. Although PFC exhibits a strong influence on the flows through neighboring systems, these PFCs are not operated in a coordinated way. As such, negative interactions between PFCs, both during steady-state operation and through dynamic interactions may occur. This problem is especially important when the controllable devices are operated by different system operators. Therefore, it is of substantial importance to deal with this issue when designing the controllers and during power system operation [53, 55].

Most PFCs are currently operated toward a certain objective which suits a single party. Coordination regarding PFC control is often limited to predefined rules and requires slow (often via telephone) interactions between the different participants. Coordination is limited for more than technical reasons: as the PFC is approved, installed, and paid by a local entity, this entity will use that asset to its optimum, without considering the "bigger picture." Using the devices to actively create capacity which can be used in the market environment is not possible in the regulatory framework. However, on the long run, it seems to be inevitable that controlling flows in the system will be come an ancillary service, which comes at a cost.

While there is significant potential in coordination of power flow coordinating devices, this potential is largely untapped at this moment. However, it is important to recognize the limitations in the system rather than searching for novel controller techniques that find no practical implementation.

4.5 Secure Operation of the Pan-European Power System

4.5.1 A New Reliability Concept is Needed

Modern society is critically dependent on a reliable electricity supply to cover basic needs such as food and water supply, residential heating, and ICT services. In the near future, electric mobility, distributed generation, distributed electricity storage, and other new uses will increase this dependency. Failing to provide a reliable electricity supply has far-reaching consequences for people, society, and the economy.

A power system's vulnerability is composed of its susceptibility and coping capacity [11, 20, 27]. A power system is susceptible to a threat if the realized threat leads to an unwanted event in the power system. The coping capacity describes the ability of the operator and the power system itself to cope with an unwanted event, limit negative effects, and restore the power system's function to a normal state. Thus, a power system is vulnerable if:

- at its intended function, it is susceptible to fail or operate with a significant loss of capacity,
- the power system is unable to cope with unwanted events and unable to quickly recover to normal function and full capacity.

The reliability of a power system is a combination of its vulnerability to external threats that can lead to failure modes and the implied consequences of such failure modes for the end-users (generators, consumers, and various facilitators, such as traders and suppliers). As such, reliability management is composed of two main subtasks: (1) reliability assessment and (2) reliability control. The overall objective is to ensure an adequate level of reliability while minimizing capital and operating expenditures. In reality, the reliability management is broken down into three main time domains in which the reliability management is performed, namely system development, asset management, and system operation, corresponding, respectively, to decision-making horizons of long term (years to decades), mid term (weeks to months/a few years), and short term (minutes to days). The $N-1$ criterion was designed to set the reliability management target within the different activities and within the different control zones of interconnected power systems at transmission level. This criterion prescribes that the system should be able to withstand at all times the loss of any one of its main elements (lines, transformers, generators, etc.) without significant degradation of service quality. The two main intrinsic limitations of the $N-1$ criterion are as follows:

- Strictly applied, it is an approach which does not take the cost of outages into account. This results in overinvestments in some instances and underinvestments in others, and hence a non-optimal social welfare.
- The $N-1$ criterion is a "binary" criterion. The system is secure, or it is not.

With the rising uncertainty in electric power systems due to the introduction of renewables and the reduced observability and controllability which are a direct consequence of the liberalization of the energy market, this causes the $N-1$ criterion to be increasingly difficult to use. The system needs to be overdesigned or underused to unacceptable levels, while the methodology does not allow to take adequate measures into account which rely on intelligent operation of the system using operator intervention or system flexibility. This also influences social welfare in a negative manner. These limitations are well known since the inception of the $N-1$ criterion. Thus, for several decades, researchers and engineers have investigated alternative formulations to measure and enhance power system reliability. These are most notably probabilistic approaches which explicitly take into account both the probabilities of the external threats and the actual socioeconomic impact of service interruptions, to more systematically balance between the different decisions to be taken from the long term to the short term. Nevertheless, the $N-1$ criterion is still massively used today, and this because the method was easy to understand, transparent, straightforward to implement while the new proposed methods are computational heavy, the electric power system was initially overdesigned, different implementations of the $N-1$ criterion were used in practice, and most importantly, the method resulted in an acceptable level of reliability for the power system. This situation, however, is rapidly changing: The operating environment of the electric power system is gradually becoming less predictable. The low social acceptance of overhead power lines leads to system operation closer to its limits and to using more complex and, sometimes, more expensive solutions (e.g., underground cables, HVDC links, new conductors, FACTS). Technology has progressed very significantly, offering new opportunities to evaluate and control electric power system reliability. At the same time, the liberalization of the energy market and the consequent unbundling of the energy system have resulted in a multi-stakeholder business, where delivering the energy with the right quality and at the correct cost is increasingly important. System operators regularly need to take decisions which influence multiple facets of the energy supply at the same time. For instance, freeing more capacity to the market might have a negative effect on the system reliability: which level of transmission capacity is appropriate, and this two days ahead, day ahead or intraday? Investing in a new transmission line can be done using HVDC, AC cable, or AC overhead line. Each of them has different failure rates, repair times, electrical characteristics and consequently influences the system behavior in a different manner. Decisions taken in one decision time frame (long term, mid term, or short term) will influence those in the others: investments versus maintenance and making transmission capacity available to the market day ahead versus redispatch in real time if needed. It is not clear how to compare these different reliability related decisions: what is the correct metric and how to evaluate it.

Such a new reliability criterion will require a system infrastructure that transcends the current one, with higher information requirements, not only technical data such as generation and load data or component failure data, but also weather forecast data and societal data. Furthermore, the new methodologies will also require tools which up to now are not available to the community.

4.5.2 Power System Calculation Tools and Methodologies for the Pan-European Power System

The European power system is the largest system engineered by men. At this moment, no adequate tools exist to adequately model and compute the entire system in a detailed and time-efficient manner, or that make fully use of the available new technologies to make full use of them. New tools must be developed to enable the system operators to model and correctly control the pan-European network in the different time frames: from the millisecond range up to the operational planning range (days).

5 Conclusion

The challenges faced by the electric power industry are overwhelming, and it is clear from the discussions in this paper that a "reality check" on current research practices is necessary; particularly if future power systems are going to hinge on the design of technologies and procedures emerging from the smart grid "hype." To this extent, if smart grids at the transmission level are to become a reality, there needs to be an alignment in the current research practices. This alignment should consider the climacteric boundary interactions between policies, the regulatory background, technology maturity and, socially responsible and farsighted investment. Although we have not covered all possible aspects, we have highlighted the key challenges and potential pitfalls in the field of STGs research.

The different time frames in which the transmission system is managed are pivotal in the appreciation of smart grids. "Smarter" grids require adjustments in the operations and the operational planning phase. At the operational side, an increased use of advanced metering infrastructure, for instance from PMU, expected to bring considerable advantages to the power system. Nevertheless, challenges such as standardization, big data management, and ICT requirements remain. If these challenges can be met, the measurement data can be applied to improve system operations, for instance through the provision of improved control and protection functions. The use of new ICT technologies and especially the possibility of visualization allow the operators to become better aware of the system state take the appropriate actions.

From the operational planning perspective, system operators are setting the first steps toward a more dynamic grid operation, but additional research is needed. More specifically on the actual use of flexibility in the power system, including the stakeholder interactions and the coordination at the international level. However, the main challenge may lie in development of new reliability concepts which can be implemented in realistic power systems and that provide a maximum social welfare to the users, and for which adequate tools are still under development.

References

1. Allen E (2010) Use of COMTRADE for exchange of PMU data. In: NASPI workgroup meeting, Vancouver, Canada
2. Bakken D, Bose A, Hauser C III, Edmond OS, Whitehead D, Zweigle G (2010) Smart generation and transmission with coherent, real-time data. Technical report TR-GS-015, Washington State University, Pullman, WA
3. Beyea J (2010) The smart electricity grid and scientific research. Sci Mag 328(5981): 979–980
4. Bobba R, Dagle J, Heine E, Khurana H, Sanders WH, Sauer P, Yardley T (2012). Enhancing grid measurements: Wide area measurement systems, NASPInet, and security. IEEE Power Energy Mag 10(1):67–73
5. Bose A (2010) Smart transmission grid applications and their supporting infrastructure. IEEE Trans Smart Grid 1(1):11–19
6. Brown R (2008) Impact of smart grid on distribution system design. In: 2008 IEEE power and energy society general meeting—conversion and delivery of electrical energy in the 21st century
7. Chakrabarti S, Kyriakides E, Bi T, Cai D, Terzija V (2009) Measurements get together. IEEE Power Energy Mag 7(1):41–49
8. Council of European Energy Regulators (CEER) (2011) CEER status review of regulatory approaches to smart electricity grids. Technical Report C11-EQS-45-04, ERGEG
9. Danielson C, Vanfretti L, Almas M, Choompoobutrgool Y, Gjerde J (2013) Analysis of communication network challenges for synchrophasor-based wide-area applications. In: 2013 IREP symposium bulk power system dynamics and control IX
10. Department of Energy (DOE) (2009) The smart grid: an introduction. Technical report, DoE. http://energy.gov/oe/technology-development/smart-grid
11. Doorman G, Uhlen K, Kjolle G, Huse E (2006) Vulnerability analysis of the nordic power system. IEEE Trans Power Syst 21(1):402–410

12. Elia System Operator (2013) Evolution of ancillary services needs to balance the belgian control area towards 2018. Technical report, Elia
13. Elmqvist H, Otter M, Mattsson SE (2012). Fundamentals of synchronous control in Modelica. In: Proceedings of the 9th international Modelica conference
14. ETSI TR 101 190 V.1.3.2 (2011)
15. European Commission (2007) Explanatory memorandum for the 3rd energy package. Technical report, European Commission (Last checked on 07 Nov 2008)
16. European Commission (2010) Energy infrastructure priorities for 2020 and beyond—a blueprint for an integrated European energy network. Technical report COM(2010) 677 final, EC, Brussels
17. European Electricity Grid Initiative (EEGI) (2010) Roadmap 2010–18 and detailed implementation plan 2010–12. Technical report, EEGI
18. European Network of Transmission Operators for Electricity (ENTSO-E) (2013a) ENTSO-E R&D implementation plan. Available online: https://www.entsoe.eu/
19. European Network of Transmission Operators for Electricity (ENTSO-E) (2013b) Research & Development Roadmap 2013–2022. Available on-line: https://www.entsoe.eu/
20. Gjerde O, Kjolle G, Detlefsen N, Bronmo G (2011) Risk and vulnerability analysis of power systems including extraordinary events. In: PowerTech, 2011 IEEE Trondheim, pp 1–5
21. Gjermundrod H, Bakken D, Hauser C, Bose A (2009) GridStat: a flexible QoS-managed data dissemination framework for the power grid. IEEE Trans Power Deliv 24(1):136–143
22. Group TH (2013) What is HDF5?
23. Guha Thakurta P, Van Hertem D, Maeght J, Belmans R (2013) Increasing transmission grid flexibility of CWE to integrate more wind energy sources while maintaining system security. Under preparation for submission a special issue of the IEEE Transactions on Power Systems
24. Guha Thakurtha P, Nguyen H-M, Antoine O, Maeght J, Dejong A, D'Hoker J, Godemann M, Van Hertem D, Schell P, Skivee F, Godard B, Doutreloup S, Warichet J, Lambin J-J, Maun J-C, Belmans R, Lilien J-L (2013) D7.3: final report on NETFLEX demo. Technical report, Twenties project
25. Hellstrom B, Danielson M, Olsson B, Lindgren P (2012) Eliminating GPS dependency for real-time wide-area synchrophasor applications. Net Insight AB, Stockholm, Sweden
26. Heydt GT (2010) The next generation of power distribution systems. IEEE Trans Smart Grid 1(3):225–235
27. Hofmann M, Kjolle GH, Gjerde O (2012) Development of indicators to monitor vulnerabilities in power systems. In: 11th Probabilistic security assessment and management and ESREL 2012 conference, Helsinki
28. IEEE PC37.242/D12 (2013) IEEE guide for synchronization, calibration, testing, and installation of phasor measurement units (PMUs) for power system protection and control
29. IEEE PC37.244 (2013) IEEE draft guide for phasor data concentrator requirements for power system protection, control, and monitoring
30. IEEE Standard C37.118-2005 (Revision of IEEE Std 1344-1995) (2005) IEEE Standard for synchrophasors for power systems
31. IEEE Std C37.118.1-2011 (Revision of IEEE Std C37.118-2005) (2011) IEEE Standard for synchrophasors for power systems
32. IEEE Std C37.118.2-2011 (Revision of IEEE Std C37.118-2005) (2011) IEEE standard for synchrophasor data transfer for power systems
33. Kezunovic M (2010) Intelligent design: substation data integration for enhanced asset management. IEEE Power Energ Mag 8(6):37–44
34. Lambert E (2008) Common information based on CIM approach: is it a dream or a reality? Remaining challenges based on concrete experience. In: IEEE Power and Energy Society general meeting—conversion and delivery of electrical energy in the 21st century, pp 1–6
35. Leelaruji R, Vanfretti L (2012) State-of-the-art in the industrial implementation of protective relay functions, communication mechanism and synchronized phasor capabilities for electric power systems protection. Renew Sustain Energy Rev 16(7):4385–4395

36. Milano F, Zhou M, Hou G (2009) Open model for exchanging power system data. In: IEEE Power Energy Society general meeting, pp 1–7
48. Moraes R, Hu Y, Stenbakken G, Martin K, Alves J, Phadke A, Volskis H, Centeno V (2012) PMU interoperability, steady-state and dynamic performance tests. IEEE Trans Smart Grid 3(4):1660–1669
49. Myrda P, Taft J, Donner P (2012) Recommended approach to a NASPInet architecture. In: 45th Hawaii international conference on system science (HICSS), pp 2072–2081
39. Net Insight (2013) Flexible multiservice MSR and IP Media Gateway-Nimbra 380. Available online: http://tinyurl.com/Nimbra-380. Net Insight AB, Stockholm
40. North American Electric Reliability Corporation (NERC) (2006) Standard PRC-002-1 define regional disturbance monitoring and reporting requirements
41. North American Synchrophasor Initiative (NASPI) (2010) NASPI phasor data NDAs
42. Otter M, Thiele B, Elmqvist H (2012) A library for synchronous control systems in Modelica. In: Proceedings of the 9th international Modelica conference
43. Panciatici P, Bareux G, Wehenkel L (2012) Operating in the fog: security management under uncertainty. IEEE Power Energy Mag 10(5):40–49
44. Patel M (RAPIR Chair) (2010) Real-time application of synchrophasors improving reliability. Technical report, North American Electricity Reliability Corporation, Princeton, NJ
45. Patterson J (2010) Hadoop as the platform for the Smartgrid at TVA. Available online: http://tinyurl.com/tva-hadoop
46. Phadke A, de Moraes R (2008) The wide world of wide-area measurement. IEEE Power Energ Mag 6(5):52–65
47. Regulators Group for Electricity & Gas (ERGEG) (2009) Position paper on smart grids—an ERGEG public consultation paper. Technical report E09-EQS-30-04, ERGEG
37. SmartGrids European Technology Platform (2010) Strategic deployment document for Europe's electricity networks of the future. Technical report, SmartGrids ETP
38. SmartGrids European Technology Platform (2012) SmartGrids SRA 2035 strategic research agenda: Update of the SmartGrids SRA 2007 for the needs by the year 2035. Technical report, SmartGrids ETP
50. Stifter M, Widl E, Andren F, Elsheikh A, Strasser T, Palensky P (2012) Co-simulation of components, controls and power systems based on open source software. In: IEEE PES general meeting 2013
51. Task Force Smart Grids Expert Group 2 (2010) Regulatory recommendations for data safety, data handling and data protection. Technical report, EC, DG Energy, Direction for Security of Supply and Energy Markets
52. Trachian P (2010) Machine learning and windowed subsecond event detection on PMU data via Hadoop and the openPDC. In: IEEE Power and Energy Society general meeting, pp 1–5
53. Van Hertem D et al (2010). Coordination of multiple power flow controlling devices in transmission systems. In: IET ACDC conference, London
54. Vanfretti L et al (2012) SmarTS Lab: a laboratory for developing applications for WAMPAC systems. In: IEEE PES general meeting
55. Van Hertem D, Rimez J, Belmans R (2013) Power flow controlling devices as a smart and independent grid investment for flexible grid operations: Belgian case study. IEEE Trans Smart Grid PP(99):1–9
56. Vanfretti L, Aarstrand V, Almas M, Peric V, Gjerde J (2013) A software development toolkit for real-time synchrophasor applications. In: 2013 IEEE PowerTech
57. Vanfretti L, Baudette M, Al-Khatib I, Almas M, Gjerde J (2013) Testing and validation of a fast real-time oscillation detection PMU-based application for wind-farm monitoring. In: Proceedings of the first international Black Sea conference on communications and networking 2013 (BlackSeaCom 2103)
58. Vanfretti L, Chow J, Sarawgi S, Fardanesh B (2011) A phasor-data- based state estimator incorporating phase bias correction. IEEE Trans Power Syst 26(1):111–119

59. Vanfretti L, Dosiek L, Pierre JW, Trudnowski D, Chow JH, Garćıa-Valle R, Aliyu U (2011) Application of ambient analysis techniques for the estimation of electromechanical oscillations from measured PMU data in four different power systems. Eur Trans Electr Power 21(4):1640–1656
60. Varaiya P, Wu F, Bialek J (2011) Smart operation of smart grid: risk-limiting dispatch. Proc IEEE 99(1):40–57

Comparison of Smart Grid Technologies and Progress in the USA and Europe

M. Godoy Simões, R. Roche, E. Kyriakides, S. Suryanarayanan, B. Blunier, K. McBee, P. Nguyen, P. Ribeiro and A. Miraoui

Abstract This work discusses historical and technical events in USA and Europe over the last few years that are aimed at modernizing the electric power grid. The US federal government has ratified the "Smart Grid Initiative" as the official policy for modernizing the electricity grid including unprecedented provisions for timely information and control options to consumers and deployment of "smart" technologies. European countries are unified in researching and developing related technologies through various structures supported by the European Union. This chapter presents the development of smart grids and an analysis of the methodologies, milestones and expected evolutions of grid technologies that will transform society in the near future.

Keywords Control · Distributed generation · Power electronics · Power systems · Smart grid · Smart metering · Storage · Calibration · Intelligent sensors · Nuclear measurements · Production facilities · Protocols · Sensor phenomena and characterization · Sensor systems · Temperature measurement · Temperature sensors · Wireless sensor networks

M. Godoy Simões (✉) · K. McBee
Colorado School of Mines, Colorado, USA
e-mail: msimoes@mines.edu

R. Roche · B. Blunier · A. Miraoui
Université de Technologie de Belfort-Montbéliard, Belfort, France

E. Kyriakides
University of Cyprus, Nicosia, Cyprus

S. Suryanarayanan
Colorado State University, Colorado, USA

P. Nguyen · P. Ribeiro
Technische Universiteit Eindhoven, Eindhoven, The Netherlands

D. Mah et al. (eds.), *Smart Grid Applications and Developments*,
Green Energy and Technology, DOI: 10.1007/978-1-4471-6281-0_11,

1 Introduction

The present-day electric grid was built about a century ago in most industrialized countries and has been growing in size and capacity ever since. Transmission lines connect large centralized power sources to the grid and have been technologically updated with automation and human monitoring over the last few decades. The electrification of our society has empowered countless advances in other fields such that the US National Academy of Engineering ranked it as the greatest engineering achievement of the last century [41]. However, this transformation was mostly in the transmission realm and not in the distribution milieu since the latter has traditionally been considered as user endpoints of service, where power was delivered to traditional loads. The last two decades saw a steady growth of distributed generation, with plans for higher penetration of renewable energy sources, and the policies on electricity distribution have been supporting needs for a "smart grid" for many reasons that will be discussed in this chapter. Centralized power plants have enormous economic constraints and benefits, and utilities have been trying to use their assets more efficiently. As an example, typically, 20 % of the US generation capacity is used only for 5 % of a year to meet peak demand and is based on coal and gas power plants, which cause environmental concerns and greenhouse gases (GHG) emissions. A paradigm shift in distribution engineering is viewed as the next frontier of advancement in electric power systems, and the smart grid is expected to introduce unprecedented changes in the distribution systems worldwide. Both the US and European economies have taken the lead in establishing some early concepts and policies for realizing the smart grid. In this chapter, a comparison of the various smart grid technologies and the path of progress in the USA and in Europe are presented.

2 Evolution of the Smart Grid in the USA and Europe

Conceptualization of techniques to improve the intelligent interaction of distributed assets for smart grids emerged in the 1980s as a call to modernize the grid, allowing deeper penetration of alternative and renewable energy sources. The first references to the term *smart grid* were provided around 2004 by Amin [5], Amin and Wollenberg [7]. While some common characteristics of a smart grid exist, Europe and USA have been following different paths to make their respective grids smarter.

2.1 Trajectory in the USA

The electric grid in the USA is composed of approximately 15,000 generators operating in 10,000 power plants, accounting for approximately 3.95 million MWh (as of 2009), with approximately 160,000 miles of high-voltage transmission lines (at voltages typically above 280 kV) [4, 13]. The electric grid in the USA evolved from small-rated centralized power stations, such as the historical Pearl Street Station in Manhattan, NY (such power stations supplied DC electricity to relatively smaller-rated loads in the late nineteenth century) to the interconnected AC system of present day that crisscrosses North America [26]. The present-day massive North American grid consists of three large asynchronous interconnections—the western electricity coordinating council (WECC) system, the eastern interconnection, and Texas (ERCOT)—and has been referred to as the "most complex man-made machine" [35]. The evolution of the smart grid in the USA may be traced to several innovations in the transmission grid, such as the wide-area measurement, fast controls, the installation of power system stabilizers, phase shifting transformers, flexible AC transmission system devices (FACTS), and phasor measurement units (PMUs). Additionally, the advent of advanced control room visualization has heralded the template for the smart grid. Public awareness and concomitant push for more renewable energy sources in the grid have also provided some impetus to the integration of newer technologies in the grid.

However, the modernization of the electricity grid has been largely restricted to the transmission systems and did not penetrate the distribution system level as much. This may be attributed to factors such as higher variability in system scenarios at the latter level and a relatively low economy of scale when compared to the transmission systems. Several legislative mandates have provided various opportunities for the modernization of the electric grid in the USA. Figure 1 provides a timeline of some events related to the electricity grid in the USA that have served as harbingers to important changes via mandates and legislations.

2.2 Trajectory in Europe

The unification of the European grid was achieved in parallel to the economical unification of European countries. This process has been slow, due to the high cost of building new infrastructures and to historical events such as the two World Wars and the political separation of Eastern and Western Europe for more than four decades.

The idea of a pan-European grid was first discussed by the League of Nations in the 1920s. This became a reality only in 1951 with creation of the Union for the co-ordination of production and transmission of electricity (UCPTE, which eventually became UCTE), which was aimed at interconnecting the grids of France, Germany, and Switzerland. Similar associations were created in other regions of Europe, such as NORDEL or SUDEL. But during the Cold War, a real interconnection of the Western and Eastern parts of Europe was nearly impossible, and European countries reorganized such interests only after 1990, with the fall of the Iron Curtain. In 2008, the regional associations ETSO, ATSOI/UKTSOA (Ireland, Great Britain), NORDEL (Finland, Sweden, Norway, and Eastern

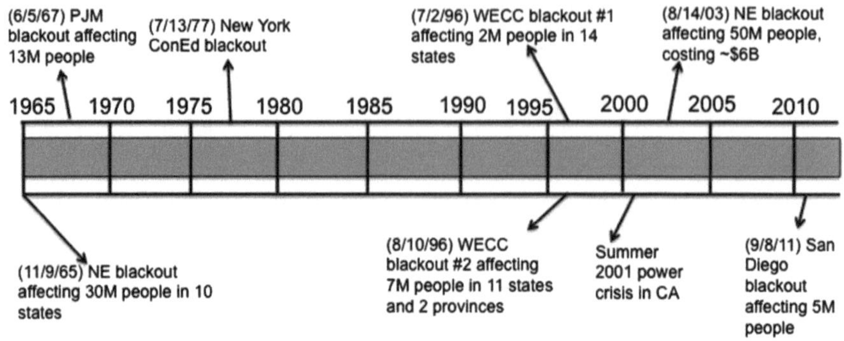

Fig. 1 Timeline of major events in the US electric grid

Denmark), UCTE (23 continental European countries), and BALTSO (Baltic countries) merged into the European Network of Transmission System Operators for Electricity (ENTSO-E), which now coordinates 41 transmission system operators (TSOs) from 34 countries [14]. Smart grid policies in Europe are relatively new, while at the same time the European grid is becoming more interconnected and sees investments decrease. However, the sense of ownership and contribution of each individual country to the whole grid is different from how the USA directs initiatives through its Department of Energy (DOE). Some recent EU initiatives are as follows:

- The European energy program for recovery (2009), which has some similarities to the US Stimulus fund. This program was aimed at speeding up and securing investments for projects in the energy sector.
- An overall energy efficiency action plan (2007–2012) establishes a firm objective of 20 % improvement.
- The European energy infrastructure package identifies smart grids as the key infrastructure for energy modernization in Europe.
- A Competitiveness and Innovation Framework Program (CIP) proposes an Intelligent Energy for Europe program.
- The European Energy Research Alliance (ERRA) aims at accelerating the development of new energy technologies by maximizing funding sources, facilities, and complementarities among institutes in participating countries.

3 Governing Bodies in Smart Grid Development

The US Smart Grid Initiative is the official policy of grid modernization in the USA as formalized by the 2007 Energy Independence and Security Act (EISA07) [3]. Under the purview of this legislative mandate, the US smart grid is characterized by:

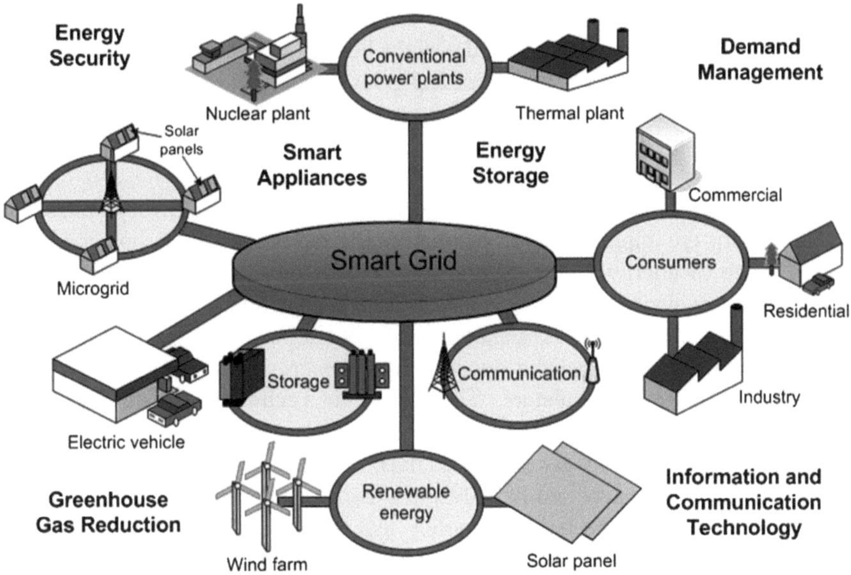

Fig. 2 The future electric smart grid

- Increased digital information and controls.
- Dynamic optimization of grid operations, including cyber security.
- Deployment of distributed resources including renewables.
- Incorporation of demand-side resources and demand response.
- Deployment of "smart" technologies and integration of "smart" appliances and consumer devices.
- Deployment of storage and peak-shaving technology including PHEV.
- Provision of timely information and control options to consumers.
- Standard development for communication and interoperability of equipment.
- Identification and lowering of unreasonable barriers to adopt smart grid technology, practices, and services [3].

An illustration of an implementation of the smart grid is depicted in Fig. 2 The National Institute of Standards and Technology (NIST) provides a conceptual model that defines seven important domains: bulk generation, transmission, distribution, customers, service providers, operations, and markets [34].

In the EU, the smart grid strategy is motivated by concepts of innovation with regard to social and environmental reforms for an interactive economy. The European energy policy relies on [21]: (1) security of supply, (2) sustainability, and (3) market efficiency. In addition, six goals have been set for the EU energy strategy to: (1) achieve the highest levels of safety and security, (2) achieve an energy-efficient Europe by improving buildings, transportations, and distribution grids, (3) extend Europe's leadership in energy technology and innovation, (4) empower consumers, (5) build a European integrated energy market, and (6) strengthen the external dimension of the EU energy market. The European Strategic Energy Technology Plan includes eight European Industrial Initiatives (EII) in the field of energy. The EII on electrical grids is called the European Electricity Grid Initiative (EEGI) and has a budget estimated to €2 billion over a period of 10 years with guidelines and activities for R&D and a program with 20 large-scale demonstration projects [22].

The EU established the Third Energy Package in 2007 with objectives and regulations for the implementation of smart grids stating that all European citizens have a range of energy-related rights, such as consumer choice and fair prices. Dispositions are also taken to favor international energy (electricity and gas) trade, collaboration and investment, separation of generation and supply from transmission networks, decentralized generation and energy efficiency, smart meters and effective national regulators. The European Technology Platforms (ETP) began to operate in 2005 [17] in formulating and promoting a vision for the development of smart grids for year 2020 in compliance with EU policy. A "Smart Grids Task Force" was created in 2009 and is composed of European Commission officials, experts from the industry, policy makers, and academia [19]. The first projects related to smart grids were grouped within the IRED (Integration of Renewable Energy Sources and Distributed Generation into the European Electricity Grid) cluster. Over 60 projects in the fields of smart grids have been supported by the 6th Framework Program (FP6) and correspond to an investment of about €190 million. FP7, the current program, will run until 2013 and has a total budget of €51 billion, with 7 % of it dedicated to energy-related projects [16]. Currently the related EU associations for promoting smart grid projects are as follows:

- The European Regulators Group for Electricity and Gas (ERGEG) and the Council of European Energy Regulators (CEER), which allow national regulators to cooperate.
- The new agency for the Cooperation of Energy Regulators (ACER), a complement to the two previous organizations and ENTSO-E.
- The European Distribution System Operators Association for Smart Grids (EDSO-SG).
- The Union of the Electricity Industry (EURELECTRIC), which represents the common interests of the electricity industry at the European level.

4 Enabling Technologies

Several technologies have to mature in order to make the smart grid a reality [40, 42]:

4.1 Distributed Generation

Distributed generation (DG) (also referred to as embedded generation or dispersed generation) refers to small rating electricity sources that are typically decentralized and located close to end-user locations on the distribution side of the electric grid. These may include conventional as well as renewable energy sources. The interconnection of DG to the grid provides a variety of advantages including on-demand power quality of supply, enhanced reliability, deferrals in transmission investment, and avenues for meeting renewable mandates in the face of growing disinvestments in transmission assets—all of which cater to the smart grid philosophy. However, the interconnection of DG is a challenge due to the safety, control, and protection issues associated with bidirectional flows of electricity. In the USA, the Energy Policy Act of 2005, enacted by the 109th Congress, recognized the IEEE 1547 Standard as the national technical standard for interconnection of DG to the electric grid [2]. Further information about DG is available in [24].

4.2 Energy Storage

Electricity is a highly perishable commodity that must be consumed within a very short span of production and cannot be easily stored, especially in high quantities. Alternatively, it may be converted into other forms such as mechanical or electrochemical energy. Storage technologies enable these processes and are among the desired features for the smart grid. Multiple existing technologies are compared in Fig. 3. Storage, which can be distributed in the grid, (i) makes the grid more efficient [36]; (ii) enables load leveling and peak shaving, while it reduces dependency on spinning reserve; (iii) improves grid reliability and power quality; (iv) provides ancillary services, supplying reactive power for voltage regulation; and (v) supports T&D investment deferring. Energy storage with power electronics interfaced units can create virtual rotational inertia, the so-called virtual synchronous generators (VSG), which can reduce the rate of change of frequency and frequency deviations [12].

4.3 Power Electronics

Power electronics is fundamental in the development of smart grids because a deeper penetration of renewable and alternative energy sources requires sophisticated power converter systems. Typically, a power converter is an interface between the smart grid and local power sources [9]. Solar PV and wind play a significant role as alternative sources for integration in smart grids and are

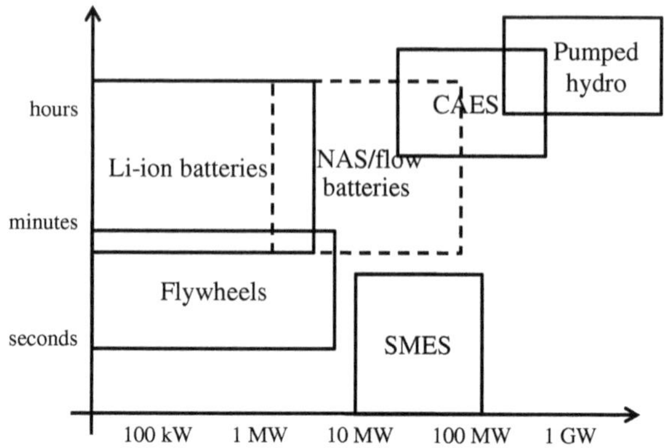

Fig. 3 Comparison of discharge duration versus rated power for some grid energy storage technologies

increasingly being installed in residential and commercial locations (typically with a power range of a few kW) as well at high rating in the high-voltage transmission grid. The intermittent nature of these sources affects the output characteristics of generator and converter sets. A power electronics converter is deemed necessary to smooth the output to desired characteristics and to allow energy storage during surplus of input power and compensation in case of lack of input power. The following characteristics are important for power electronics systems for smart grids:

- High efficiency: Only a negligible part of the power should be dissipated during conversion stages;
- Optimal energy transfer: All renewable energy sources are energy constrained and, as such, need algorithms to achieve the maximum power point which must be considered in the design of the power electronics interface;
- Bidirectional power flow: Power converters have to be able to supply the local load and/or the grid;
- High reliability: Continuity of service is a major issue when delivering energy;
- Synchronization capabilities: All power sources connected to the grid have to be fully synchronized, thus ensuring high efficiency and eliminating failures; therefore, standards such as IEEE 1547 should be incorporated in power electronics interfaces [28];
- Electromagnetic interference (EMI) filtering: The quality of the energy injected into the grid must adhere to apt electromagnetic compatibility (EMC) standards;
- Smart metering: The interface between the local source/load and the grid must be capable of tracking the power consumed by the load or injected into the grid;
- Real-time information must be passed to an automatic billing system capable of taking into account parameters such as energy bought/sold in real time, informing end users of all required pricing parameters;

- Communications: The intelligent functioning of the smart grid depends on the capability to support communications layer in tandem with an energy delivery layer in the grid; and
- Fault tolerance/Self-healing: A key issue is a built-in ability to minimize the propagation of failures and resilience against such local failures. This capability should be incorporated with monitoring, communication, and reconfiguration features of power electronics systems. Additionally, power electronic interfaces must be configured to avoid nuisance trips.

4.4 Control, Automation, and Monitoring

A smart grid is a highly complex, nonlinear dynamic network of distributed energy assets with bidirectional flow of power and information that presents many theoretical and practical challenges. Monitoring and control are key issues that need to be addressed to make it more intelligent and equip it with self-healing, self-organizing, and self-configuring capabilities. This requires much more sophisticated control, sensing, and computer-oriented monitoring than in the contemporary grid, where grid operations are rather reactive, with a number of critical tasks performed by human operators. Therefore, some modern control techniques have been claimed to be the best fit for smart grids, for example, agent-oriented programming, implementing computational intelligence into distributed systems operation [15]; however, most of these are yet to transcend the research domain into large-scale deployment. A combination of agent-based control techniques and power electronics possess the potential to create an intelligent and flexible interface between consumers, storage, DGs, and the network, as well as among network areas. Figure 4 illustrates different power electronics layers to integrate a cluster of prosumers (an entity in the future grid capable of both producing and consuming electric power) into the grid.

The two-way communication ability of smart meters allows the transmission of delivered and generated energy data along with actionable commands to customers. With technologies such as WIFI, ZigBee, and home area network (HAN) communication systems, smart meters can now act as interfaces for energy management entities, customers, and utilities to control a number of appliances within a residential home based on price signals [32].

Power quality analyzing capabilities of smart meters may improve the ability to identify system and customer voltage deficiencies, harmonic distortions, and onset of equipment failure. Daily energy usage and generation profiles may be recorded for forecasting-relevant system parameters. Threshold voltage events can trigger communication with utilities for providing alerts to disturbances, prior to customer equipment failure or discomfort. These functionalities may also improve the ability of utilities to locate the source of system events, which is a difficult and complex task on the existing electric distribution system, but at may introduce the

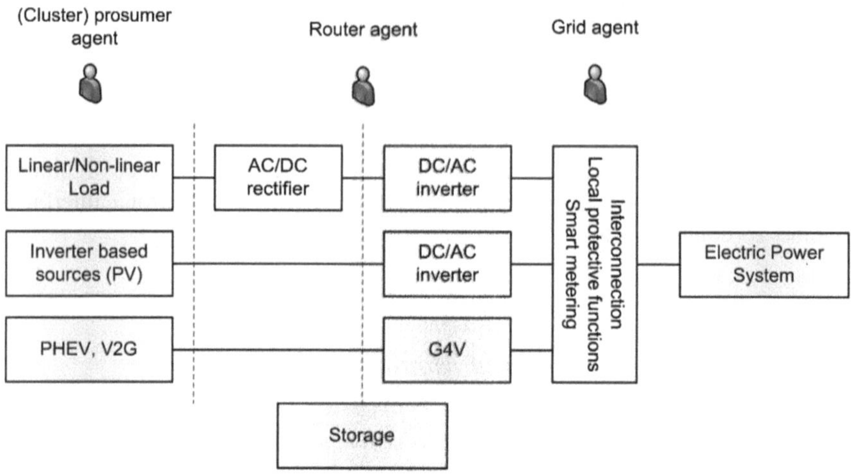

Fig. 4 Intelligence-based control structure for power electronics in smart grids

risk of increased susceptibility to cyber intrusions by malicious agents [11]. Several utilities in the USA and Europe have already seen improved power quality due to the installation of smart meters [31].

4.5 Demand-Side Management

Demand-side management (DSM) refers to the ability to change energy consumption patterns and characteristics via structured programs. Historically, most DSM programs aimed at achieving targets in energy efficiency, while some conservation programs aimed at deferring investments in new assets (including generating facilities, power purchases, and transmission and distribution capacity additions). However, with the advent of the smart grid, DSM may provide paradigm shifts in the normal operation of the electricity market or from government-mandated energy efficiency standards. In the last few years, there has been an increased interest in dynamic pricing, i.e., a time-varying pricing at the end-user level, different from the state of the art of tariffs. Dynamic pricing, which is available in the US bulk power (transmission) markets since mid-90s following deregulation of the industry, has enabled economic efficiency, fostered investments in technological innovations, and for the most part removed the ills of market power and monopoly [30]. Such a differentiated or tiered rate of electricity in the distribution system is viewed as an enabler for the smart grid. When enabled with such information on dynamic pricing of electricity, customers and utilities will need to interact via DSM structures aimed at increased energy efficiency, lowered cost of engaging inefficient and costly generators at peak periods, coordinated charging of plug-in hybrid electric vehicles (PHEVs), or other similar objectives. Several dynamic pricing models have been proposed in order to better reflect the actual cost of producing energy on the specific day and time and provide incentives to customers to become more active in controlling their electricity consumption [33, 37]. The following time-varying pricing methodologies are viewed as possible enablers of DSM:

- Time-of-use (TOU) electricity rates have been shown to be effective in promoting customer participation in demand management [37]. TOU rates use predetermined time intervals during which electricity use is recorded. Each time interval has a fixed price that is proportional to the electricity availability during that same time interval.
- Real-time pricing is a methodology whereby the customer is informed a time period ahead of the electricity price so as to make rational decisions regarding the consumption of electric energy. The retail price of electricity in this pricing methodology floats based on the actual cost of electricity. If the time period is 1 h, then real-time pricing is known as hour-ahead pricing.
- Critical peak pricing is used to "force" customers to avoid consuming electric energy during specific peak periods. These periods are well defined by the utility, and the cost of electricity during these periods is increased significantly.
- The peak time rebate has a similar concept as the critical peak pricing. Instead of a penalty, the conscious customer receives a reward for reducing their electricity consumption below a certain baseline. If the customer consumes more than the baseline, there is no penalty imposed.

Data management is critical for the widespread operation of the smart grid in the near future, because vast raw data will have to be processed, aggregated, validated, and transmitted for further processing and analysis.

4.6 Distribution Automation and Protection

Smart grid technology supports a wide range of applications in power systems, such as protection and automation of the distribution system and security. It is possible to design self-healing protection systems using the capabilities of the advanced distribution automation (ADA) [25]. The protection of the power system is a critical necessity in order to achieve continuity, reliability, and security of supply. Protection deals with the detection and clearance of abnormal system conditions such as faults and overloads. High penetrations of DG can compromise existing protection schemes, which are based on single sources supplying unidirectional power through radial distributed lines. In the smart grid, where bidirectional flow of electricity through partially networked systems in the distribution milieu will be prevalent, protection mechanisms are adaptive and incorporate intelligent automated functions. Recent activities in protection engineering focus on developing microprocessor-based devices called intelligent electronic devices

(IEDs). Such devices may be smart distribution switches or integrated solutions for substations. Standard IEC 61850 (Communication Networks and Systems in Substations) provides authoritative information relevant to the design of substation automation [29].

4.7 Communication Systems

Self-healing systems have been sought to be incorporated into power systems, especially as the complexity and interactions of several market players significantly increase the risk for large-scale failures. Reconfiguring the system in islanded mode may require a hitherto unknown rate and amount of data exchange, two-way communication links, and advanced central computing facilities. Decentralized intelligent control could enable islands to accommodate their native load and generation in a more reliable and efficient manner. Local controllers may ensure that each island is operating within the security limits, safeguarding the electricity supply to its customers [6]. A self-healing system should be based on a wide-area monitoring network that incorporates a variety of sensors, such as PMUs that obtain phasor measurements by synchronizing with each other through the global positioning service (GPS) [6, 10].

Measurements and signals obtained from sensors may be used either by local (distributed) or centralized controllers to enable the self-healing of the system under disturbance or fault conditions. These measurements and signals may be submitted for processing to a single controller, despite the fact that they may be originating from different proprietary networks [8]. IEEE Standard 1451.4 requires analog sensors to have a transducer electronic data sheet (TEDS) to provide calibration information to the data acquisition system [27, 43]. Figure 5 shows how several communication technologies can be applied for such data, according to their characteristics. Ranges of operation and bandwidth of these technologies vary significantly, and some of them may be chosen for HAN, while others may be used for longer distances as from houses to concentrators or between substations [31].

Two-way communication enabled smart appliances, smart meters for control of sources, and loads, and storage must be implemented in a platform that allows both digital information and electric energy to flow through a two-way smart infrastructure. The requirements for these communication infrastructures are reliability and resilience, bandwidth, interoperability and costs. Several communication protocols and media are currently under various stages of research and development R&D for implementation in smart grids. Examples include: broadband over power line (BPL) and power line communications (PLC), which use existing power lines to transmit information; Ethernet, DSL, and optic fiber, already in use for the Internet; ZigBee, and WIFI, which are already used for HAN applications; WiMaX, a "super-WIFI", with a much higher range; and 3G, LTE/4G, and other mobile telephone communication protocols.

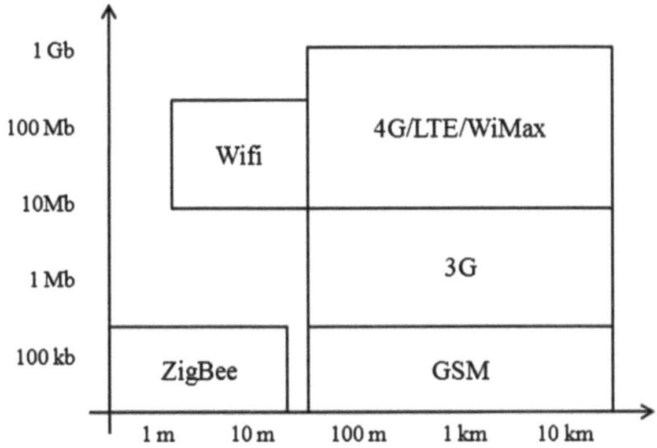

Fig. 5 Characteristics of some wireless communication technologies: bandwidth versus transmission range

5 Comparative Metrics for USA Versus Europe

Making the grid smarter requires considering all aspects of smart grids as part of the decision-making process. This section compares the practices in the USA and the EU on several topics for the development of smart grid technologies.

5.1 Legislation in the USA

The legislation that led to the present-day US Smart Grid Initiative might be traced back to the 70s, when deregulation was initially introduced as a direct result of the Arabic oil embargo that escalated a nationwide energy crisis. In 1977, the Federal Power Commission (FPC) was restructured to form the Federal Energy Regulatory Commission (FERC), with the objective of regulating energy transactions and transmission across different states in the Union. The National Energy Act of 1978 was passed to conserve energy and increases efficiency by judicious use of resources and amenities by utilities and introduced the Public Utilities Regulatory Policies Act (PURPA), which advocated the need for small power productions, for cogeneration and for renewable energy sources that could compete as independent power producers (IPPs) in the electricity market with other utilities. The 102nd Congress of the USA passed the Energy Policy Act of 1992, which included provisions for alternative fuels, electric motor vehicles, energy efficiency improvement, and energy conservation techniques [1]. FERC mandated open transmission access, with the mandated open-access transmission tariff (OATT) requiring that all investor owned utilities (IOU) make separate functionalities of generation, distribution, transmission, and marketing services. It also mandated the creation of independent system operators (ISO), eventually making possible the creation of an open-access same-time information system (OASIS)—a web-based secure database of transmission system-related information. These orders were the

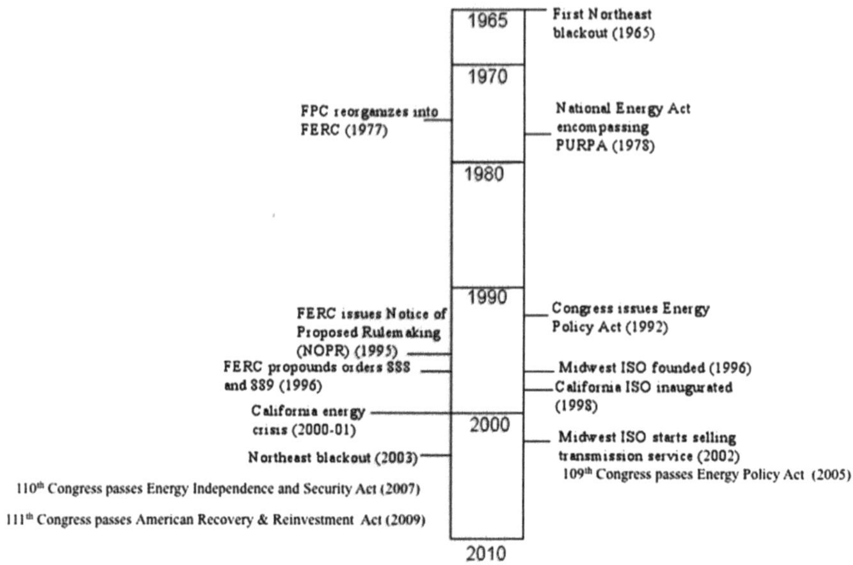

Fig. 6 Key US legislative events vis-à-vis the smart grid

template for the evolution of the deregulated electricity market structure that is currently in place in the USA.

Following the Northeast blackout of August 2003 in the USA, the 109th US Congress passed the 2005 Energy Policy Act, with provisions for tax incentives and subsidies for renewable energy integration and energy efficiency technologies. Figure 6 depicts a timeline of some key legislative events in US history vis-à-vis the smart grid. The 110th US Congress is credited with passing the EISA07, which explicitly characterized the smart grid through the Smart Grid Initiative in Title XIII. The other highlights of this act included electrifying the transportation fleet, reductions in fossil fuel usage in certain sectors, and carbon sequestration. This was followed by the American Recovery and Reinvestment Act of 2009 (ARRA09), passed by the 111th US Congress, which included provisions for energy infrastructure improvements via the implementation of the smart grid. Table 1 lists some details of selected smart grid projects in USA supported under ARRA09 [38]. The Smart Grid Clearing House is a web resource that lists all ARRA09-funded smart grid projects geographically as well as according to their technical focus.

5.2 Legislation in Europe

Energy needs are responsible for 80 % of all European GHG emissions [17]. Therefore, climate change legislation and energy policy have been intimately

Table 1 Selected US smart grid projects from Smart Grid Clearing House [38]

Project type	Number of projects
Automated meter infrastructure (AMI)	81
Customer systems	8
Distribution systems	15
Equipment manufacturing	2
Integrated systems	54
Regional demonstration	16
Storage demonstration	16
Transmission system	10
Total number as of June 2011	~202

linked with a strong impact on investment decisions from private companies. As a consequence of the Kyoto protocol, European leaders made a unilateral commitment to reach three legally binding objectives by 2020, known as the "20–20–20" targets [18]:

- Reach a 20 % share of energy consumption coming from renewable sources.
- Achieve a reduction of 20 % of primary energy use through energy efficiency measures.
- To reduce GHG emissions by at least 20 % below 1990 levels.

With the Third Energy Package, additional requirements were introduced, such as the encouragement to roll out 80 % smart meters in Europe by 2020 [23]. GHG target reductions may be increased to 30 % if other major emitting countries set themselves ambitious objectives. A proposal to cut emissions by 80–95 % by 2050 has also been suggested. Table 2 shows a list of selected smart grid projects in Europe supported by EU funding in the last few years [20].

5.3 Barriers

Several barriers may slow down the development of smart grids on both continents, as depicted next. A first issue is related to financing. During 2009, stimulus plans were used for funding dozens of projects. However, as more and more governments are taking austerity measures, this funding is expected to decrease, or not be renewed, and will need to be either replaced or supplemented by private funding sources. This raises the question of the real interests of the many stakeholders in the smart grid. The cost to modernize distribution networks is high, and utilities may consider if the benefits will overweigh the costs. Moreover, the smart grid requires utilities to make significant changes to their present business models (e.g., reducing demand is contradictory with present-day models). Regulators are expected to balance costs in order to ensure that each player finds an acceptable ratio between the costs and returns on investments. They should also enable dynamic electricity pricing, a requirement for demand response actions, and DSM

Table 2 Selected EU smart grid projects from European Commission CORDIS [20]

Project	Focus
More-microgrids (FP6)	DER, microgrids
FENIX (FP6)	RES integration
EU-DEEP (FP6)	Business models
ADDRESS (FP7)	DER integration, demand management
SmartHouse/SmartGrid (FP7)	Smart buildings
MERGE, G4V (FP7)	Impact of electric vehicles
Green eMotion (FP7)	Business models for electric vehicles

programs to achieve success. The acceptance of consumers regarding smart metering and changes in general is a challenge. In some US states, consumers raised concerns to the installation of smart meters, regarding an increase in the electricity bill, or the privacy of information transmitted to the utility. Technology maturity and availability presents another challenge, especially regarding DER integration and control. The operating security is also critical, as shown by the Stuxnet case [39]. The establishment of standards for smart grids is also a crucial step. By allowing components to interact with each other, and ultimately to reduce costs, standards will enable true interoperability between assets produced by various companies. The US NIST, IEEE, IEC, and other organizations have been working on several standardization activities.

6 Path Forward

The EU and the USA have different approaches in fostering smart grid technology. Europe has been influenced by concerns derived from the diversity and evolution of power grids across European countries, while the USA needs to increase security and to respond to the predicted growth in demand for a long-term vision. It is expected that such technologies will have widespread growth subject to economies of scale. Distribution networks will dramatically change in the near future, and energy storage is expected to become increasingly available, even at the distribution level, in order to compensate the intermittent nature of renewable energy sources that will penetrate the distribution sector. Communication and user interfaces will be pervasive, and the integration with the new web of things (WoT) will allow individual home and business electric devices to be controlled and operated from remote locations. Power distribution will be more controllable and dispatchable, and new distribution systems will have massive automation and sensing systems that will allow any user to interact with any other one on the electrical network. A smart grid is expected to emerge in the USA and in Europe in the next decade and to evolve thereafter; notwithstanding the avatar of this smart grid, which will be a function of the policies shaping this evolution, the desired characteristics of resilience, sustainability, increased energy efficiency, engaging highly dispersed assets with temporal and spatial stochastics, and breeding a new class of informed customers who engage in the grid operations are expected to be achieved.

Dedication We dedicate this chapter to the memory of our friend and co-author, Dr. Benjamin Blunier, formerly an Associate Professor at the University of Technology of Belfort-Montbéliard (UTBM), who passed away on february 23, 2012.

Acknowledgment This chapter is an updated version of the paper "A Comparison of Smart Grid Technologies and Progresses in Europe and the USA" by Simoes, M.G., Roche, R., Kyriakides, E., Suryanarayanan, S., Blunier, B., McBee, K.D., Nguyen, P.H., Ribeiro, P.F., Miraoui, A., Industry Applications, IEEE Transactions on 48 (4), 1154–1162. The copyright permission for reusing the paper has been granted by IEEE.

References

1. 102nd US Congress (Oct 1992) 1992 Energy Policy Act (EPAct92)
2. 109th US Congress (Jul 2005) 2005 Energy Policy Act (EPAct05)
3. 110th US Congress (Dec 2007) 2007 Energy Independence and Security Act (EISA07)
4. American Society of Civil Engineers (2005) Report card for american's infrastructure: US electric power grid. http://www.asce.org/reportcard/2005/page.cfm?id=25. Accessed Jun 2011
5. Amin SM (2004) Balancing market priorities with security issues. IEEE Power Energ Mag 2(4):30–38
6. Amin SM (2008) For the good of the grid. IEEE Power Energ Mag 6(6):48–59
7. Amin SM, Wollenberg BF (2005) Toward a smart grid: power delivery for the twenty first century IEEE Power Energ Mag 3(5)
8. Betts B (2006) Smart sensors. IEEE Spectrum, 43(4), 50–53. doi:10.1109/MSPEC.2006.1611760, http://ieeexplore.ieee.org/stamp/stamp.jsp?tp=&arnumber=1611760&isnumber=33851. Accessed April 2006
9. Carnieletto R, Brandão DI, Suryanarayanan S, Farret F, Simões MG (2011) Smart grid initiative. IEEE Ind Appl Mag, 27–35
10. Chakrabarti S, Kyriakides E, Eliades DG (2009) Placement of synchronized measurements for power system observability. IEEE Trans Power Deliv 24(1):12–19
11. Clements S, Kirkham H (2010) Cyber-security considerations for the smart grid. Paper presented at the IEEE power and energy society general meeting, Minneapolis, USA
12. ECN (2011). VSYNC project. http://www.vsync.eu/ Accessed Jun 2011
13. Energy Information Administration (2011). Electric power annual report revision. http://www.eia.doe.gov/cneaf/electricity/epa/epa_noticerev2.html Accessed Jun 2011
14. ENTSO-E (2010) Annual report 2009. https://ww.entsoe.eu/fileadmin/user_upload/_library/publications/entsoe/Annual_Report/100610_ENTSO-E_Annual_Report_2009.pdf
15. Ericsen T, Khersonsky Y, Schugart P, Steimer P PEBB-power electronics building blocks, from concept to reality. In: 3rd IET international conference on power electronics, machines and drivers, pp 12–16)
16. European Commission (2006) FP7 Tomorrow's answers start today. http://ec.europa.eu/research/fp7/pdf/fp7-factsheets_en.pdf
17. European Commission (2006) Vision and strategy for Europe's electricity networks of the future. Smart Grids European Technology Platform

18. European Commission (2008) Combating climate change: The EU leads the way
19. European Commission (2010) Task force smart grids—vision and work programme. http://ec.europa.eu/energy/gas_electricity/smartgrids/doc/work_programme.pdf
20. European Commission CORDIS (2011). http://cordis.europa.eu Accessed Jun 2011
21. European Electricity Grids Initiative (2010) Energy 2020: a strategy for competitive, sustainable and secure energy. COM (2010) 639 final
22. European Electricity Grids Initiative (2010) Roadmap 2010–18 and detailed implementation plan 2010–2012
23. European Parliament (2009) Texts adopted—part II, at the sitting of Wednesday 22 April 2009 (Vol PE 425.401)
24. Farret FA, Simões MG (2006) Integration of alternative sources of energy. IEEE, Wiley, Hoboken
25. Higgins N, Vyatkin V, Nair NC, Schwarz K (2011) Distributed power system automation with IEC 61850, IEC 61499, and intelligent control. IEEE Trans Syst Man Cybern Part C Appl Rev 41:81–92
26. IEEE Global History Network (2008) Pearl street station. http://www.ieeeghn.org/wiki/index.php/Pearl_Street_Station. Accessed Jun 2011
27. IEEE Standard (Dec 2004) IEEE standard for a smart transducer interface for sensors and actuators—mixed-mode communication protocols and transducer electronic data sheet (TEDS) formats (Vol 1451.4)
28. Institute of Electrical and Electronics Engineers (IEEE) (2003) IEEE 1547 Standard for interconnecting distributed resources with electric power systems
29. International Electrotechnical Committee Standard (2003) Communication networks and systems in substations (Vol IEC61850)
30. Kirschen D, Strbac G (2004) Fundamentals of power systems economics. Wiley, West Sussex
31. McBee K, Simões MG (Accepted for Publication) Utilizing a smart grid monitoring system to improve customer voltage quality. IEEE Trans Smart Grid
32. Metke AR, Ekl RL (2010) Security technology for smart grid networks. IEEE Trans Smart Grid 1(1):99–107
33. Mohsenian-Rad AH, Leon-Garcia A (2010) Optimal residential load control with price prediction in real-time electricity pricing environments. IEEE Trans Smart Grid 1(2):120–133
34. NIST (2010) Framework and roadmap for smart grid interoperability standards. Release 1.0
35. Overbye TJ, Weber JD (2000) Visualizing the electric grid. IEEE Spectr 38(2):52–58
36. Rodriguez GD (2010) A utility perspective of the role of energy storage in the smart grid. In: IEEE power energy society general meeting, Minneapolis, USA, pp 1–2
37. Shao S, Zhang T, Pipattanasomporn M, Rahman S (2010) Impact of TOU rates on distribution load shapes in a smart grid with PHEV penetration. Paper presented at the 2010 IEEE PES transmission and distribution conference and exposition, New Orleans, USA
38. Smart Grid Clearing House (2011). http://www.sgiclearinghouse.com Accessed Jun 2011
39. Stromback J, Dromacque C (2010) Evaluation of residential smart meter policies, WEC-ADEME Case studies on Energy Efficiency Measures and Policies. http://www.ffydd.org/documents/ee_case_study__smart_meters.pdf. Accessed Nov 2011
40. Suryanarayanan S, Mitra J (2009) Enabling technologies for the customer-driven microgrid. In: IEEE power and energy society general meeting, Calgary, Canada, pp 1–3)
41. Wolf WA (2000) Great achievements and grand challenges. Nat Acadamy Eng Bridge 30(3, 4):6
42. Xinghuo Y, Cecati C, Dillon T, Simões MG (2011) The new frontier of smart grids. IEEE Ind Electron Mag 5(3):49–63
43. Zou Q, Qin L (2010) Integrated communications in smart distribution grid. In: International conference on power system technology (POWERCON), Oct 2010, pp 1–6)

Towards Sustainable Energy Systems Through Deploying Smart Grids: The Japanese Case

Amy Poh Ai Ling

Abstract The deployment of smart grids has a major role to play in Japan's aspiration to achieve sustainable energy systems. A smart grid in Japan is designed to have an intelligent monitoring system, which not only keeps track of all the energy coming in from diverse sources but also can detect where energy is needed through a two-way communication system that collects data about how and when consumers use power. This chapter examines the vision, concepts, and elements of Japan's smart grid development. Smart grids are a part of the Japanese government's 'go green' effort to bring Japan to a leadership position in environmental and energy sustainability. This analysis provides an overview of Japan's major initiatives in deploying smart grids, including smart communities, large-scale smart grid pilot projects in four major cities, as well as overseas collaboration. Major smart grid awareness promotion bodies in Japan are discussed along with their important initiatives for influencing and shaping policy, architecture, standards, and traditional utility operations. Implementing a smart grid will not happen quickly because when Japan does adopt one, it will continue to undergo transformation and be updated to support new technologies and functionality.

Keywords Japanese · Smart grid · Smart communities · Pilots · Collaboration

1 Introduction

Culture encroachment happens through the interplay of technology and everyday life. The emergence of the smart grid created a drastic increase in the demand for the smart supply of energy flows [27]. For Japan, the evolution of smart grids in

A. P. A. Ling (✉)
The Japan Society for the Promotion of Science (JSPS), Graduate School of Engineering, Department of Technology Management for Innovations, University of Tokyo, Tokyo, Japan
e-mail: amypoh@tmi.t.u-tokyo.ac.jp

D. Mah et al. (eds.), *Smart Grid Applications and Developments*,
Green Energy and Technology, DOI: 10.1007/978-1-4471-6281-0_12,

the country could be said to have begun after the oil shock in the 1970s. Japan then started the development of renewable energy just as Denmark did. The ministry known today as the Ministry of Economy, Trade and Industry (METI), and the New Energy and Industrial Technology Development Organization (NEDO) launched new Sunshine Project and started exploring the possibility of making a major shift to alternative energy such as wind power, solar power, geothermal power, and energy from biomass [32].

The Japanese version of the 'smart city' is envisaged for the post-fossil fuel world. Alternative energy sources are harnessed in mass quantities [19]. In Japan, 'smart grid' implies energy transmission and distribution to promote the stability of the electric power supply by using information and communication technology while introducing a high level of renewable energy [39]. The focus will be on how to stabilize power supplies nationwide as large amounts of wind and solar power start entering the grid. This is because unlike conventional power sources, such as hydro, thermal, and nuclear power, solar and wind energies are prone to the vagaries of the weather. People in Japan are still not familiar with the smart grid concept because the system has yet to gain currency. According to a nationwide survey released in December 2010 by the advertising agency Hakuhodo Inc., only 36.4 % of about 400 respondents aged from 20 to 70 years said they understood, or had heard of, a smart grid [36].

To address the likely impact of the smart grid on customers, utilities, and society as a whole, it may be necessary to conduct a pilot study [9, 29]. It is an attempt of Japanese government to implement a fully smart grid city in the future.

2 Energy Context of Japan

Japan is a highly industrialized country in Asia, and its GDP is the third highest in the world in 2011, only after the USA and China [41]. From 2000 to 2010, Japan's electricity consumption increased by 8 % to 8,250 kWh per capita in 2010 (Statistics Bureau [40]. The increasing trend of CO_2 emissions since 1990 stopped in 2007 at 1,296 million tons and thereafter has been declining—only 4 % higher than the pre-industrial level in 2010 [40]. Indigenous energy in Japan is very limited, and its energy sources are mainly imported and fossil fuel dependent. About 40 % of energy supply in Japan comes from oil, 22 % from coal, 19 % from natural gas, 11 % from nuclear, and 7 % from renewables [40]. The majority of oil and natural gas (97 %) are imported, in which half of these primary energy sources are converted to electric power [7].

Unlike most other industrial countries such as the USA, Canada, and Russia, Japan does not have a single national grid, but instead has separate eastern and western grids. The four main power generation sources in Japan are hydroelectric power, thermal power, nuclear power, and solar photovoltaic power (PV). Hydroelectric power is one of the few self-sufficient energy resources in Japan, which is expected to provide stable supply and generation cost over the long term [11]. Most

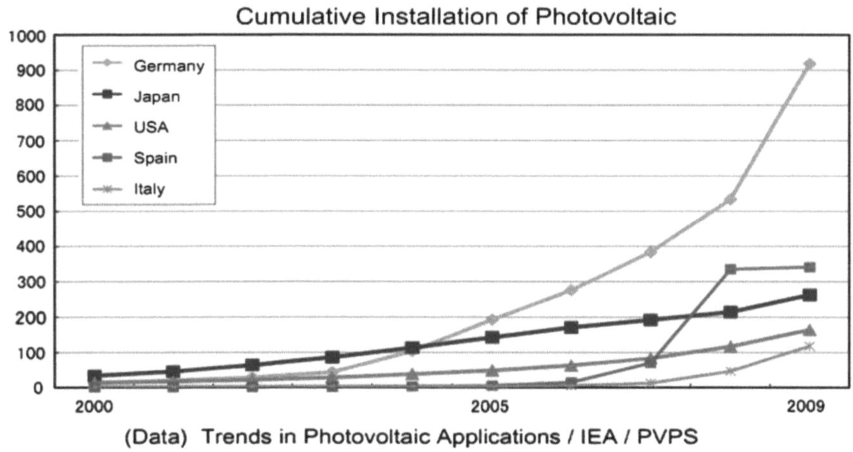

Fig. 1 Cumulative installation of PVs in Japan, Germany, the USA, Spain, and Italy (*Source* IEA PPSP [20])

of the ideal sites for hydropower have been developed, so the recent growth of this source is mainly on small-scale hydro. Pump storage stations are commonly used to cope with excessive supply during off-peak hours, for example, in Okumino and Arimine Daiichi Hydroelectric Power Plant [13].

A diverse range of fuels including coal, oil, and liquid natural gas (LNG) are utilized for thermal power generation. Recent development focuses on LNG-fired plants and combined-cycle generating plants, aiming to generate thermal power with lower emissions and higher efficiency. In addition, Japan has the third largest geothermal energy potential in the world after the USA and Indonesia, which accounts for 0.1 % of its total electricity generation [19].

As one of the major sources of electricity, nuclear has accounted for more than 30 % of Japan's electricity generation [40]. As of January 19, 2011, Japan had 54 reactors operating around the country [8]. By 2018, the nuclear output share is expected to reach 40 %. Currently, there are three plants under construction and another 10 in the advanced planning stages [4, 43]. The Fukushima accident in 2011 caused radical changes in the nuclear energy industry. Nuclear became unavailable and unstable after the earthquake and tsunami, which led to power outages to 2.4 million households [4, 43]. From this incident, the government realized the vulnerability of nuclear power and the need for a more resilient energy plan.

Solar PV is another growing source of electricity generation—the cumulative installation in Japan is the third largest in the world as of 2009 (see Fig. 1) [21]. The Japanese government introduced a new policy on PV systems in 2010, aiming to increase independent energy supplies by 32 % and emission-neutral electricity generation by 36 % by 2030 [44]. There are ten existing PV communities in Japan (including Kasukabe and Yoshikawa in Saitama), and efforts in promoting

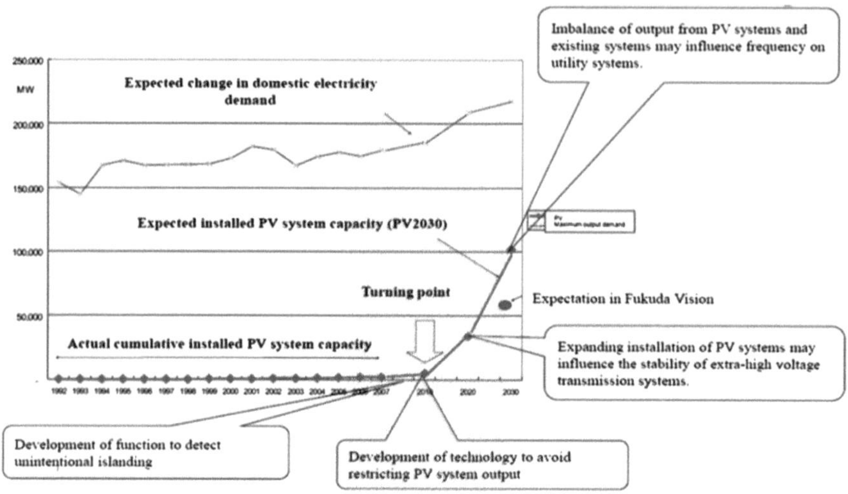

Fig. 2 Estimation of Japanese PV penetration by 2030 (*Source* Japan Guide [23])

residential PV appear to continue even if the subsidy program by METI is no longer available. As shown in Fig. 2, solar PV capacity is expected to rise, following a number of technological breakthroughs, with the forecasted spike in electricity demand between 2010 and 2030. With such expansion, imbalance between output from PV systems and existing systems is predicted to affect on utility systems.

3 Smart Grids in Japan: Key Concepts and Elements

The main objectives of adopting smart grid technology differ by region. Japan's main objective is to achieve a total shift from fossil fuels to renewable energy [22], generating a low-carbon society. Reducing carbon dioxide (CO_2) emissions is by no means easy and requires a large number of combined measures. One such measure is the utilization of renewable energy as one of the key elements of smart grid development in Japan.

In June 2010, the Japanese cabinet adopted a new Basic Energy Plan. This was the third such plan that the government had approved since the passage of the Basic Act on Energy Policy in 2002, and it represented the most significant statement of Japanese energy policy in more than 4 years since the publication of the New National Energy Strategy in 2006. Among the targets are a doubling of Japan's energy independence ratio, a doubling of the percentage of electricity generated by renewable sources and nuclear power, and a 30 % reduction in energy-related CO_2 emissions, all by 2030 [6].

3.1 Low-Carbon Society

It was believed that a low-carbon society would not be realized without a fundamental shift in energy source use. The Japanese smart grid concept aims to make the best use of local renewable energy with a view to maximizing total efficiency.

The Japanese government is aiming to increase the reliability of the grid system by introducing sensor networks and to reduce losses by introducing smart meters. The introduction of the smart grid will promote the use of renewable energy by introducing a demand response system. By focusing on electric vehicle (EV) technology, Japan is moving toward introducing charging infrastructure for electric cars [11]. Recently, increasing numbers of PV and wind power plants have been installed across the country as clean energy sources that emit no CO_2 [11].

3.2 Key Concepts

There is a wide recognition in Japan that there is an urgent need to develop and adopt products, processes, and behavior that will contribute toward more sustainable use of natural resources. In developing new green technologies and approaches, engineers in Japan are inspired by natural products and systems. Since 2010, the Tokyo Electric Power Company (TEPCO) and the Kansai Electric Power Company (KEPCO) have been testing the effects of smart meters on load leveling under the Agency for Natural Resources and Energy project.

Figure 3 illustrates the concepts of the smart grid in Japan. As noted, the Japanese government aims to increase the reliability of the grid system by introducing sensor networks and reduce losses by smart metering. The introduction of the smart grid will promote the use of renewable energy by introducing a demand-responsive system. Increasing numbers of PV and wind power plants have been installed across the country [29].

There are six major smart grid elements—smart office, smart school, smart house, smart store, smart factory, and national grid system. The smart office refers to intelligent building design involving cabling, information services, and environmental controls and envisages a desire for architecture with permanent capacity for EVs for office energy backup. In order to accelerate grid modernization in schools, the Japanese government started to develop a vision of schools that operate by drawing energy from PVs; this government program assists in the identification and resolution of energy supply issues and promotes the testing of integrated suites of technologies for schools. The smart house projects relate to smart houses interacting with smart grids to achieve next-generation energy efficiency and sustainability [26], and information and communication technology-enabled collaborative aggregations of smart houses that can achieve maximum energy efficiency. The potential benefits of smart houses include cheaper power and cleaner power, a more efficient and resilient grid, improved system reliability,

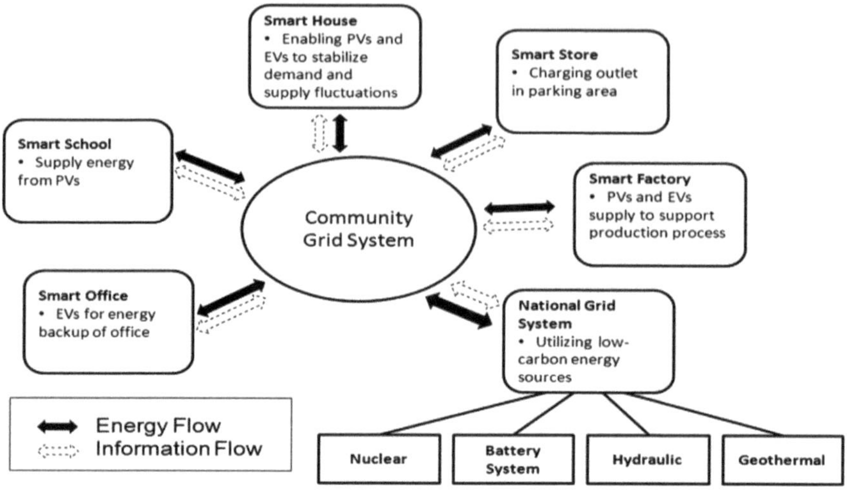

Fig. 3 Key concepts of smart grid in Japan

and increased conservation and energy efficiency. A smart house enables PVs and EVs to stabilize demand and supply fluctuations. Smart factory systems enable full factory integration of the PV cell into the bitumen membrane. The PVs and EVs in a smart factory will be supplied to support its production process. Smart stores refer to the charging outlets in parking areas and the deployment of public charging stations for EVs. The arrows shown in Fig. 3 indicate that there exist real-time energy flows and information flows within smart grids. Japan's national grid system will be enhanced by utilizing low-energy sources such as geothermal, hydraulic, battery systems, and nuclear.

The natural grids are expected to play a vital role in making effective use of renewable energy and providing a stable supply of power by controlling the balance between electricity supply and demand through telecommunications technology.

4 Major Developments of Smart Grids in Japan

Implementation of smart grid tends to differ between countries, as do the timing and adoption of these technologies [33]. Japan is pushing for advanced integrated control, including demand-side control, so as to be ready to combine unstable power (that is, reliant on weather conditions), such as solar, with its strong foundation of a highly reliable grid, as shown in Table 1.

With regard to Japan's nuclear contribution to energy supply, as of 2010, Japan is currently operating 54 commercial nuclear reactors with a total generation capacity of 48,847 MW, and about 26 % of electricity comes from nuclear power [12].

Table 1 Japanese smart grid efforts

Description	Japan
Nuclear as % of all energy sources	26 %
No. of electric power companies	10 electric power companies
Design	Vertically integrated in each region
Energy supply	0.7–25 million customers
Aim	A low-carbon society
Reliability	• Japan already has a highly reliable grid • Going for advanced integrated control, including demand-side control, to accommodate unstable renewable power
Smart grid focus	• More than ¥100 billion investment in the 1990s to upgrade generation, transmission, and SCADA network • Last mile and demand-side management (DSM) • Home solar power
Cyber security research	Protect smart meters, mutual monitoring, privacy in cloud computing

Japan's ten electric power companies are monopolies, being electric giants vertically integrated in each region. Japan depends greatly on nuclear power as an energy source to achieve a low-carbon society. With a highly reliable grid, Japan is developing its smart grid at a steady pace and has already been investing in grid projects for almost 20 years; over this period, there have been many developments. With proper security controls, smart grids can prevent or minimize the negative impact of attacks by hackers and thus increase the reliability of the grid, thereby gaining the trust and meeting the satisfaction of users [28].

Since smart grid cyber security is significantly more complex than the traditional IT security world, Japan focuses on areas of security beyond smart metering, such as mutual monitoring and privacy in cloud computing. In conclusion, Japan is striving to move toward a low-carbon society by developing the smart grid system.

5 Japan's Smart Communities

Japan's smart community initiative is based on a systemic approach. There are five identified action items: sharing vision and strategy for smart communities; social experiments for development and demonstration; standardization and interoperability; data-driven innovation and privacy protection; and smart communities for development. To address simultaneously the three Es (environment, energy security, and economy) requires the right mix and match of power sources through renewable and reusable energy (RE) utilizing storage.

Table 2 illustrates the future social system at which Japan is aiming, concentrating on the regional energy management system (EMS) and lifestyle changes under such an energy supply structure.

Table 2 Japan's future social system (*Source* NEDO)

	Current period to 2020	Period from 2020 to 2030	2030 onward
Renewable energy	Solar panel prices will decrease significantly because of the large-scale introduction of panels to houses and commercial buildings	Because of a decline in PV prices, more PV systems will be installed at houses	Cost competitiveness of RE will improve as fossil fuel prices increase more than twofold. Use of RE will be prioritized, and nuclear power will be used as a base
	Measures will be introduced to maintain the quality of electricity supply while the large-scale introduction of PV systems is conducted mainly for the grid side. Storage cells will be installed at substations	A regional EMS, which contributes to the effective use of RE generated at houses, will become more important	
Energy management system (EMS)	As the regional EMS is further demonstrated, technology and know-how will be accumulated	A regional EMS will be achieved as storage cells become cheaper and are further disseminated	An EMS that can provide an optimized balance in terms of economy and security between regional EMS and the grid will be established
	The cost of storage cells will decrease because of technology development and demonstration	Distribution and transmission networks that enable two-way communication between the demand side and the grid side will be actively established	An EMS that creates demand by charging EVs at the time of excessive RE reliance, and supplies energy to the grid at times of high demand, will be used
Smart houses	Remote reading using smart meters will start	The home EMS and the regional EMS will be integrated. All power generated at houses will be used optimally	A fully automated home EMS will be achieved
	The home EMS will be disseminated. Some houses will install home servers. Demand response demonstration will start	Various services using home servers will be disseminated	
Electric vehicles (EV)	Demonstration of EVs will start	EVs will be used for power storage as well	

The development of Japan's smart community is divided into three stages: the first is the development plan for the current period up to 2020; the second is the development plan for 2020 to 2030; and the third is the development plan for 2030 onward. With regard to the relation between the regional EMS and the grid, the decrease in solar panel prices following their large-scale introduction will cause more PV systems to be installed at houses, which is expected to create cost competitiveness in RE. The EMS will become more important and will provide an optimized balance in terms of economy and security. When EMS technology and know-how has accumulated and the cost of storage cells has fallen because of technology development, the distribution and transmission networks that enable two-way communication between the demand and grid sides will be actively established. By that time, the EMS that creates demand by charging EVs at times of excessive reliance on RE, and supplies energy to the grid at times of high demand, will be used.

As for the development of houses, the remote reading of smart meters will start when the home EMS and the regional EMS are integrated and all power generated at houses will be used optimally. In addition, zero-energy buildings (ZEBs) will be introduced from 2020 to 2030, initially for new public buildings. The introduction of ZEBs is expected to significantly reduce emissions for all new buildings as a group.

The aims of Japan's smart community are as follows: (a) to cut energy costs through competitive advantage and establish a new social system to reduce CO_2 emissions; (b) to introduce widespread use of RE; and (c) to facilitate the diversification of power supplier services [28] This approach is also directed at helping Japanese customers see how cutting their energy costs can give them a competitive advantage.

Smart community efforts also extend an international collaboration. A consortium of Japanese companies is preparing a report on the feasibility of smart community development projects in Gujarat, India. According to preliminary estimates by Japanese experts, of the 6.23 million tons of hazardous waste generated in India annually, 22 % comes from Gujarat. This report prompted the Gujarat government to sign a memorandum of understanding for developing 'Surat' as an eco-town along the lines of 'Kitakyushu Eco-Town' in Japan and for developing Dahej along the lines of the 'reduce, reuse, and recycle'-oriented environmentally smart community development concepts prevalent in Japan [31].

6 Japan's Smart Grid Pilot Projects

On April 8, 2010, four sites were selected from four cities in Japan to run large-scale, cutting-edge pilot projects on the smart grid and smart community (budget request for FY2011: 18.2 billion yen) [4]. The four cities are Kyoto, Kitakyushu, Yokohama, and Toyota City. The community EMS will be achieved based on a combination of the home EMS, the building EMS, EVs, PVs, and batteries. Not

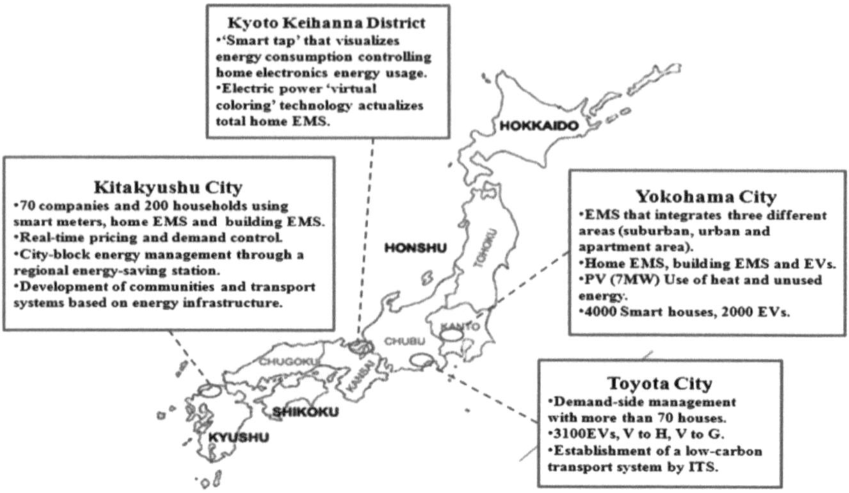

Fig. 4 Japan's pilot projects on smart grid and smart community (*Source* METI [35])

only METI's smart grid-related projects, but also projects in other ministries, such as communications, environment, agriculture, and forestry, will be implemented at these four sites (Fig. 4).

Apart from these four major smart grid community pilot projects, there are two other pilots that are particularly important to note. These are the demonstration project in Rokkasho Village, in the Aomori Prefecture and the Hawaii–Okinawa Partnership on Clean and Efficient Energy Development and Deployment in Okinawa.

6.1 *Smart Grid Pilot Project in Yokohama City (Large Urban Area)*

There were 900 PV systems installed in the so-called progressive city of Yokohama in 2009, and the Japanese government plans to install about 2,000 more 10 years hence [25]. The aim of the Yokohama Smart City Project is to build a low-carbon society in a large urban area, involving 4,000 smart houses. This project is a 5 year pilot program being undertaken by a consortium of seven Japanese companies: the Nissan Motor Co., Panasonic Corp., Toshiba Corp., TEPCO, the Tokyo Gas Co., Accenture's Japan unit, and Meidensha Corp. The project focuses on the development of the EMS, which integrates the home EMS, the building EMS, and EVs. It is expected to generate PVs with a capacity of 27,000 KW. The EMS, which integrates suburban, urban, and apartment areas, will have PV use of heat and unused energy.

6.2 Smart Grid Pilot Project in the Kyoto Keihanna District (R&D Focus)

The Smart Grid Pilot Project in the Kyoto Keihanna District involves the Kyoto Prefecture, Kansai Electric Power, Osaka Gas, Kansai Science City, Kyoto University, Doshisha, Yamate, the Sustainable Urban City Council, and other local governments and utilities. It makes use of the 'smart tap,' which visualizes energy consumption controlling home electronics energy usage. It is also a pilot project to test 'electric power virtual coloring' technology, which actualizes the overall home EMS. This project calls for the installation of PVs in 1,000 houses and an EV car-sharing system. It also studies nanogrid management of PVs and fuel cells in houses and buildings on the visualization of demand. This project grants 'Kyoto eco-points' for the usage of green energy.

6.3 Smart Grid Pilot Project in Toyota City (Metropolitan Area)

The Toyota Rokkasho Village in the Aomori Prefecture began to experiment with the smart grid in September 2010. The key feature of the project is the pursuit of optimal energy use in living spaces at the community level at the same time as achieving compatibility between environmental preservation and resident satisfaction. This project involved Toyota City and companies including Toyota Motors, Chubu Electric Power, Toho Gas, Utilities, Denso, Sharp, Fujitsu, Toshiba, KDDI, Circle K Sunkus, Mitsubishi Heavy Industries, and Dream Incubator. The project focuses on the use of heat and unused energy as well as electricity. It has a demand response with more than 70 homes and 3,100 EVs. Through this project, houses that contain an IT network of electrical appliances and other household equipment, solar panels, household storage batteries, onboard automobile storage batteries, and other devices, can develop household power leveling and optimized energy usage. As of June 2011, model smart houses in the Higashiyama and Takahashi districts of Toyota City for testing EMSs had been completed successfully and had begun trial operations under the Verification Project for the Establishment of a Household and Community-Based Low-Carbon City in Toyota City, Aichi Prefecture [1].

6.4 Smart Grid Pilot Project in Kitakyushu City (Industrial City)

Located in this industrial city, this project involves 46 companies and organizations, including the Kitakyushu City Government, GE, Nippon Steel, IBM Japan, and Fuji Electric Systems. It focuses on real-time management in 70 companies and 200 houses. Energy management will be controlled by the home EMS and the building EMS. The pilot project study of the energy system integrates demand-side management and the high-energy systems. The Kitakyushu Hibikinada area is promoting low carbon emissions, recycling, and nature coexistence in a balanced manner. This project implements various demonstrations, including communications, urban planning, a transportation system, and lifestyle, with an emphasis on the demonstration of energy projects such as electric power. Implementation will take place in five years, from FY2010 to FY2014, involving the operation of 38 projects, and is worth 16.3 billion yen [37]. The Kitakyushu Smart Community Project focuses on the development of technologies and systems related to a smart grid with an eye on international standardization and the expansion of international business, and the presentation of new urban planning for a smart city, by developing various human-friendly social systems compatible with next-generation traffic systems and an aging society. In addition, a recycling community will be constructed in which all sorts of waste will be used as raw materials for other industrial fields so as to eliminate waste and move toward a zero-emissions community.

6.5 Smart Grid Demonstration Project in Rokkasho Village

In September 2010, the Japan Wind Development Company, the Toyota Motor Corporation, the Panasonic Electric Works Company, and Hitachi Ltd. started a 2 year smart grid demonstration project in Rokkasho Village, in Aomori Prefecture, where wind power stations with large-capacity batteries were built several years ago [22]. This project aims to verify technologies that allow for the efficient use of energy for the achievement of a low-carbon society [30]. It involves a so-called smart grid village composed of six 'smart houses' equipped with automatic electricity control systems, eight Toyota Prius plug-in hybrid vehicles, and a battery system, all powered exclusively by renewable energy sources and detached from the national electricity grid. Families of the employees of the corporations participating in the project reside in these smart houses, where they go about their normal lives [3]. In addition, an experimental situation has been created in isolation from the external power grid, where approximately eight kilometers of private distribution line has been laid between the Rokkasho-mura Futama Wind Power Station and the smart houses. The station is outfitted with 34 units of 1,500 KW windmills with a total capacity of 51,000 KW and is equipped with large-capacity network-attached storage batteries of 34,000 KW. The objective of the experiment is to examine such factors as changes in electricity usage in different seasons and at different times of day and to investigate trends in electricity usage based on different family configurations. The experiment will help create a system that efficiently balances electricity supply and demand.

6.6 *Smart Grid Trial Project in Okinawa*

Six months after the launch of the demonstration project in Rokkasho Village, the Hawaii–Okinawa Partnership on Clean and Efficient Energy Development and Deployment began. This pilot aims to help the two island regions switch from thermal power to renewable energy systems, which are considered crucial for reducing CO_2 emissions but whose power supply is unstable [34].

The Okinawa Electric Power Company has begun operating a smart grid to control the supply of renewable-energy-derived electricity for the 55,000 strong population of the remote Okinawa Prefecture island Miyako-jima. The Hawaii–Okinawa Partnership on Clean and Efficient Energy Development and Deployment is an agreement between the US Department of Energy, METI, the State of Hawaii, and the Prefecture of Okinawa, which was signed in June 2010. The partnership is intended to foster the development of clean and energy-efficient technologies needed to achieve global energy security and meet climate change challenges. Japan and the USA designated Hawaii and Okinawa as the representatives for this groundbreaking partnership because of their demonstrated leadership and experience in clean energy and energy efficiency. The trials started in October 2010. The infrastructure links the existing power grid to a 4-MW solar power plant and a sodium sulfide battery complex capable of storing 4 MW of power. Some lithium-ion batteries have also been installed. In addition, the system also controls power from existing 4.2-MW wind farms situated on Miyako-jima. Okinawa Electric spent 6.15 billion yen (US$75.8 million) on the infrastructure, two-thirds of which was subsidized by the national government [34]. The programs will help the two island regions switch from thermal power to renewable energy systems.

Each of the smart grid pilots and demonstration projects has its own challenges. Building smart grids requires meeting the requirements for electricity supply, including power sources and transmission lines, and the communications infrastructure of each specific country and region, as well as introducing such elements as renewable energy generation facilities, EVs and plug-in hybrid vehicles, storage batteries, EcoCute electric water heating and supply systems, and heat storage units. The Japanese government is strongly promoting the efficient use of energy by developing smart grid pilot projects and the promotion of the smart community.

7 Overseas Collaborations on Smart Grid Projects

Japan has proactively developed extensive collaboration with other countries to facilitate the development of smart grids.

7.1 Smart Grid Pilot Project in the USA

To such pilots are located in the southwestern US states of New Mexico and one in Hawaii. Toshiba, Kyocera, Shimizu, the Tokyo Gas Company, and Mitsubishi Heavy Industries will spend $33.4 million on a smart grid project at Los Alamos and Albuquerque, New Mexico [24]. NEDO will participate in research at Los Alamos and Albuquerque and in collective research on the overall project. Toshiba will install a 1-MW storage battery at the Los Alamos site, while Kyocera and Sharp will test smart homes, energy management, and load control technologies.

7.1.1 Los Alamos

The Microgrid Demonstration Project in Los Alamos which began operating in August 2012 involves concentrated PV energy generation and the installation of power storage cells on distribution lines of 2–5 MW. In addition, absorption experiments on PV output fluctuations will be conducted using PV-induced efficiencies obtained by changing grid formation, and a distribution network with high operability will be installed and demonstrated by introducing smart distribution equipment. The smart house is intended to maximize demand response by using a home EMS, and a demonstration will be carried out to verify its effectiveness relative to an ordinary house. In Los Alamos, with a population of 20,000, NEDO offers help to optimize the amount of stored electricity by controlling battery systems and monitoring electricity demand of household and business.

7.1.2 Albuquerque

The microgrid demonstration is located in commercial areas of Albuquerque the largest city in New Mexico with a population of 480,000. This pilot was kicked off in May 2013, focusing on demonstrating the demand response by using facilities in industrial and commercial buildings. The move is prompted by the aim of catching up with the USA, which has taken the lead in developing technological global standards [36]. It is also intended to evaluate smart grid technology from Japan and the USA based on research results obtained at the demonstration sites of the New Mexico project [42]. In Albuquerque's project, NEDO helped build an urban smart city with ecological buildings. The project includes the installation of a solar power generator, a cogeneration system, fuel cells, and a building and energy management system in the Mesa Del Sol Aperture Center, the model building for the project.

7.1.3 Smart Grid Project in Hawaii

A project supported by Japan's NEDO, in cooperation with the State of Hawaii, the Hawaiian Electric Company, the University of Hawaii, and Pacific Northwest

National Laboratory, whose involvement is based on the Japan–US Clean Energy Technologies Action Plan, was started in November 2009 [18]. A feasibility study was completed in September 2011 [15]. The project includes the installation of a computer system dubbed the electric vehicle management system, that optimizes a way to charge electric vehicles by using oversupplied electricity at a time of low demand.

While appearing to be 'state-of-the-art' technological marvels, smart grids expose energy production and distribution to higher levels of security vulnerability than ever before. In considering this matter, the project will incorporate the installation of smart controls in the Kihei area on Maui at the regional and neighborhood levels to improve the integration of variable renewable energy resources, such as PV systems [2]. Installation of the smart grid technology began operational in 2013. The demonstration project is scheduled to run from 2013 to 2015. This project may be useful for the design of future microgrids that will provide secure backup and UPS services to distributed energy residences and light industry. Independent, distributed energy appliances in homes and businesses guarantee the highest possible level of security and reliability in a national power system.

7.2 Smart Grid Pilot Project on the Island of Jeju, Korea

Large electronics conglomerates in Japan and Korea, such as Sharp, Panasonic, Samsung, LG, SK Telecom, and KT, are building a domestic smart grid pilot on the island of Jeju, which is south of Seoul in South Korea [10]. In March 2011, IBM announced that two new utilities had joined its Global IUN Coalition, a group of utility companies designed to further the adoption of smarter energy grids around the world: TEPCO from Japan and KEPCO from Korea. These two companies are in charge of the world's largest comprehensive smart grid test bed in Jeju Island, which brings together smart technologies in the areas of generation, power grids, electrical services, buildings, and transportation [16].

7.3 ZEBs in Lyon in France

NEDO has been exploring the possibility of collaborating with Grand Lyon, the second largest city in France, to introduce Japanese leading-edge technologies for ZEBs in France and to establish an EV-charging infrastructure coinciding with the Lyon Confluence urban development project [42]. This project has started in phases in 2012 [14]. Lyon is in southeastern France, and the area has a population of more than 1 million. In line with the European Union's effort to raise the ratio of renewable energy to 20 %, France requires buildings erected in 2020 or later to generate more electricity than they consume, an initiative called positive energy building (PEB).

7.4 Smart Grid Collaboration in Malaga, Spain

In Malaga, Spain, NEDO will help install a system to manage the use of electric vehicles and infrastructure for charging station of EVs. The Malaga project starts in March 2013. In Spain, 40 % of energy consumption is in transportation, and the source of the energy is mostly thermal. In order to meet the EU's ecological goal of raising ratio of renewable energy to 20 % by 2020, Spain aims to increase the number of EVs to 250,000 by 2014.

8 Major Smart Grid Awareness and Activities Promotion Bodies in Japan

Apart from the above-mentioned smart grid pilot projects in Japan and the overseas collaborative programs, another major initiative that has been undertaken in Japan is smart grid awareness activities. Major initiatives have been launched to engage stakeholders in smart grid promotional activities that complement activities in existing organizations and groups that currently influence and shape policy, architecture, standards, and traditional utility operations [38]. A number of major smart grid awareness and activities promotion bodies and associations affirm the successful implementation of smart grids in Japan and raise people's awareness through education.

8.1 Ministry of Economy, Trade, and Industry

METI play a key role in overseeing smart grid projects, policies, and research. Some of their major initiatives are the four smart community pilot projects, the *International Standardization Roadmap for Smart Grid*, establishment of the Working Group on International Standardisation of Smart Grid, and a number of research institutes [30]. METI is also reviewing closely Japan's PV/CSP technologies and programs to support smart grid development.

8.2 NEDO

NEDO corporates with local government and companies around the world in creating smart communities. NEDO called it oversea projects verification, meaning it develops technology in Japan and verifies if the technology is commercially viable oversea.

NEDO is committed to contributing to the resolution of energy and global environmental problems and further enhancing Japan's industrial competitiveness, and strongly supports numerous smart grid research and development projects.

8.3 The Japan Smart Community Alliance

The Japan Smart Community Alliance (JSCA) is a member of the Global Smart Grid Federation [17], which aims to promote public–private cooperative activities relating to the development of smart communities by tackling common issues such as dissemination, deployment, and research on smart grid standardization. The JSCA has members from the electric power, gas, automobile, information and communications, electrical machinery, construction, and trading industries as well as from the public sector and academia. Four working groups have been established at a practical level for discussion and deliberation in order to facilitate JSCA's activities. The four working groups are the International Strategy Working Group, the International Standardization Working Group, the Roadmap Working Group, and the Smart House Working Group.

8.4 Democratic Party of Japan

Generally, the Democratic Party of Japan (DPJ) is perceived to be more proactive on environmental regulations than the Liberal Democratic Party of Japan (LDP) [45]. The DPJ is promoting the development and diffusion of smart electricity grid technologies. In its 2013 Manifesto, DPJ states its commitment to an efficient and decentralized energy system. It aims to encourage renewables and smart grid development by creating special economic zones and reform electric grids through improving interconnection, market deregulation, and unbundling [5].

9 Conclusion

Smart grid technology helps to convert the power grid from a static infrastructure that is operated as designed to a flexible and environmentally friendly infrastructure that is operated proactively. This is consistent with the Japanese government's goal of creating a low-carbon society, and maximizing the use of renewable energy sources, such as photovoltaics and wind power. Nevertheless, public–private sector cooperation is necessary to establish smart communities.

This chapter provides an understanding of the Japanese concepts of smart grid deployment. It also discussed the major government strategies, major initiatives, and their achievements in this important area.

Japan is currently focusing on demand-side management and home solar power. Researchers have started to address challenges caused by large-scale solar power generation connected to the power grid as well as information security issues. Because the smart grid remains a novel field of study in Japan, it has great potential for further research.

Acknowledgments This chapter is an updated republication of the paper 'The Japanese Smart Grid Initiatives, Investment, and Collaborations' by Amy Poh Ai Ling, Sugihara Kokichi, and Mukaidono Masao, International Journal of Advanced Computer Science and Applications 3(7), 44–45. The copyright permission for reusing the paper has been granted by SAI Organization Inc. The author was supported by the MIMS Ph. D. Program of the Meiji Institute for Advanced Study of Mathematical Sciences (MIMS) and DC2 Scholarship of the Japan Society for the Promotion of Science (JSPS).

References

1. Asahi (2011) Toyota city low-carbon project model homes completed. Asahi, June
2. Brighter Energy News (2011) US and Japan work on Maui Smart Grid for electric vehicles. Brighter Energy News. Retrieved 5 Aug 2013, from http://www.brighterenergy.org/22835/news/transport/us-and-japan-work-on-maui-smart-grid-for-electric-vehicles/
3. Cabinet Office (2011) Living Off the Grid. Tokyo: Cabinet Office, Government of Japan. Retrieved 5 Aug 2013, from http://www.gov-online.go.jp/pdf/hlj/20110201/20110201all.pdf
4. Carson P (2011) Smart grid more attractive, post-Japan? Intell Util Daily, Mar 15. Retrieved 2 Aug 2013, from http://www.intelligentutility.com/article/11/03/smart-grid-more-attractive-post-japan
5. DPJ (2013) The DPJ manifesto 2013: helping to protect lives and livelihoods. The democratic party of Japan, Tokyo. Retrieved 5 Aug 2013, from http://www.dpj.or.jp/english/manifesto/manifesto2013.pdf
6. Duffield JS, Woodall B (2011) Japan's new basic energy plan. Energy Policy 39(6): 3741–3749. doi:http://dx.doi.org/10.1016/j.enpol.2011.04.002
7. ECCJ (2005) Japan energy conservation handbook 2004/2005. The Energy Conservation Center, Japan. Retrieved 30 July, 2013, from http://www.eccj.or.jp/databook/2004-2005e/index.html
8. European Nuclear Society (2013) Nuclear power plants, world-wide. Retrieved 2 Aug 2013, from http://www.euronuclear.org/info/encyclopedia/n/nuclear-power-plant-world-wide.htm
9. Faruqui A, Hledik R, Sergici S (2009) Piloting the smart grid. Electri J 22(7):55–69. doi:http://dx.doi.org/10.1016/j.tej.2009.06.012
10. Fehrenbacher K (2010) The new smart grid players: Korea, Japan, China, Oh My! Blog Retrieved from http://gigaom.com/2010/02/01/the-new-smart-grid-players-korea-japan-china-oh-my/
11. FEPC (2011) The federation of electric power companies of Japan, Tokyo. Electr Rev Japan. Retrieved 2 Aug 2013, from http://www.fepc.or.jp/english/library/electricity_eview_japan/__icsFiles/afieldfile/2011/01/28/ERJ2011_full.pdf
12. FEPC (2013) Nuclear Power Plants in Japan. Nuclear Power Generation. Retrieved 2 Aug 2013, from http://www.fepc.or.jp/english/nuclear/power_generation/plants/
13. FEPC (2013) Profile of Japan's major power generation sources, energy and electricity. Information Plaza of Electricity. Retrieved 2 Aug 2013, from http://www.fepc.or.jp/english/energy_electricity/electric_power_sources/index.html

14. Grand Lyon (2012) Lyon smart community: media kit. Grand Lyon. Retrieved 5 Aug 2013, from http://www.economie.grandlyon.com/fileadmin/user_upload/fichiers/site_eco/20121121_gl_lyon_smart_community_dp_en.pdf
15. Green Car Congress (2011) US and Japan collaborating on smart grid project in Hawaii; EV operation and charging, including grid-balancing services. Retrieved 5 Aug 2013, from http://www.greencarcongress.com/2011/05/hitachi-20110518.html
16. Greenbang (2011) Japanese, Korean utilities join smart-grid coalition. Retrieved 5 Aug 2013, from http://www.greenbang.com/japanese-korean-utilities-join-smart-grid-coalition_16971.html
17. GSGF (2012) Global smart grid federation: Japan Smart Community Alliance. Retrieved 5 Aug 2013, from http://www.globalsmartgridfederation.org/japan.html
18. Hitachi Ltd., Cyber Defense Institute Inc., JFE Engineering Corporation, Sharp Corporation, Hewlett-Packard Japan Ltd., Mizuho Corporate Bank Ltd. (2011) U.S. and Japan companies collaborate on smart grid project in Hawaii. Retrieved 5 Aug 2013, from http://sharp-world.com/corporate/news/110517.html
19. Hosaka TA (2010) Japan creating 'smart city' of the future. San Francisco Chronicle, 11 Oct 2010. Retrieved 2 Aug 2013, from http://www.sfgate.com/business/article/Japan-creating-smart-city-of-the-future-3171405.php
20. IEA PPSP (2012) Trends in photovoltaic applications. International energy agency photovoltaic power systems programme, France. Retrieved 29 Sept 2013, from http://www.iea-pvps.org/index.php?id=92&no_cache=1&tx_damfrontend_pi1[showUid]=1239&tx_damfrontend_pi1[backPid]=92
21. IEA PVPS (2010) Trends in photovoltaic applications: survey report of selected IEA countries between 1992 and 2009. International energy agency photovoltaic power systems programme, Switzerland. Retrieved 2 Aug 2013, from http://www.iea-pvps.org/fileadmin/dam/public/report/statistics/tr_2009_neu.pdf
22. Ito S (2009) Japan's initiative on smart grid–a proposal of 'nature grid'. Information Economy Division, Ministry of Economy Trade and Industry, Tokyo. Retrieved 2 Aug 2013, from http://documents.eu-japan.eu/seminars/europe/other/smart_grid/presentation_ogawa.pdf
23. Japan Guide (2013) Electricity in Japan. Retrieved 2 Aug 2013, from http://www.japan-guide.com/e/e2225.html
24. Japan Today (2010) 4 companies start smart grid demonstration project in Rokkasho. Japan Today. Retrieved 5 Aug 2013, from http://www.japantoday.com/category/technology/view/4-companies-start-smart-grid-demonstration-project-in-rokkasho
25. Japanese Nikkan Kogyo Shimbun (2010) Smart grid of the future urban development. Japanese Nikkan Kogyo Shimbun, June 2010
26. Kok K, Warmer C, Venekamp G, Weidlich A, Karnouskos S, Silva PD, Liolou V (2011) SmartHouse/SmartGrid: D2.5 final architecture for SmartHouse/SmartGrid. European Commission, Brussels. Retrieved 2 Aug 2013, from http://www.smartgrid.gov/sites/default/files/doc/files/SmartHouse_SmartGrid_D25_Final_Architecture_for_SmartHouse_201103.pdf
27. Ling APA, Masao M (2011) Grid information security functional requirement—fulfilling information security of a smart grid system. Int J Grid Comput Appl 2(2):1–19
28. Ling APA, Masao M (2011) Selection of model in developing information security criteria on smart grid security system. Paper presented at ninth IEEE international symposium on the parallel and distributed processing with applications workshops (ISPAW), 26–28 May 2011. Retrieved 2 Au 2013
29. Ling APA, Mukaidono M, Kokichi S (2011) The natural grid concept and the strategy of Asia's energy-balance pioneer. Paper presented at the smart grid electronics forum, 24–25 Oct 2011, San Jose
30. Mah DN, Wu YY, Ip JC, Hills PR (2013) The role of the state in sustainable energy transitions: a case study of large smart grid demonstration projects in Japan. Working paper. The Kadoorie Institute, The University of Hong Kong. Hong Kong. Retrieved 28 Aug 2013, from http://www.kadinst.hku.hk/PDF_file/KI_working_paper_18_revised_on_3Jul2013.pdf

31. Matsuoka T (2011) Kitakyushu City's challenge toward a low-carbon society (Kitakyushu smart community creation project). Paper presented at the international symposium: learning the way to building low carbon society (Mar 10), Shinagawa. Retrieved 2 Aug 2013, from http://www.challenge25.go.jp/roadmap/media/s5_en.pdf
32. Matsutani M (2012) 'Smart grid/community' special. Japan Times, 16th Year. No. 40,739, 14 Oct 2012. Retrieved 2 Aug 2013, from http://info.japantimes.co.jp/ads/pdf/1014z01-08.pdf
33. McGuire K (2010) The smart grid movement: Japan vs. U.S. Retrieved 2 Aug 2013, from http://smart-grid.tmcnet.com/topics/smart-grid/articles/73301-smart-grid-movement-japan-vs-us.htm
34. METI (2011) Hawaii–Okinawa partnership on clean and efficient energy development and deployment. Tokyo: Ministry of Economy, Trade and Industry Retrieved 5 Aug 2013, from http://www.meti.go.jp/english/press/2011/0603_01.html
35. METI (2013) Smart community. Retrieved 5 Aug 2013, from http://www.meti.go.jp/english/policy/energy_environment/smart_community/index.html, 5 Aug 2013
36. Nakata H (2011) Smart grid pursuit slow off mark. Japan Times, 11 Jan 2011. Retrieved 2 Aug 2013, from http://www.japantimes.co.jp/news/2011/01/11/reference/smart-grid-pursuit-slow-off-mark/#.UftQI9JkOQ0
37. Rediff (2011) Japan to help Gujarat in smart community project. Retrieved 5 Aug 2013, from http://www.rediff.com/business/report/japan-to-help-gujurat-in-smart-community-project/20110113.htm
38. Renesas (2010) Special Interview—part 5 of 5: Guido Bartels, Chairman of GridWise Alliance, talks about 'Smart Grid'. Renesas EDGE. Retrieved 5 Aug 2013, from http://sg.renesas.com/media/edge_ol/feature/05/r70pf0028ej0101_edge.pdf
39. Shinkawa T (2010) Smart grid and beyond. Paper presented at the 2010 BNEF energy smart technologies leadership forum, 27–28 Oct 2010, New York. Retrieved 2 Aug 2013, from http://about.bnef.com/presentations/smart-grid-and-beyond/
40. Statistics Bureau, Japan (2012) Japan statistical yearbook 2013. Japan: Statistics Bureau, from http://www.stat.go.jp/english/data/nenkan/index.htm
41. The World Bank (2013) World development indicators—GDP (current US$). Retrieved 8 Feb 2013 http://data.worldbank.org/indicator/NY.GDP.MKTP.CD
42. Watanabe H (2010) Smart community activities in Japan. Paper presented at the Korea Smart Grid Week, 10 Nov 2010, Jeju Island. Retrieved 5 Aug 2013, from http://www.ksgw.or.kr/down/pr/KSGW_11_HiroshiWatanabe(101110).pdf
43. World Nuclear News (2011) Rolling blackouts as Japanese efforts continue. World Nuclear News. Retrieved 2 Aug 2013, from http://www.world-nuclear-news.org/RS-Rolling_blackouts_as_Japanese_efforts_continue-1403117.html
44. Yasui M (2010) Japanese policy on photovoltaic systems. Paper presented at the he nature photonics technology conference (19–21 Oct 2010), Tokyo. Retrieved 5 Aug 2013, from http://www.natureasia.com/pdf/ja-jp/events/photonics/yasui_opening.pdf
45. Zipangu (2009) Japan's world-class energy-saving and environmental technologies: part 2. Zipangu, 2nd edn, pp 13–16. Retrieved Aug 31, from http://www.iri-japan.co.jp/en/zipangu_en/pdf2/ZIPANGU2%20P13-16.pdf

Governing the Transition of Socio-technical Systems: A Case Study of the Development of Smart Grids in Korea

Daphne Mah, Johannes Marinus van der Vleuten, Jasper Chi-man Ip and Peter Hills

Abstract This chapter examines the motivations, processes, and outcomes of the development of smart grids in South Korea through the perspectives of governance and innovation systems. Drawing on desktop research and semi-structured interviews, this chapter has two major findings. First, the development of smart grids in Korea has been shaped by various factors including macroeconomic policy, the role of the government, and experimentation. The complex interactions between these factors at the landscape, regime, and niche levels have impacted on the development of smart grids. Second, while Korea's government-led approach has its strengths in driving change, it has also exposed weaknesses in the country's ability to mobilise the private sector and consumer participation. Major obstacles including partial electricity market reform and public distrust exist. A systemic perspective is needed for policy in order to accommodate the changes required for smart grid development. Regulatory reforms, particularly price-setting mechanisms, and consumer engagement are priority areas for policy change.

1 Introduction

Smart grids are electricity networks that utilise information technology to enhance the reliability, security, and efficiency of power systems [38]. A distinctive feature of smart grids is the ability to integrate the actions of all users including generators and consumers [24]. Smart grids have a major role to play in a low-carbon future: they can be instrumental in energy saving and accommodating a broad range of

---

D. Mah (✉) · J. C. Ip
Department of Geography, Hong Kong Baptist University, Kowloon, Hong Kong
e-mail: daphnemah@hkbu.edu.hk

D. Mah · J. M. van der Vleuten · P. Hills
The Kadoorie Institute, The University of Hong Kong, Pokfulam, Hong Kong

D. Mah et al. (eds.), *Smart Grid Applications and Developments*,
Green Energy and Technology, DOI: 10.1007/978-1-4471-6281-0_13,

generation and storage options including renewable energy and thus are a key to both demand- and supply-side management of energy systems.

The potential opportunities and benefits of smart grids could be substantial. The Electric Power Research Institute (EPRI) estimated that an investment of $338 billion to $476 billion for a fully functional smart grid could result in benefits up to $2 trillion in the USA [12]. Faruqui et al. [14] also estimated that 67 billion euros for building and running peak infrastructure could be avoided in the EU if dynamic pricing which is based on smart grids can be adopted.

The USA, EU, Japan, and Korea have been among the first movers in the development of smart grids. These economies have adopted different pathways with varying levels of achievement. In the USA, for example, the emphasis is on smart metering and grid modernisation [13]. In contrast, Europe places emphasis on decentralised systems with active participation from end-users who can sell surplus electricity that they generate from microgeneration technologies such as small wind turbines at household and community levels [52].

The transition from large-scale carbon/nuclear-based electricity grid systems to smart grid systems, however, is a difficult and complex process that goes beyond technological challenges. Smart grids look very different from today's grid systems, involving a shift from centralised, fossil fuel/nuclear-based and non-participatory power systems to one which can accommodate a wide range of energy sources including both centralised energy systems and decentralised renewable energy such as wind and solar energy [43]. Smart grids are also characterised by two-way relationships with well-informed and actively involved end-users. Dynamic pricing, which is the charging of different electricity rates at different times of the day and year to reflect the time-varying cost of supplying electricity [16], and smart meter roll-outs would need to be introduced to enable effective consumer engagement in demand-side management [24]. The challenges of smart grid development therefore are numerous, including realigning the interests of business, government, and electricity consumers to overcome resistance to change [13, 24]. However, little is known about how these challenges can be overcome.

This chapter examines governing processes for the transition of socio-technical systems, applied through a case study of smart grids in South Korea (hereafter Korea). The chapter examines the motivations, processes, and outcomes of the development of smart grids in Korea using the concepts of governance and innovation systems as an analytical framework.

Korea merits study because its government-led and export-oriented approach to developing smart grids appears to differ in many interesting ways when compared with other countries. A number of recent policy developments, most notably the national smart grid roadmap and a major demonstration project known as the Smart Grid Testbed on Jeju Island, can provide useful information for analysis.

The analysis presented here draws on data and information derived from desktop research, semi-structured interviews, and field observations. The interview data consist of seven in-depth interviews with stakeholders conducted in Korea in April 2011, two follow-up email correspondences, and four telephone interviews.

The richness of the information derived from our face-to-face interviews has the strength to reveal the critical interactions of complex social phenomena [42]. However, qualitative case studies may suffer from what Miles and Huberman [42] have termed the "limitations of interpretivism"—they may be a "person-specific, artistic, private/interpretive act that no one else can viably verify or replicate it".

Our study adopted several measures to overcome these limitations. First, the interviewees were carefully selected informants who occupy roles or positions in an organisation, social networks, and communities of a political system and are therefore knowledgeable about the issues studied [26]. They came from the government, energy companies, universities, and research institutes (see Appendix 1). Second, we used semi-structured questionnaires which were developed on the basis of our literature review as a way to facilitate systematic interviews across interviewees. Third, email correspondence and follow-up telephone interviews were conducted to collect supplementary information and to clarify data. Fourth, the interviews were recorded and transcribed to reduce inaccuracies due to poor recall. Fifth, as far as such information is accessible to us, we have used data we collected from publications to corroborate data provided by our interviewees.

The following section provides an overview of smart grid developments in Korea. This is followed by a discussion on the theoretical framework that integrates the key concepts of governance and innovation systems. The framework is then used to inform our analysis of the Korean case.

2 Smart Grid Development in Korea: An Overview

South Korea, officially the Republic of Korea, has a geographical area of 99,900 km^2 and a population of 49.41 million in 2011 [34]. It is a major developed country in Asia which ranks 14th globally by GDP [59]. Korea was the world's 10th largest energy consumer in 2008 [10]. As a country that has no oil, no high quality coal, and produces only 1.5 % of the natural gas it requires, Korea is dependent on imports to meet almost all of its energy needs [9]. The electricity system in Korea is fossil fuel based. Coal, natural gas, and nuclear amounted to about 30.9, 25.4, and 23.6 % of its electricity generation capacity, respectively, while renewable energy amounted to about 5.7 % (2011) (Table 1). Electricity consumption increased by 52 % between 2001 and 2009 and reached 433.6 TWh in 2009 [35]. Climate change is a key policy issue in Korea [43]. CO_2 emissions have been rising since the 1990 s with carbon intensity reaching 0.67 kg CO_2 per US$2,000 of GDP in 2008 [23].

The government's rationale for smart grids centred around President Lee Myung Bak's "Low Carbon, Green Growth" vision announced in 2008. The vision aspires to use green technology and green industries including smart grids as new engines for growth [29]. Detailed studies quantifying the benefits and costs of the deployment of smart grids in Korea are not publicly accessible. However, official data show that smart grids are expected to bring Korea economic benefits that

Table 1 Overview of the electricity generation in Korea (2011)

Fuel mix	
Coal (%)	30.9
Oil (%)	6.4
Natural gas (LNG) (%)	25.4
Nuclear (%)	23.6
Renewable and community energy services (%)	5.7
Hydro (%)	8.1
Generation capacity (MW)	79,342
Peak demand (MW)	71,137
Reserve margin (%)	10.3

Source [37]

could outweigh investment. It has been estimated that by 2030, smart grids would bring a range of benefits that include new global and domestic markets for smart grid technology that worth approximately 49 trillion won and 74 trillion won, respectively, the creation of 50,000 new jobs annually, saving 47 trillion won of energy imports, cost savings of 3.2 trillion won by avoiding building new power plants, and a reduction of 230 million tonnes in GHG emissions [44]. These benefits would outweigh the investment which is estimated to be 27.5 trillion won, in which 90 % would come from the private sector [44].

To realise the potential benefits of smart grids, the Korean government has introduced three strategies: the announcement of the national smart grid vision in 2009, the release of a national smart grid roadmap in 2010, and the launch of the Smart Grid Testbed on Jeju Island in 2009. The roadmap sets out a work plan for implementing the smart grid vision in five key areas, namely smart power grids, smart consumers, smart transportation, smart renewables, and smart electricity services. The Testbed is a large-scale demonstration project for testing technologies for the global market and developing business models. The Testbed involves 12 consortia (involving about 170 corporations) and 2,000 participating households [38, 44, 54]. Stage 2 of the Jeju Testbed was recently launched on 1 June 2011 following the completion of Stage 1 in end of May. According to the roadmap, the Testbed will be scaled up to a citywide level through the pilot of a Smart Grid City by 2012 and to the national scale by 2030.

Smart grid developments in Korea are currently still at an early stage and are mostly in the area of R&D while regulatory and policy frameworks are being strengthened. Following the enactment of the Smart Grid Act in May 2011, the corresponding decree and rule which are critical for the implementation and enforcement of the act were promulgated in November 2011 [45–47]. Dynamic pricing and smart meter roll-outs—which have been regarded as key enablers of smart grids—have been deployed but on a limited scale. Dynamic pricing has been recently piloted in the Jeju Testbed, but the participation rate is low. A total of 2.5 million smart meters have been installed in the industrial, household, and other consumer sectors, and the penetration rate is expected to reach 100 % with a total of 25 million smart meters to be installed by 2020 (Interview: 12/2011; [44]).

3 Smart Grid in Theoretical Perspective

A scanning of the literature suggests that two substantive bodies of theory are instructive in helping to analyse the development of smart grids: governance and innovation systems studies.

3.1 Governance Perspective

Central to the concept of governance is the move away from government to governance [51]. The perspective of governance highlights the importance of new approaches to enhancing governing capacity in which governments reach outwards and downwards to localities, engage with markets, and move out to civil society [51]. This multi-level and multi-actor approach therefore relies more on collaboration, networking, and learning [19, 41] (Gouldson et al. 2008).

Governance is a relevant perspective to analyse Korea's smart grid developments for a number of reasons. Smart grids require a transition in electricity systems from a fossil fuel-based and centralised model to a more decentralised form. This transition requires the involvement of a larger number of actors in more open electricity systems in which both established actors (such as the established electricity generation companies) and newcomers (such as renewable energy developers and well-informed consumers) interact. The traditional producer–consumer relationship changes to one in which well-informed consumers can play a much more active role in energy saving and even function as a "co-provider" of electricity supply [48]. These changes also raise a number of key governance issues, such as the changing role of the state, power asymmetries, conflict of interests, regulatory governance, participatory governance, and trust [2, 22].

3.2 Innovation Systems Perspective

Since the late 1980s, the idea of systems of innovation has become a major theme in science and technology studies in the western literature [8]. The notions of "socio-technical" system and the multi-level perspective are particularly useful to highlight the complexity and dynamics of the transition of energy systems to accommodate smart grid technologies.

The notion of socio-technical systems emphasises that transitions of electricity systems are embedded in a broader context that goes beyond technological change. According to Geels et al. [18], "such system innovations not only involve new technological artefacts, but also new markets, user practices, regulations, infrastructures, and cultural meanings". This notion emphasises the importance of co-evolution of technological, social, and environmental systems [30].

The multi-level scheme of Rip and Kemp [53] distinguishes three inter-related levels of changes: the landscape, regime, and niche levels of socio-technical systems. The landscape consists of a range of contextual factors that influence technological developments [17]. Regimes refer to rules and institutions that are built up around an established technology [17]. Niches are "protected" space in which innovation takes place [17]. This literature is particularly instructive in highlighting the drivers of change and what interactions should be created at the three inter-related levels to drive changes.

Coenen et al. [6] and Watson et al. [57, 58] on the other hand have shed light on how government policies, business incentives, and consumers' motivations have to converge to overcome barriers such as the problems of "lock-in" and the lack of a level playing field.

The complementary insights of the perspectives of governance and innovation systems provide us with a general framework for guiding and evaluating changes in smart grid-related innovation systems. This framework suggests that governments, business, and consumers are all key players who interact in new kinds of relationships at the three inter-related levels of landscape, regime, and niche and that such relationships can drive changes in socio-technical systems. Government has an important role to play in formulating coherent policy, fiscal and regulatory frameworks, articulating expectations, creating a level playing field, and enhancing market certainty for innovation processes [5, 58]. The business sector needs to collaborate with government to develop new business models such as energy service contracts that can put business incentives and consumers' motivations in place to drive towards changes in energy infrastructure [58]. Consumers who have access to better information through smart meters can play an active role in microgeneration investment as well as energy saving [24]. This framework suggests that the interactions of these key actors would create important forces for change at and between landscape, regime, and niche levels. Such forces for change, including visioning, expectation articulation, institutional arrangements, social networking, niche experimentation, second-order learning, and feedback are critical for accelerating the mainstreaming of smart grids [6, 53, 58].

The literature, however, is limited in illuminating how system innovations occur in the specific context of smart grids and is particularly so in the Asian context. Our study therefore addresses the following key questions in the case of Korea: Who were the key actors and what were their motivations? How did key actors interact at the landscape, regime, and niche levels in the socio-technical system for smart grids? And how did such interactions facilitate or constrain the development of smart grids?

4 Factors Underpinning Korea's Smart Grid Development

Korea's socio-technical system for smart grids possesses a number of characteristics which appeared to create opportunities as well as barriers for the development of smart grids. These factors can be found at the landscape, regime, and niche levels of the socio-technical system.

4.1 Landscape Level

At the *landscape* level, macroeconomic policies, a tradition of government-led growth strategies, pre-existing strengths in information technology, and world views on GHG reduction are key factors in Korea. Korea's macroeconomic policies, particularly the "Low Carbon, Green Growth" vision has played a significant role in the emergence of smart grids. The Green Growth vision introduced by President Lee Myung Bak in the wake of the financial crisis in 2008 was to develop green energy technologies, including smart grids as "new growth engines" [29].

Another landscape factor is a tradition of government-led growth strategies. The government has had a central role in spearheading the rapid development of strategic industries since the 1950s [49], including the heavy and chemical industries, petrochemicals, and shipbuilding in the 1970s and 1980s, and more recently, the growth of the electronics and information and communication technology during the 1990s [28]. Korea's recent move into the emerging industry of smart grids is an extension of this government-led growth strategy into the energy sector.

The pre-existing strengths of Korea in IT and other smart grid-related industries are another key factor at the landscape level. This high-tech industrialised economy [5] is particularly well placed to enter the global smart grid market because this industry can be built upon Korea's existing IT infrastructure and research networks.

World views on GHG reduction are another important landscape factor. Although Korea is a non-Annex 1 country of the Kyoto Protocol and is not obliged to commit to mandatory emissions reduction targets, it has voluntarily committed to reducing CO_2 emissions by 30 % by 2020 based on a "business as usual" baseline, implying a 4 % cut from 2005 levels [27]. Smart grid development has been regarded as a key strategy to meet this commitment [44].

Another landscape factor is Korea's long-standing reliance on government-funded institutes as a bridge between government and industry. A key institute for smart grids is the Korea Smart Grid Institute (KSGI). The Ministry of Knowledge and Economy (MKE) is the government agency which is responsible for formulating and implementing the national smart grid vision and policies. KSGI is in effect the executive arm of MKE. Fully funded by the government, KSGI is designated as the secretariat for smart grid initiatives, and it is responsible for implementing the national smart grid roadmap, managing the Jeju Smart Grid Testbed, and coordinating and managing R&D funding.

4.2 Regime Level

At the *regime* level, a number of dominant practices, rules, and shared assumptions [30] in Korea's electricity system have shaped the setting in which smart grids have evolved. These include Korea's partial electricity market reform, a government-led approach to energy policy making and management, limited market competition, structural problems of Korea Electric Power Corporation (KEPCO), distorted pricing systems, and public distrust.

A defining feature of Korea's electricity sector is the presence of partial electricity market reforms. The reforms, which were first launched in 1999, introduced competition into the generation industry in 2001 by dividing KEPCO's generation capacity into six power generation companies (GenCos) as KEPCO's subsidiaries [39, 56]. The reforms also created a new institution, the Korea Power Exchange (KPX), as an independent NGO which coordinates the flow of electricity in all regions of Korea [36]. Plans for further reforms, including the privatisation of five of the six state-owned GenCos, however, stalled in 2004 as a result of serious opposition from labour unions for concerns relating to price fluctuations and supply reliability [39, 56].

As a result of the partial market reform, considerable market distortion including state monopolies, energy subsidies, and distorted pricing systems remained. To date, the government has retained its commanding position throughout the entire electricity sector through the state-owned KEPCO. KEPCO, which owns 92 % of the total electricity generation capacity in Korea [50], has remained vertically integrated in nature. It is the sole transmission, distribution, and retail company [50] (Fig. 1). Even in the generation industry where competition has been introduced, the daily operations of five of the six KEPCO-owned GenCos are being managed by two government agencies, namely the MKE and the Ministry of Strategy and Finance (MOSF) (Fig. 1). As such, the government manages the daily operation of the five GenCos rather than adopting an arm's length approach.

Market distortions have given rise to another key regime factor—a highly distorted tariff system. On the one hand, the tariff system has been distorted by the cross-subsidy policy from residential and commercial to industrial and agricultural sub-sectors [39]. At present, the household tariff is approximately 114 KRW/kWh, which is 54 % higher than the industrial tariff (which is approximately 74 KRW/kWh) and 37 % higher than the average retail price to all consumers (Table 2). On the other hand, tariffs have been modulated by the government and have remained at relatively low levels to control inflation [39].

These features of Korea's electricity sector have created opportunities as well as barriers for smart grid development. The vertically integrated nature of KEPCO has ensured it has ready access to the planning and management in grid facilities that are critical components of smart grid infrastructure. In addition, Korea's government-led approach to energy planning and management [56] has created a political obligation for KEPCO to implement the national smart grid vision. KEPCO has become a first mover into the emerging smart grid business and has committed to making a US$7.18 billion investment by 2030 [4].

However, the vertically integrated nature of KEPCO has posed limitations for smart grid developments. New entrants including independent power producers (IPPs) such as renewable entities can broaden energy options in smart grids. However, although five of the six GenCos are in competition under the cost-based pool system ([3, 39]), market competition between the five GenCos and the IPPs

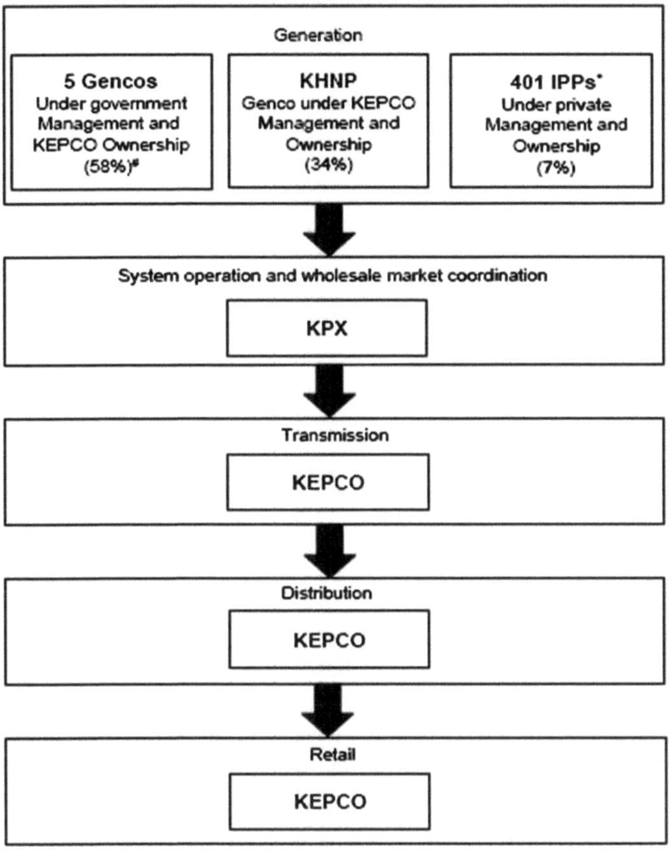

Fig. 1 The electricity sector in Korea. *Hash* (#) numbers in brackets are percentage of generation capacity in Korea in 2009 (*Source* Authors; data from [35, 50])

has remained limited. The 401 IPPs are numerous. They, however, contributed only 7 % of the total generation capacity in 2009 [35, 50]. Similarly, the 353 renewable companies represented only 2 % of the total installed capacity as of April 2011 [50].

Furthermore, the distorted pricing system and the associated public distrust of the government and KEPCO appear to limit the opportunity for introducing dynamic pricing systems. Dynamic pricing, which generally imposes higher price during peak periods and offering lower prices during off-peak periods, is a key to effective demand response programmes and provides an alternative to the distorted system in Korea (Interviews: 05/2011; 06/2011). However, the public has been highly sceptical about the government's motives in changing tariff levels. Although dynamic pricing has the potential to deliver price reductions [1], the public would regard introducing new dynamic pricing systems for smart grid as simply the government and KEPCO disguising tariff increases (Interviews: 06/2011; 07/2011).

Table 2 Overview of the electricity consumption and electricity rates in Korea (2009)

Total electricity demand (GWh)	433,604
Residential (%)	19.9
Industrial (%)	52.5
Public and service (%)	22.7
Agricultural (%)	2.5
Educational (%)	1.6
Street lighting (%)	0.7
Retail electricity rates (KRW/kWh)	
Residential	114.45
Industrial	73.69
Public and service	98.50
Agricultural	42.13
Educational	83.56
Street lighting	76.65

Sources compiled by authors; data from [35]
The data presented in this table exclude electricity that was generated by private generation companies which amounted to approximately 13 % of the total installed capacity in Korea (2009)

Long-standing public distrust is rooted in a sense of inequity that has emerged from the cross-subsidy policy (Interviews: 04/2011; 06/2011; 07/2011).

Another regime factor is the presence of three structural problems in Korea's electricity sector—reliance on imports, peak load problems, and financial losses. The peak load problem has threatened the reliability of the electricity system with potential blackouts ([11]; Interview: 08/2011). The reserve margin against peak load has been decreasing since 2003 (except 2008) and reached a record low of 6.2 % in 2010 [36]. KEPCO also faced financial problems. It has suffered from financial losses amounting to a total of 3 trillion won for the years between 2008 and 2010 [31].

These problems have motivated the government and KEPCO to explore smart grids as a potential solution. Dynamic pricing which could be introduced as a component of the smart grid policy is perceived by KEPCO as an opportunity to introduce changes into the current tariff system (Interview: 03/2011). The opportunities to introduce dynamic pricing also provide an opportunity to manage peak load problems more effectively through price-responsive demand. Studies elsewhere show that dynamic pricing has the potential to reduce peak loads by up to 16 % (see e.g. [15]).

4.3 Niche Level

At the *niche* level, a major development has been the establishment of the Smart Grid Testbed on Jeju Island in 2009. The Testbed is a government-led large-scale niche experimentation project for domestic companies to test and demonstrate their technologies for global markets.

The Testbed, with a geographical area of 185 km^2, is located in a remote community in the rural, north-eastern part of Jeju Island. Jeju won a national competition and became the hosting province for the Testbed in June 2009 [38]. Jeju was selected for good reasons. Jeju, as Korea's only autonomous province, has the flexibility in institutional, regulatory, and legal arrangements and therefore is particularly well placed to pioneer innovative incentives for R&D investment and to experiment with new policy ideas (e.g., dynamic pricing) which are politically sensitive and would be difficult to implement on a nationwide scale.

The Testbed has two distinctive features. The first is its emphasis on collaboration between the central government, industries, and the Jeju local government. Twelve consortia, involving approximately 170 companies, have been formed in the Testbed. Those companies came from diverse sectors that range from energy to information technology, to steel manufacturing, electric vehicles, and home appliance manufacturing. The involvement of well-known companies including KEPCO, KPX, Samsung, LG, and Hyundai has enhanced the prominence of the Testbed.

The second key feature is the involvement of local residents. Approximately 2,000 households, about one-third of the total households within the Testbed geographical boundary, are participating in the Testbed on a voluntary basis.

To sum up, three features of the socio-technical system are noteworthy. First, as Fig. 2 shows, the transition of this socio-technical system involved a wide range of factors that extend beyond technological one. The system is influenced by factors that range from world views to macroeconomic policies, a distorted tariff system, the presence of public distrust, and to experimentation in a remote island. Second, a broad range of actors including the government, established actors in the electricity sector (such as KEPCO and KPX), new entrants (such as the IPPs), industries, institutes, consumers, and global companies interacted and shaped the path and scale of the smart grid developments. Third, the interactions of the actors are taking place across various governing levels in a multi-level system that comprises of the landscape level at the macro level, the regime level at the meso level, and the niche level at the micro level.

5 Discussion: The Strengths and Weaknesses of the Government-led Approach

The emergence of the Korean model, which is distinguished by a government-led and export-oriented approach, poses a number of important questions: To what extent has this model driven changes in the innovation system? Who are driving (or creating barriers) for change and how? Before discussing our findings, we must

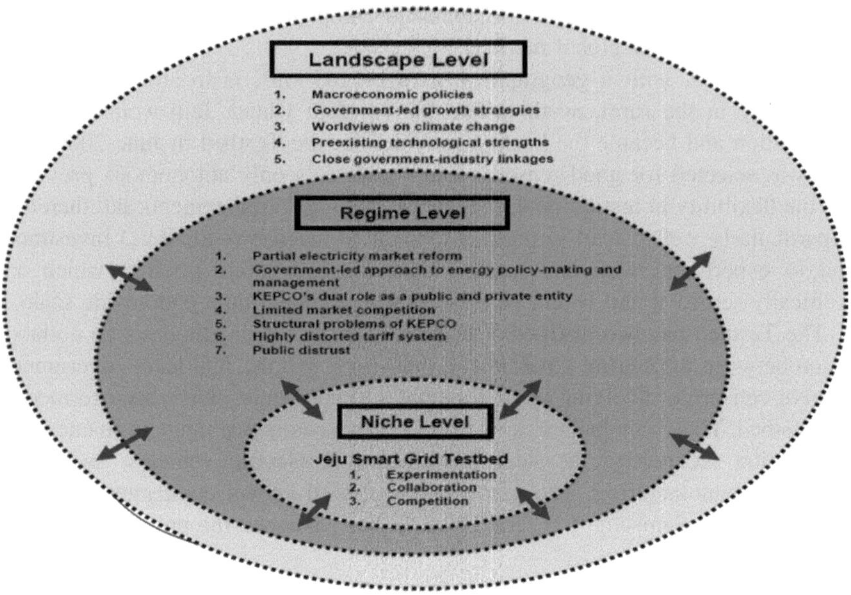

Fig. 2 An innovation system approach to understanding the development of smart grid in Korea (*Source* authors)

begin by acknowledging that our observations should be interpreted with caution. In electricity sectors where large existing investments in fossil fuel-based and nuclear infrastructure have been made, the lock-in effect as a result of sunk investments in infrastructure of established energy technologies tends to make short-term transitions towards smart grids difficult to achieve [55]. In addition, many of the smart grid initiatives, notably the Testbed is at the pilot stage and is ongoing. It would be premature to provide an evaluation of the successes and failures of the Korean model. Our observations about causal connections are therefore tentative.

Despite these limitations to our data and observations, our analysis leads us to make some observations relating to the strengths and weaknesses of Korea's government-led approach. The Korean government has been motivated by a number of factors and has played a pivotal role in driving changes throughout the landscape, regime, and niche levels in the socio-technical system. It has initiated, incubated, and set the pace for the development of smart grids through the announcement of the national smart grid vision, the formulation of the national Smart Grid Roadmap, the creation of the Smart Grid Institute as an executive arm of the government to implement the roadmap, and the launch of the Smart Grid Testbed to facilitate experimentation through government–industry–consumer collaboration for technological innovation.

Korea's government-led and export-oriented model, however, has also suffered from some weaknesses. In contrast to our framework that suggests government policies, business incentives and consumer motivations have to converge to drive changes in socio-technical systems, our analysis has found that those desirable interactions have been limited in three aspects.

The first weakness is related to the government's regulatory and policy capacity. Our framework suggests that the government has an important role to play in articulating expectations and reducing uncertainty in innovation processes by formulating a coherent regulatory and policy framework [6]. In Korea, however, a strong regulatory and policy framework is still lacking. According to some industrial experts, the policy support for smart grid has been clouded as possible changes in leadership in the upcoming presidential election and National Assembly election in 2012 may lead to a withdrawal of policy support (Interviews: 05/2011; 06/2011). The enactment of the Smart Grid Act in April 2011 could have ensured policy consistency across presidential terms. However, the passage of this law itself was delayed five months from December 2010 to April 2011 as consensus between various stakeholders was difficult to achieve (Interviews: 05/2011; 06/2011). The recent promulgation of the corresponding decree and rule has been regarded as a key to strengthen the implementation and enforcement of the Act. However, there are concerns that the current regulatory framework is still not sufficient to drive major stakeholders, particularly utilities, to break the inertia and move away from the "lock-in" situation (Interview 13/2012).

The second weakness is related to the business sector. Smart grid developments require a new producer–consumer relationship in which the business sector and the government would need to collaborate and develop new market rules, user practices, and energy infrastructure to enable the more active participation of consumers [48]. Business models such as energy service contracts therefore are a critical element for accelerating smart meter roll-outs and major changes in energy infrastructure [57, 58]. However, although business models have been identified as one of the strategic pilot areas in Stage 1 of the Testbed, experimentation on business models have been negligible, if any (Interviews 5/2011; 6/2011). Driven by the aspiration to access global markets, the Testbed has a rather narrow focus on R&D with a much higher priority given to demonstrate Korea's technological capacity.

The third weakness relates to consumer engagement, in particular the upscaling of experimentation. At present, Korea adopts a flat rate electricity pricing system. Dynamic pricing presents an opportunity for Korea to address the problems of energy subsidies and other structural problems of the electricity sector such as KEPCO's long-standing financial deficit. Elsewhere, countries such as Japan and China have also been contemplating dynamic pricing as an alternative which is more politically feasible to changing the tariff level [25, 40].

Dynamic pricing can take place in various forms, and the three main forms are real-time pricing (RTP), critical peak pricing, and peak time rebate [7]. In Korea, a rebate-based RTP pilot has been introduced in Stage 2 of the Testbed since June 2011. In order to secure participation, the system is designed in such a way that the participating households would not suffer loss. Consumers who sign up to the RTP would receive a rebate for the electricity saved. The rebate system caps the bill to existing rates so that even if a household would have to pay more under the RTP system it would not be charged more than the existing rate. Furthermore, households participating in the Testbed were offered facilities and equipments such as smart meters and PV panels at no cost (Interview 09/2011).

Evaluations of the Testbed are ongoing and are expected to be released by mid-2013 (Interview: 11/2011). Although it is premature to assess the achievements of the Testbed, this pilot—the first of its kind in Korea—has exposed weaknesses in two critical areas. The first weakness is the low participation rate. This rebate-based pilot has been tested in some 2,000 rural households who reside within the geographical boundary of the Tested and who volunteered to participate. These 2,000 households, however, represent only one-third of the total household residing within the Testbed area, about 1 % of the total number of households on Jeju [32], and just 0.01 % of the total households of the country [33].

Although experiences elsewhere with dynamic pricing have reported positive results (see e.g. [16]), the Jeju pilot was not the case. The low participation rate is particularly a concern as the actual reduction in electricity consumption achieved appears to be minimal. According to a senior executive of KEPCO who is a core member managing the pilot, a preliminary analysis on 105 consumer bills collected in September 2011 shows that only 27 consumers reacted to the RTP and the reduction of electricity bills ranged from 3 to 5 % on average (Interview 12/2011). Evaluations on the changes in energy consumption and peak load have been ongoing and are not publicly accessible as of October 2011 (Interview 12/2011). Preliminary observations from industrial experts suggest that KEPCO's lack of marketing knowledge and skills appears to be a key factor for the low participation rate and poor consumer responses in a Korean context in which electricity price has been modulated at a relatively low level that make dynamic pricing difficult to be effective (Interview 13/2012). Whether other barriers identified in the western experiences such as a lack of economic incentives and inertia can also apply in this pilot is an important area that needs to be better understood [7, 21].

The second weakness is related to the programme design. While the rebate element and the provision of free facilities can be regarded as a pragmatic and transitional strategy to attract consumer participation, a major drawback of this approach is its limits in achieving second-order learning. Second-order learning emphasises that learning for innovation should extend from technological advancement to testing actual changes in user practices, new markets, regulator performance, and energy infrastructure [6]. In contrast to our framework which highlights the critical role of demonstration projects as a protected and "niche" area in which new ideas can be tested and second-order learning can take place, this pilot is not able to explore critical issues such as consumer acceptability of dynamic pricing and how feedback information from smart meters can change consumption behaviour.

Those weaknesses may limit the potential of the upscaling this experimentation. On the one hand, if this rebate-based system is scaled up to a city or national scale without introducing major changes in the incentive system, such a system would be unlikely to build up policy legitimacy. Instead, it may crowd out the adoption of dynamic pricing options. On the other hand, this pilot, which is characterised by its voluntary and virtually zero-risk nature, appears to be designed to reflect many of the unique contexts of the Jeju pilot. The results of this pilot may therefore not be generalisable from this rural community to other parts of Korea and elsewhere in the world where the socio-political and economic contexts can be substantially different.

6 Conclusion

Driven by its "Green Growth Vision", Korea embarked on its smart grid initiatives in 2009. Within just 3 years, the country has made some important progresses in the development of smart grids although problems still exist. This chapter adopted the perspectives of governance and innovation systems as a general analytical framework for understanding and critically examining the recent evolution of smart grids in Korea.

We have two major findings. First, we have revealed the complexity of the socio-technical system by highlighting the breadth and depth of those factors which have influenced smart grid development in Korea. Our findings reinforced the view that the convergence of policies, business incentives, and consumer motivations is critical to drive changes in socio-technical systems (see e.g. [57, 58]). We have demonstrated how the government, business, and consumers interacted at and across the landscape, regime, and niche levels, and how such interactions facilitated or constrained smart grid development. Various factors that span from macroeconomic policies and global views, to partial electricity market reform and public distrust, and to experimentation in a demonstration project are found to be crucial in the socio-technical system for smart grids in Korea.

Second, our analysis has shed light on the mechanisms for change in socio-technical systems. It is important to understand how socio-technical systems manage to change. However, the barriers to changes must also be understood. We found that the presence of partial electricity market reform and public distrust has created barriers for Korea to develop some of the favourable conditions for change. Favourable conditions such as policy consistency, second-order learning, and the development of financially viable business models (see e.g. [57, 58]) are still lacking in Korea. These observations can provide a better understanding of the "lock-in" phenomenon (see e.g. [6]).

Our findings have some policy implications. A systemic perspective is needed for policy in order to accommodate the changes required for smart grid development. Regulatory reforms, particularly price-setting mechanisms, and consumer engagement are priority areas for policy change. Experiences elsewhere suggest that smart grid literacy programmes may be useful in enhancing the public understanding of such complex issues relating to dynamic pricing and for restoring public trust [13, 54].

Our findings are country- and sector-specific. We recognise that there are limits to the generalisability of our findings to other countries and other socio-technical systems in non-electricity sectors. The highly regulated electricity sector in which market signals and market competition play a much lesser role than administrative regulation has set Korea apart from other economies such as the USA in which electricity sectors are liberalised. However, further research may generate useful results if, for example, it investigates the transferability of the findings to China—which has also been pursuing smart grid initiatives. China's electricity sector also remains highly regulated as a result of partial electricity market reforms.

Acknowledgments This chapter is an updated republication of the paper "Governing the transition of socio-technical systems: A case study of the development of smart grids in Korea" by Daphne Ngar-yin Mah, Johannes Marinus van der Vleuten, Jasper Chi-man Ip, Peter Ronald Hills, Energy Policy 45, 133–141. The copyright permission for reusing the paper has been granted by Elsevier. We would like to acknowledge appreciation to our anonymous interviewees in Korea for contributing their insights and providing useful information. We gratefully acknowledge the funding of this research by the University of Hong Kong through the Initiative on Clean Energy and Environment, and the Hong Kong Baptist University through the Faculty Research Grant of Faculty of Social Sciences (FRG2/12-13/057 and FRG1/13-14/051). We remain solely responsible for any errors and omissions in the findings and interpretations expressed in this paper.

A.1 7 Appendix 1: List of Interviewees

As some interviewees agreed to be interviewed anonymously, this study indicates interviews by number. The first two digits indicate the interview numbers and that followed by the year of interviews. The interview formats included face-to-face interview (FI), telephone interview (TI), and email correspondence (EC).

Code	Interviewees background	Type of interview	Date of interview
01/2011	A senior executive, Korea Smart Grid Institute (KSGI)	EC	27 May 2011
02/2011	A senior executive, Korea Power Exchange (KPX)	TI	31 May 2011
03/2011	Same interviewee as 01/2011, KSGI	FI	25 April 2011
04/2011	An associate professor, Graduate School of Environmental Studies, Seoul National University	FI	26 April 2011
05/2011	Same interviewee as 02/2011, KPX	FI	26 April 2011
06/2011	A senior executive, Korea Electric Power Corporation (KEPCO)	FI	26 April 2011
07/2011	A postgraduate student, Graduate School of Environmental Studies, Seoul National University	FI	25 April 2011
08/2011	A senior executive, Total Operation Centre, KPX	FI	28 April 2011
09/2011	Same interviewee as in 06/2011, KEPCO	EC	16 June 2011
10/2011	A government official of Smart Grid Division, Jeju Special Self-Governing Province	FI	28 April 2011

(continued)

(continued)

Code	Interviewees background	Type of interview	Date of interview
11/2011	Same interviewee as 02/2011, KPX	TI	27 October 2011
12/2011	Same interviewee as in 06/2011, KEPCO	TI	27 October 2011
13/2012	Same interviewee as 02/2011, KPX	TI	30 January 2012

References

1. Albadi MH, El-Saadany EF (2008) A summary of demand response in electricity markets. Electr Power Syst Res 78:1989–1996
2. Beierle T, Cayford J (2002) Democracy in practice: public participation in environmental decisions. Resources for the Future, Washington, DC
3. Byrne J, Glover L, Lee H, Wang YD, Yu JM (2004) Electricity reform at a crossroads: problems in South Korea's power liberalization strategy. Pac. Aff. 77(3):493–516
4. Cho MY (2011) S. Korea KEPCO to invest $7.2 billion in smart grid by 2030. http://www.reuters.com/article/2011/02/18/kepco-korea-idUSTOE71H00R20110218. Accessed 20 May 2011
5. CIA (2011) The World Factbook, South Korea. Central Intelligence Agency. https://www.cia.gov/library/publications/the-world-factbook/geos/ks.html. Accessed 20 June 2011
6. Coenen L, Raven R, Verbong G (2010) Local niche experimentation in energy transitions: a theoretical and empirical exploration of proximity advantages and disadvantages. Technol Soc 32(4):295–302
7. DOE (2008). What the smart grid means to Americans. The United States Department of Energy, Washington, DC. http://energy.gov/sites/prod/files/oeprod/DocumentsandMedia/ConsumerAdvocates.pdf. Accessed 10 Jan 2012
8. Edquist C (1997) Systems of innovation approaches—their emergence and characteristics. In: Edquist C (ed) Systems of innovation: technologies, institutions and organizations. Pinter, London, pp 1–35
9. EIA (2010) Country analysis briefs, South Korea. Energy Information Administration. http://www.eia.doe.gov. File://Z:/NewCABs/V6/South_Korea/Full.html. Accessed 20 June 2011
10. EIA (2011) International energy statistics. Energy Information Administration. http://www.eia.gov/cfapps/ipdbproject/IEDIndex3.cfm?tid=44&pid=44&aid=2. Accessed 25 May 2011
11. EMSC and KPX (2009) 2009 annual report: electricity market trends & analysis. Electricity Market Surveillance Committee, Korea Power Exchange, Seoul
12. EPRI (2011) EPRI analysis estimates costs-benefits of fully developing smart grid. http://my.epri.com/portal/server.pt/gateway/PTARGS_0_234325_317_205_776_43/http%3B/uspalecp604%3B7087/publishedcontent/publish/epri_analysis_estimates_costs_benefits_of_fully_developing_smart_grid_da_777189.html. Accessed 20 Jun 2011
13. Executive Office (2011) A policy framework for the 21st century grid: enabling our secure energy future. Executive Office of the President of the United State, Washington, DC
14. Faruqui A, Harris D, Hledik R (2010) Unlocking the 53 billion Euro savings from smart meters in the EU: How increasing the adoption of dynamic tariffs could make or break the EU's smart grid investment. Energy Policy 38:6222–6231
15. Faruqui A, Hledik R, Sergici S (2010) Rethinking prices: the changing architecture of demand response in America. Public Utilities Fortnightly 30–39
16. Faruqui A, Palmer J (2011) Dynamic pricing and its discontents. Regulation, Fall 16–22

17. Geels F (2001) Technological transitions as evolutionary reconfiguration processes: a multi-level perspective and a case-study. Paper presented at the Nelson and Winter Conference (12–15 June 2001). Aalborg, Denmark. Organised by DRUID (Danish Research Unit for Industrial Dynamics), Research Policy and Corporate and Industrial Change
18. Geels F, Elzen B, Green K (2004) General introduction: system innovation and transitions to sustainability. In: Elzen B, Geels F, Green K (eds) System innovation and the transition to sustainability: theory, evidence and Policy. Edward Elgar, Cheltenham, pp 1–18
19. Goodwin M, Painter J (1996) Local governance, the crises of Fordism and the changing geographies of regulation. Trans Inst Br Geogr 21(4):635–648
20. Gouldson A, Hills P, Welford R (2008) Ecological modernisation and policy learning in Hong Kong. Geoforum, 39(1):319–330
21. Hargreaves T et al (2010) Making energy visible: a qualitative field study of how householders interact with feedback from smart energy monitors. Energy Policy 38(10):6111–6119
22. Hood C, James O, Scott C (2000) Regulation of government: has it increased, is it increasing, should it be diminished? Public Adm 78(2):283–304
23. IEA (2010) CO_2 emissions from fuel combustion highlights, 2010 edn. International Energy Agency, Paris. http://www.iea.org/co2highlights/co2highlights.pdf
24. IEA (2011) Technology roadmap: smart grids. International Energy Agency, Paris. http://www.iea.org/papers/2011/smartgrids_roadmap.pdf
25. Japan FS (2011) Kitakyushu to test dynamic pricing for electricity in FY2012, 17 Oct 2011. http://www.japanfs.org/en/pages/031319.html. Accessed 25 Oct 2011
26. Johnson J (1990) Selecting ethnographic informants. Sage Publications, Newbury Park
27. Jones RS, Yoo B (2010) Korea's green growth strategy: Mitigating climate change and developing new growth engines. OECD economics department working papers, No. 798, ECO/WKP 54, OECD Publishing, Paris
28. Kang MH (1996) The Korean business conglomerate: Chaebol then and now. Korea Research Monograph. Southgate Publishers, Devon, UK
29. Kang MJ, Park H (2011) Impact of experience on government policy toward acceptance of hydrogen fuel cell vehicles in Korea. Energy Policy 39(6):3465–3475
30. Kemp R, Rotmans J (2005) The management of the co-evolution of technical, environmental and social systems. In: Weber M, Hemmelskamp J (eds) Towards environmental innovation systems. Springer, Berlin, pp 33–54
31. KEPCO (2011) Dividends payment section. http://www.kepco.co.kr. Accessed 25 May 2011
32. KOSTAT (2011) Preliminary results of the 2010 population and housing census. Statistic Korea, Ministry of Strategy and Finance, the Republic of Korea. http://kostat.go.kr/portal/english/news/1/1/index.board?bmode=read&aSeq=245048. Accessed 07 Nov 2011
33. KOSTAT (2011) Results of the 2010 population and housing census (Households and Housing Units.) Statistic Korea, Ministry of Strategy and Finance, the Republic of Korea. http://kostat.go.kr/portal/english/news/1/9/index.board?bmode=read&bSeq=&aSeq=249192&pageNo=1&rowNum=10&navCount=10&currPg=&sTarget=title&sTxt. Accessed 07 Nov 2011
34. KOSTAT (2012) Explore Korea through statistics 2012. Statistics Korea. http://kostat.go.kr/Explore12/EBook.htm. Accessed 26 Aug 2013
35. KPX (2010) Handbook of electricity statistics in Korea (as of Dec 2009). Korea Power Exchange, Seoul
36. KPX (2011) Korea power exchange website: about KPX. http://www.kpx.or.kr/english. Accessed 18 May 2011
37. KPX (2012) Korea power exchange website: statistics. http://www.kpx.or.kr/english/. Accessed 26 Aug 2013
38. KSGI (2011) Korea smart grid institute website. http://www.smartgrid.or.kr/10eng2-4.php. Accessed 18 June 2011
39. Lee BH, Ahn HH (2006) Electricity industry restructuring revisited: the case of Korea. Energy Policy 34(10):1115–1126

40. Lin B, Jiang Z (2012) Designation and influence of household increasing block electricity tariffs in China. Energy Policy 42(March):164–173
41. Mah D, Hills P (2009) Collaborative governance for sustainable development: wind resource assessment in Xinjiang and Guangdong Provinces, China, Sustainable Development, In Press, Corrected Proof, Available online 1 June 2010. doi:10.1002/sd.466
42. Miles M, Huberman A (1994) Qualitative data analysis: An expanded sourcebook. Thousand Oaks, Sage, California
43. MKE and KSGI (2009) For the World's Green Growth—Jeju smart grid test-bed in Korea. Ministry of Knowledge Economy and Korea Smart Grid Institute. Seoul. http://www.smartgrid.or.kr/Ebook/JejuTestBed.pdf
44. MKE and KSGI (2010) Korea's Smart Grid Roadmap 2030—Laying the Foundation for Low Carbon, Green Growth by 2030. Ministry of Knowledge Economy, Korea Smart Grid Institute, Seoul. http://www.smartgrid.or.kr/10eng4-3.php
45. MOLEG (2011) Act on the creation and facilitation of use of smart grids. Ministry of Government Legislation, Seoul, Republic of Korea. http://www.moleg.go.kr/english/searchListR.mo?q=smartgrid&x=31&y=11. Accessed 10 Jan 2012
46. MOLEG (2011) Enforcement decree of act on the creation and facilitation of use of smart grids. Ministry of Government Legislation, Seoul, Republic of Korea. http://www.moleg.go.kr/english/searchListR.mo?q=smartgrid&x=31&y=11. Accessed 10 Jan 2012
47. MOLEG (2011) Enforcement rule of the act on the creation and facilitation of use of smart grids. Ministry of Government Legislation, Seoul, Republic of Korea. http://www.moleg.go.kr/english/searchListR.mo?q=smartgrid&x=31&y=11. Accessed 10 Jan 2012
48. Nye M, Whitmarsh L, Foxon T (2010) Sociopsychological perspectives on the active roles of domestic actors in transition to a lower carbon electricity economy. Environ Plann A 42:697–714
49. Consulting Ovum (2009) Broadband policy development in the Republic of Korea: a report for the Global Information and Communications Technologies Department of the World Bank. Ovum Consulting, London
50. Park KS (2011) The Korean Electric Power Industry & Corporate Governance of KEPCO. Paper presented at the 6th meeting of the OECD network on Corporate Governance of State-Owned Enterprises in Asia, Seoul, 17–18 May 2011. http://www.oecd.org/dataoecd/35/60/48049493.pdf
51. Pierre J, Peter G (2000) Governance, politics and the state. Macmillan Press, Basingstoke
52. Ragwitz M, Held A, Stricker E, Krechting A, Resch G, Panzer C (2010) Recent experiences with feed-in tariff systems in the EU—A research paper for the International Feed-in Cooperation: a report commissioned by the Ministry of the Environment, Nature Conservation and Nuclear Safety (BMU)
53. Rip A, Kemp R (1998) Technological change. In: Rayner S, Malone EL (eds) Human choice and climate change, vol 2. Battelle Press, Ohio, pp 327–399
54. SGA (2011) Smart Grid Australia: maximising consumer benefits. Smart Grid Australia, Melbourne. http://www.smartgridaustralia.com.au/uploads/documents/VIC060411_Consumer_Working_Group.pdf
55. UNEP (2006) Global environment outlook year book 2006. Nairobi, United Nations Environment Program, Kenya
56. Vine E, Rhee CH, Lee KD (2006) Measurement and evaluation of energy efficiency programs: California and South Korea. Energy 31(6–7):1100–1113
57. Watson J, Sauter R, Bahaj B, Patrick J, Myers L, Gann D et al (2007) Integrating micro-generation into energy networks and buildings : full research report. ESRC End of Award Report, RES-338-25-0003. ESRC, Swindon
58. Watson J, Sauter R, Bahaj B, Patrick J, Myers L, Wing R (2006) Unlocking the power house: policy and system change for domestic micro-generation in the UK. Science and Technology Policy Research, The Freeman Centre, University of Sussex, Brighton
59. World Bank (2011) World development indicators database. http://siteresources.worldbank.org/DATASTATISTICS/Resources/POP.pdf. Accessed 01 Jul 2011

Developing Super Smart Grids in China: Perspective of Socio-technical Systems Transition

Yuan Jia-Hai

Abstract This chapter adopts the framework of socio-technical systems transition to discuss the development of smart grid (SG) in China. Multi-level perspective (MLP) analysis, namely landscape, regime, and niches, is presented to set the institutional background of China's power sector. Then, a scenario of power supply and demand is compiled, and its main features are elaborated to justify the necessity of developing SGs. The overall transition pathways of China's power sector are drafted, and then, the road map for developing SGs in China is proposed. Key policy implications are as follows: first, working out clear national strategy and road map for SG development and carrying out integrated translation research and strong front-end support; second, restructuring power sector in parallel with SG development; and third, reforming pricing mechanism and empowering the customers with rights to choose service/enter power market.

Keywords Smart grids · Multi-level perspective · Transition pathways · China

1 Introduction

During the 18th National Congress of CPC of November 2012, ecological civilization [Sheng Tai Wenming] was highlighted for the first time among the guiding principles for the completion of a well-off society [Xiaokang Shehui] in 2020. Promoting revolution in energy production and consumption, controlling total energy consumption, enhancing energy conservation, and supporting low-carbon industry and renewable energy development will set the keynotes for China's energy strategy and play essential roles to realize *Scientific Development Perspective* in China.

Y. Jia-Hai (✉)
School of Economics and Management, North China Electric Power University, Beijing, China
e-mail: yuanjh126@126.com

D. Mah et al. (eds.), *Smart Grid Applications and Developments*, Green Energy and Technology, DOI: 10.1007/978-1-4471-6281-0_14,

Power system is at the center of modern energy system. Logically, the energy revolution calls for radical revolution in power system. Considering the currently coal-based system and the enormous growth potential of power demand in the future, this issue is especially pressing and complicated in China.

Smart Grid (SG), characterized by large-scale integration of renewable energy and energy storage, distributed power generation and smart demand response, is expected to facilitate the energy revolution worldwide. In China, the first road map for SG was proposed by State Grid Corporation of China (SGCC) in May of 2009. Then in July of 2010, Southern Grid of China (SGC) issued its strategy for building "smart, high-efficient, reliable, and green power grid." In March of 2011, in "the 12th Five-Year-Plan of National Economy and Social Development" (12th FYP), SG was regarded as a key strategy to upgrade the power sector and enhance its core competency. Then in May of 2012, the Ministry of Science and Technology (MOST) issued "the 12th Special Program for Key Industrialization of Smart Grid during 12th FYP Period." The program is a clear signal that SG has been incorporated into the national energy strategy and that the Chinese Government is resolute to implement it seriously in the coming decades.

Up to now, to the best of our knowledge, primarily the "hard parts" or the technical aspects of SG are addressed. However, the transformation of current carbon-locked and dumb power system to future low carbon and smart one represents a radical system transition. The "soft parts," i.e., the organization, institution, and policy, which are essential to the transition, are seldom discussed. This chapter will address the following issues: Why is SG so important for China's low-carbon economy? What are the key features of SG in China? How can China build SG and set clear road map for it? From a perspective of socio-technical systems transition, what are the main institutional barriers for developing SG in China and what are proper policies to overcome them? Section 2 will present a brief analysis of China's power sector in a MLP; Sect. 3 will discuss a scenario for power demand and supply scenario into year 2030 and analyze the landscape and driving forces; Sect. 4 will elaborate the key features of the power scenario and discuss the unique functions of SG in China; Sect. 5 will propose a road map for developing SG in China; Sect. 6 will discuss the institutional barriers during the transition and propose countermeasures to overcome them; and Sect. 7 concludes.

2 Power Sector in China

2.1 Overview

As summarized in Table 1, the history of China's power system can be divided into four periods. At present, China's power sector consists of two national grid companies, SGCC and SGC, five national generation companies, a dozen of generation companies affiliated to provincial governments, and a few relatively smaller independent private generators.

Table 1 Brief history of the Chinese electric power sector, 1949–present

Period	Guiding policies	Description
1949–1985	Centrally planned and administered system	Vertical integrated state-run enterprise; government agencies plan, finance, manage, and operate the system; oscillations between centralized and decentralized management
1985–1997	Decentralization to provinces; opening of investment	Opening up of the sector to provincial government, private, and foreign investment; guaranteed investment return on generation investment
1997–2002	Separation of government and business	Corporatization of the sector through the creation of the State Power Corporation (SPC); ministry of electric power dissolved, functions transferred to the State Economic And Trade Commission (SETC) and the State Development And Planning Commission (SDPC), later merged into the National Development And Reform Commission (NDRC)
2002–2013	Unbundling of generators and grid companies	Dismantling of the SPC into 5 national SOE generating companies, 2 national grid companies; creation of State Electricity Regulatory Commission (SERC) and National Energy Administration (NEA)
2013-	Merge of NEA and SERC	Creation new NEA and integrating the function of SERC into it

Growth of electric power consumption has been faster than that of primary energy consumption in the past three decades. The rapid growth in power demand has resulted in tremendous growth of generation capacity expansion and rapid electrification. Measured by the ratio of electricity over final energy consumption, the electrification level has increased from 7.7 % in 1980 to 20 % in 2008 [2]. From 1980 to 2009, annual electricity demand in China grew more than 12-fold, from 300 to 3,660 TWh [5]. Demand growth of this magnitude and speed has contributed to consistently severe capacity shortages. At the same time, generation capacity in China has increased significantly to meet with the power demand. In 1980, the total generation capacity in China was only 65.8 GW, while in 2009 the installed capacity amounted to 863.6 GW, which is a 12-fold growth. Especially since 2004, annual increase in generation capacity ranges between 50 and 100 GW, creating a new record in the history of world power system growth. Only the USA maintained an annual growth of over 50 GW during the 1970s.

The coal-based generation mix results in carbon lock-in in China's power system, making it the largest contributor of CO_2 emissions as well as the main source of other serious environmental problems in China [4]. The share of thermal power units in the entire generation mix was as high as 70 %. As a result, the thermal power units consumed more than 50 % of the country's total coal production.

Although China's generation mix has been relatively stable over the past two decades, the composition of coal-fired power plants has undergone a significant shift toward larger and more efficient units. In the 1990s, most of the thermal power generators above 300 MW/unit installed in China were imported, and in 1993, the share of 300 MW units and above accounted for only 23 % of the total thermal generation capacity. In 2007, the share of 300 MW and above, and 600 MW and above accounted for 50 and 21.6 %, respectively. Then at the end of 2009, the shares of 300 MW and above rose to 69 % [5]. As a result of this push toward higher generation efficiency, the average thermal efficiency has sustained a linearly increasing trend since the 1990 s and now reportedly surpasses the average efficiency of American coal plants by a significant margin. In 2007, CO_2 emission per KWh electricity generation in coal-based plants in China was 893 grams (down by 7 % from the 1992 level), lower than world average (903 g), the US average (920 g) and Japanese average (910 g), but was still higher than European Union 27 nations average (835 g).

The imbalance between power load and resource center necessitates long-distance large-scale power transmission in China. China's physical grid is divided into four regional synchronous grids, with the Northeast-North-Central, East and Northwest regions operated by the SGCC, and the Southern grid operated by the CSG. Although basic DC interconnection among regional grids was achieved in 2005 [22], power flow among regions and even between provinces within regions remains limited.

Operation efficiency, especially in an environmental sense, is notoriously low because of the guaranteed return policy during the 1990s. Dispatch in the Chinese power sector has, since the early 1980s, operated under an "equal shares" formula whereby generators of a given type are guaranteed a roughly equal number of operating hours to ensure adequate revenues to recover their fixed costs. Economically and environmentally, this practice is inefficient, as generating units with high heat rates (i.e., low efficiency) may receive the same number of operating hours as those with low heat rates. In addition, equal shares dispatch has contributed to inefficient investment by encouraging overcapacity [13]. A related issue is that many generation services provided by natural gas units in other countries are instead provided by coal or hydropower units in China. In regions that do not have hydropower resources, coal units are used for load-following and peaking generation, requiring significant cycling of coal units and reducing the efficiency of these units.

Without a formal and transparent pricing mechanism that can link real cost and retail price, the current pricing policy poses significant barriers to transition to a more efficient power system. Wholesale generation rates in China have historically been loosely based on average costs. Since 2004, rates for thermal generators have been set using benchmark pricing, in which generators in the same technology class are given the same tariff, based on an estimate of annual output and fixed and variable costs for that class. As coal prices rose in the 2000s, China's central government developed a "comovement" mechanism that allows for some pass through of fuel cost increases. Wholesale rates for renewable generators are set using regional benchmark prices, while rates for hydropower and nuclear generators are set on a facility-by-facility basis (because of the vast initial capital investment cost of these generators). Provision of ancillary services has historically been limited in scale and scope, mandatory, and uncompensated, but plans to compensate generators for services are currently in the early stages of implementation. Because of the dominance of coal in China's electricity system, this predominant benchmark-based approach to wholesale pricing means that generation supply curves in China tend to be relatively flat.

The revenues grid companies receive for transmission and distribution (T&D) services are currently the residual between retail sales and generation costs. This residual is inherited from historical prices and is not based on a bottom-up accounting of T&D costs. Beginning in 2005, SERC developed accounting standards and reporting requirements for grid companies, but the level of detail and transparency in the disclosures required by SERC is not sufficient to assess whether costs are reasonable [3]. Moving toward cost-based T&D pricing is a continuing priority for regulators [18]. Retail electricity prices in China have historically been designed to reflect government policy and social priorities (e.g., maintaining low fuel prices to boost economic growth and affordable fuel prices for households), instead of the cost of service [10].

Developing renewable energy in power systems still faces many challenges. China has a rich endowment of renewable energy and has formulated favorable policy for renewable energy development. Given the relatively high costs for solar and biomass power, wind is and will likely continue to be the principal non-hydro renewable resource in China. Ever since the publication of the Renewable Energy Law formulated in 2005 and the Medium- and long-Term Development Planning for Renewable Energy in 2007, wind power has experienced rapid growth in China [15]. Installed wind power capacity connected to the grid increased from 1.06 GW in 2005 to 31.07 GW at the end of 2010.

Although wind power growth in China has been the most spectacular in the world in the past decade, the following factors have posed and will pose serious threats to its sustainable development. First of all, because of the intermittence of wind power generation, grid companies in China are reluctant to integrate it into the power grid. Even though the *Renewable Energy Law* stipulates the obligation of the power grid to purchase the full amount of renewable energy generated accessible to the grid, without specific operation rules, grid operators simply reject wind power access with reasons such as unpromising access conditions and operation safety requirements. The capacity factor of wind power was as low as 23.7 % in 2009, which was an evidence of the difficulties of grid integration [5]. Secondly, the incompatibility between wind power projects and power grid expansion planning has made grid access difficult. China has formulated ambitious planning for eight large-scale inland wind power bases in 10 GW scale in provinces of Xinjiang, Inner Mongolia, Hebei, Jilin and Shandong. However, without synchronous power grid expansion and proper backup plants, it is unlikely that this plan can be realized. Thirdly, the current policy to develop wind power is unsustainable because it neglects the vast investment requirement on power grids and discriminates against small projects and small investors. So it is no surprise that most wind projects are operated by the five big SOE generators and other large province-owned generators, unlike the success of wind power in Germany and the

Netherlands where various small investors enter into wind power market. Considering the long-term prospective, an optimal development of wind power should consider the solutions of both "large windmill base and long-distance transmission" and "distributed power generation," which in turn call for the entrance of small investors. Besides, there are still no concrete stipulations addressing the issues of grid cost recovery for additional transmission investments and reserve capacity when incorporating wind power into the grids.

2.2 Multi-level Perspective Analysis

2.2.1 Landscape and Regime

The MLP of [7, 16] distinguishes three interrelated levels of changes: the landscape, regime, and niche levels of socio-technical systems. This MLP literature is particularly instructive in understanding the dynamics of transition and the institution interactions during the process.

The current power regime in China for meeting industrial process, lighting, heating, and power-related services may be characterized as a centralized system. Electricity is centrally generated, largely from coal, hydropower, and a small but growing amount of nuclear and renewable sources; it is delivered to businesses and homes through large-scale T&D networks, before being used to provide power, lighting, heating, and services with the aid of end-use technologies and the buildings infrastructure. The absolute low starting point, the vast growth in power demand incurred by rapid industrialization of the Chinese Economy, and the rise of China as a World Workshop, as well as the increasing demand from households as a result of enhanced living standards and the popularization of home appliances, have made rapid growth of supply capability a policy priority in China. The strategic importance of electric power to enabling economic activity and well-being means that the system is the subject of intense policy activity, which focuses on ensuring secure and affordable supplies, and other social objectives. The meta "socio-technical landscape" of electrification in China can be characterized as follows:

- a dependency upon fossil fuel-based energy supply (mainly coal and rising supply capability of gas) and large-scale electricity generation technologies,
- a coal extraction, processing, and transportation infrastructure always unable to catch up with increasing demand,
- province-based power grids and a growing interconnected national grid to ease the geographic imbalance between supply and demand,
- historically vertical integration monopoly just being restructured into two grid operators and a set of SOE generators who are regulated by government bodies,
- a weak but expanding rail and road infrastructure,
- a traditionally centralized administration system, with initiated but rather fruitless market reform.

Key processes that influence or "drive" the power regime at the landscape level in China include the following:

- the increasing power demand from the industrial sector because of the ongoing industrialization process and a big gap in energy efficiency compared to the rest of the world,
- the increasing power demand from households because of increasing income and rapid urbanization,
- a huge population base of 1,340 million, which is still growing,
- the increasing pressure on primary energy supply, environment, and ecology because of the heavy reliance on coal,
- global concern of GHG reductions and increasing pressure on China's "measureable, reportable, and verifiable" obligation to cut GHG emissions,
- the Chinese Government's commitments to reduce emissions and reduce GDP carbon intensity by 40–45 % in 2020 compared to the 2005 level and to promote clean energy sources, increasing the share of non-fossil over primary energy by 15 % in 2020,
- concerns over security of primary energy supplies,
- external factors leading to high and/or volatile oil and gas prices and related concerns over energy affordability and fuel poverty, as well as physical disruption of external supplies (war, terrorism, foreign governments limiting supply, etc.),
- the gradual transition of Chinese Economy from centralized planning to a more market-based economy that will gradually blend into the global economic system,
- the international efforts to deliver renewable energy technologies.

Many processes that drive the power landscape in China, such as energy security concern and international factors, are the same as other countries face. However, under the macroeconomic-social transition process, the interactions of industrialization, urbanization, and energy–environment–ecology constraints will pose important impacts on the power landscape and thus influence the transition pathways. Meanwhile, the international pressure on China to assume "measureable, reportable, and verifiable" CO_2 reduction targets will intensify during China's growth process. Currently, by emphasizing that as a developing country and according to "shared but differentiated" principle, China can and will take active measures to cut CO_2 emissions so far as they do not impede the economic development. As expected in 2020, per capita GDP of China will exceed 10,000 US$, and in 2030, it is expected to exceed 15,000 US$ (2010 constant prices). At that time, with its raised share of global GHG emissions and the expected intensified international conflicts, it is very likely that Chinese Government would need to take a more positive attitude toward GHG reductions. Thus, in the future 5–10 years, the power landscape will definitely experience radical change toward decarbonization and in turn exert more pressures on the power regime to change.

2.2.2 Technical Niches

A wide range of competing energy technologies are currently being developed, reflecting not only the underlying scientific and technological base but also the perceived opportunities arising from the emerging low-carbon socio-technical regime. Shackley and Green [17] have proposed a categorization of these technologies, distinguishing between mature technologies and those that are at various stages of commercialization, demonstration, research, and development. Zhang et al. [21] also appraised the renewable energy technology progress in China and proposed their categorization. However, their categorizations mainly focus on pure technological level and neglect final consumption as a means for efficiency improvement by technology innovation in the industrial process. Based on their work and adapted to our pathway study, we propose our decarbonization categorization for power sector transition in China (Table 2). In our categorization, we classify available and potential options into five streams, namely fossil fuel, nuclear, renewable, demand side and energy carriers and storage technology, and appraise the availability of options according to typical technology innovation stages: mature (being available at hand with affordable prices in large scale); early commercialization (technically feasible but still expensive and in need of technological learning to become fully effective); development and demonstration; and finally the research stage. The time line in the table is not strict, given the uncertainty in innovation and the inaccuracy of appraisals, and is calibrated based on technology launch stage.

3 Planning of China's Power Sector into 2030

3.1 Landscape and Driving Forces

According to the objective of completion of a well-off society in 2020 indicated by President Hu Jintao's report on 18th National Congress of CPC, per capita GDP, measured in 2010 constant price and current exchange rate, will reach 10,000 US$ at that time. Then, with an annual growth rate of 4.5 %, per capita GDP will reach 15,000 US$ in 2030. According to the standard of the World Bank, at that time China will approach the lower bound of high-income group. According to socio-economic comparative study, in 2020, per capita electricity consumption will reach around 5,000 KWh, and in 2030, it will reach around 8,000 KWh.

The output structure will also exert significant impact on power demand. Currently, in the group of lower–middle income, China's output is predominantly in the secondary sector. However, during the ongoing industrialization process, the share in the secondary sector will not decrease significantly until 2020, thus posing great challenge to China's energy supply and low-carbon development.

Table 2 Niche energy technology options

Stage of (technology) option	Fossil fuel based	Nuclear	Renewables	Demand-side technologies	Energy carriers and storage technologies
Mature	USC boilers	Existing fission reactors	Wind turbines	Energy-efficient appliances	Batteries
	Combined heat and power generation (CHP)		Traditional geothermal technology	Compact fluorescent light	Pump storage
	Gas (coal bed gas) power			Industrial process innovation	
	Generation		Solar heat water system	House insulation	
			Biomass generation	Heat pump (geothermal or air)	
			Hydropower		
Early commercialization (6–15 years)	Some gasification technologies	New fission reactors	Some wind turbines	Passive solar	Fuel cells
			Biomass boilers	Low-carbon buildings	
	Some CO_2 capture technologies		Biofuels	LED lighting system	
			PV and CSP		
	CO_2 storage		Grid modification		
	Retrofitting of old boilers to USC				

(continued)

Table 2 (continued)

Stage of (technology) option	Fossil fuel based	Nuclear	Renewables	Demand-side technologies	Energy carriers and storage technologies
Development and demonstration (D&D) stage (15–20 years)	Some CCS technologies	Fourth generation reactors (high-temperature gas reactors etc.)	Wave	Intelligent buildings	Hydrogen from gas and electrolysis
	Integrated gasification combined cycle (IGCC)		Tidal	Smart meter	Fuel cells
	Underground coal gasification		Biofuels, e.g., gasification, AD, transport fuels	The Internet of things	
	Polygeneration		Grid modification		
	Fischer–Tropsch process				
Research stage (20 years and beyond)	Novel CO_2 capture technologies	Nuclear fusion	Biofuels, e.g., pyrolysis		Hydrogen generation from biomass, waste, nuclear, etc.
			Enhanced geothermal system		
			New materials for PV		

(*Source* authors' compilation using Larsson [12], Foxon et al. [6], Weiss and Bonvillian [19])

With an expected income where consumption is bound to take off, what are the most important factors that exert pressure on China's power system regime? Here, we briefly analyze four of them, namely industrialization, built environment, appliances (home and commercial), and transportation.

The leading factor is industrialization. During the past decade, the growth in consumption of energy-intensive products such as steel and cement in China has been very fast. According to Yu [20], China accounts for 70 % of the increased production in crude steel, 83 % in cement, and 118 % in steel during 2000–2008, which implies that with GHG reduction pressures in the industrialized nations, more energy-intensive products are outsourced to China. On the other hand, energy efficiency in China's manufacturing sector is significantly lower than in developed countries. Because of strong domestic and foreign demand, even though the share of heavy industry over GDP will decrease, it will nevertheless experience remarkable growth and the absolute scale will more than double in 2020 and triple in 2030 (as of 2010 level), thereby leading to vast power demand growth and posing perhaps the biggest challenge to the power system transition in China. Therefore, a well-designed industry policy to guide the development of manufacturing sector, as well as stricter energy efficiency standards to promote conservation, will be the most important strategies for China.

The second factor is urbanization. Together with the growth of income, it will exert significant impact on the power system in two ways. First, urbanization is the aggregation of population in urban areas. The urbanization rate in China increased from 36.2 to 46 % during 2000–2010 periods, equivalent to 14 million citizens moving into cities annually, and will increase to 56 % in 2020. Urbanization will be accompanied by vast housing demand, and China has set the goal of constructing new homes for 400 million people by 2017 [12]. Houses are built for the long term, on the expectation that houses will be used for 50–70 years or more. On one hand, a house is a kind of stand-alone system in that a new house could be built using the most modern and energy-efficient technologies (for example, a passive solar house) available without regard for how other houses in the surrounding area have been built. On the other hand once built, buildings will consume substantial amount of energy for lighting, ventilation, cooling, and heating purposes over its life cycle. It is reported that energy consumption per square meter in the existing 40 billion square meters of housing in China is three times that of advanced country level, while 90 % of the newly built houses are energy inefficient. Hence, to cope with the impact of urbanization and reduce power demand in buildings, it makes sense to immediately implement as many available energy-efficient house technologies as possible in the new construction projects. On the other hand, refurbishment of existing houses, by improved insulation, new heating systems such as heat pumps to replace central air-conditioning, can also improve energy efficiency remarkably.

Urbanization will also result in the expansion of cities, in turn increasing demand on regional transportation. The relationship between transportation and power sector is complicated. On the one hand, all kinds of biofuels could also be used for power generation. Considering the gigantic oil demand in the future, the inadequate domestic oil supply and the already high dependence on oil import (50 % in 2009), biofuels should take priority over bio-generation. On the other hand, the next generation vehicles, be it hydrogen/fuel cells or battery driven, will have significant impact on the power sector. Hybrid Electric Vehicle technology may be the only one mature enough in the next 10 years to contribute to energy transition in China [1]. Pending significant improvements in battery technology, plug-in hybrids could possibly start making an impact in about 10 years, while vehicles powered by fuel cells are unlikely to enter high-volume production for at least 20 years. Whatever technology evolves, power demand will be significantly pushed up by direct consumption, by batteries or indirect consumption by hydrogen/fuel production. Considering the vast growth potential of private car demand, transportation alone will vastly increase power demand in China.

With growing per capita GDP and more disposable income, electric appliances will be popularized in households. Comparing the difference of appliance inventories in rural and urban families [14], there is vast potential for the demand of appliances such as refrigerators, washing machines, and computers, which in turn will consume more electricity. On the other hand, with the expected rapid growth of the service sector in the coming two decades, the popularization of office automation and more large-scale power-consuming data centers in operation, there will be growth of power consumption in the service sector.

3.2 *Planning Scenario*

With demand growth at two digits and the intensified pressure for carbon reduction, it is unwise to simply build more thermal power plants to meet the power demand. More diversified power sources must be included in the generation mix. Similarly, it is unwise to simply increase supply without considering conservation potential on the demand side. This is especially important for China with notoriously low energy efficiency. Hu et al. [9] proposed an integrated resource strategic planning (IRSP) model for power planning, and its core is a minimized power supply planning model considering both traditional power plants (thermal power, hydropower, nuclear power, and renewables) on the supply side and efficient power plants (EPPs, including energy-saving lightings, high-efficiency motor systems, high-efficiency appliances, and others) with environment constraint. Based on the model presented in Hu et al. [9], generation capacity planning for China during 2010–2030 periods is projected in Table 3. Considering the potential of EPPs, in 2020, 142 GW of generation capacity could be avoided and save 285 TWh electricity on the demand side in 2020; in 2030, 277 GW could be avoided and save electricity by 783 TWh. Meanwhile, with a diversified generation mix, in 2020, clean capacity would amount to 593 GW, accounting 35.5 % of total generation capacity and clean generation would amount to 1,948 TWh, accounting for 25.38 % of total generation; in 2030, clean capacity would amount to 1,091 GW, accounting for 46.78 % of generation capacity, and clean generation

Table 3 Scenario planning of power generation capacity for China 2010–2030

(GW)	Coal	Hydro	Nuclear	Wind and solar	Gas	Total	Efficiency power plants
2010	671	200	10	40	12	933	**15**
2015	857	260	23	100	12	1,251	47
2020	1038	335	78	165	44	1,660	142
2025	1141	450	170	280	50	2,091	204
2030	1180	500	230	361	61	2,333	277

would amount to 3,887 TWh, 37.19 % of total electricity generation. The significant increase in clean generation will vastly avoid potential CO_2 emissions from China's power sector. Supposing that the carbon emission factor and clean generation share were constant at 2007 level, 640 and 2,114 million tons more CO_2 would be emitted in 2020 and 2030.

The preceding analysis indicates that with radical improvement in generation mix on the supply side and energy efficiency on the demand side, China's power sector could make obvious progress on decarbonization. Unfortunately, a simple though not precise calculation indicates that these efforts are not enough. Suppose that the efficiency improvement plus the installation of gas power cut the CO_2 emission factor of fossil plants to 803 g per KWh (10 % lower than 2007 level), multiplying by fossil generation share of 63 % (and supposing no emission at all for clean generation), in 2030, CO_2 emission per KWh from China's power sector would be 504 g. Then, multiplying by the expected consumption, power sector alone will emit 5,600 million tons CO_2 or 93 % of China's total emissions in 2007. Including other sectors, it would be very likely to double the 2007 emissions to 11,000–12,000 million tons, which is far beyond the IEA 450 ppm scenario for China [11] and would ruin the global effort to stabilize GHG emissions.

4 Developing Super Smart Grid in China

4.1 Key Features of China's Power Scenario

To fuel socio-economic development properly, while at the same time manage the GHG emissions at controllable scale, we need to understand the key features of China's power scenario and build SG to deliver it.

Start with power demand first. From regional perspective, the demand from developed regions such as Yangtze River Delta, Pearl River Delta, and Bohai Economic Rim will flat and saturate around 2020–2025 after a swift growth from 2010 to 2020. On the other hand, the demand from underdeveloped western and central regions will continue to increase during rapid industrialization and urbanization process. The implication is that the historical imbalance between load and resource will exacerbate until 2020. Due to the bottleneck in the transportation system, ultra-high voltage and long-distance transmission system will continue to be a priority before 2020. However, with the takeoff of economic growth in western and central regions, the local demand from these regions will explode and consume most of the local supply. Besides, with stricter environmental and ecological standard, there will be more constraints on the construction of transmission corridor. All these pressures call for more active regionalized balance of demand and supply.

From the perspective of final use, demand from the manufacturing sector is supposed to flat and saturate at around 2020 when China finishes the industrialization process. Meanwhile, demand from service, residential, and transportation will take off when tertiary sector dominates the GDP, and higher disposable personal income results in the burst of domestic consumption. One implication is that at around 2020 when demand from service, residential, and transportation sector begin to challenge the role of manufacturing sector in total consumption, the demand will become more irregular, unpredictable, and hard to be managed. The other implication is that the pattern of the new demand from service, household, and transportation sector depends on how the city, buildings, and transportation system are designed and managed during the urbanization process. The key decision making of city planning, building standard, and transportation system should incorporate energy, environment, and ecology objectives in the priority agenda and be the results of integrated planning. Market mechanism, energy efficiency standard for new buildings and electric appliances, energy management service, and technology innovations would play major roles for smart management of the demand. Retrofitting of existing buildings can contribute significant energy-saving and GHG abatement. All kinds of demand-side measures can serve as low-cost energy storage options when incorporating intermittent renewable generation in large scale.

Now we turn to supply side. For thermal power, enhancement of energy efficiency by building more new USC and/or CHP units can save coal and avoid GHG emissions. With available technology, retrofitting existing coal-burning units into USC can make another contribution. Because coal-based units will still account for half of the generation capacity in 2030, to control total GHG emissions, carbon capture and storage (CCS) technology must be put into commercial application beginning from new coal units above 300–400 MW at around 2015 after pilot projects. With growing and secured supply from both regular and shale gas, gas generation will enter into the generation mix in China as reserve units and improve the capability of the power grid to integrate more portion of renewable generation. Exploitation of hydropower in China will peak at around 2020 when technically and economically exploitable resource is used up. Nuclear power will peak at around 2020–2025 at scale of 80–120 GW installation. But next generation reactor, in cogeneration of power and hydrogen, will open a new window for reshaping power system. Large-scale integration of inland wind power and solar power will peak at around 2020 when the wind/solar power bases at scale of 1/10 GW are completed and commissioned. Then, the pressure on the construction of more transmission lines and the requirement of regionalized balance of demand and supply will call for the integration wind/solar and other forms of renewables at small scale in the form of distributed generation or microgrid, which in turn calls for a radical change in the pattern and management of power system.

4.2 Function Analysis on SSG in China

A strong power grid infrastructure is needed to manage the challenges/opportunities of the scenario and deliver the transition. According to the preceding analysis, in China, the power grid should be able to assume the following functions in the future:

- Optimize primary energy supply structure and facilitate clean generation access, which in turn requires large-scale semi-reliable and intermittent power generation units such as wind and solar power and thus poses big challenge to power system planning, operation, and control. On the other side, to promote clean generation as much as possible, generation technologies in different types and sizes, connected into distribution networks plus self-generators injecting into, or taking from the grid from time to time, should be installed to the grids (especially the distribution grids), which in turn calls for reconfiguration of distribution systems.
- Serve as an alternative system besides railway, waterway, and road systems for large-scale energy transportation, which in turn calls for renovation of the current 500 kV backbone grid with less than 800 km transmission radius to 750 kV and higher with a longer transmission radius, functioning as long-distance and large capacity resources optimization framework.
- Promote energy conservation on power demand side, as well as on supply side, which in turn calls for more consumer choices, customized power services and smart energy management.
- Control GHG emissions effectively within the power sector and leave more space for other sectors, which in turn calls for the construction of large coal power bases nearby collieries to reduce energy consumption for coal transportation and to provide the base for large-scale CCS applications, innovative dispatch rules to support carbon-free or low-carbon generation, and exploration of the demand-side energy conservation to the utmost extent.

Super smart grid (SSG) has been studied to integrate SG and super grid for energy conservation and renewable energy utilization in many countries. However, "SGs" do not enjoy a commonly accepted definition in the literature. Even its architecture and basic elements are of universal sense, its functions must be tailored to accommodate the differences in resource and demand in different countries.

Here, we would only discuss the main features of SG that could satisfy the above functions. We define super SG as the power grid system that will economically accept and transmit renewable power on the supply side and operate energy efficiency improvement on the demand side. On the transmission side, SSG can take advantage of renewable and clean power generation over a long distance and integrate huge amounts of power generation into the grid, and on the distribution side, SSG can stimulate customers to save electricity and sell their renewable power generation to the grid (Table 4).

Table 4 Key components of developing SSG in China

Aspects	Transition direction	Key SSG components
Generation	• Large-scale access-in of intermittent renewable generation in northwest resource center	• Wind power grid-access technology
	• Development of large-scale thermal and hydropower bases in western resource center	• Solar power generation and grid-access technology
	• Development of nuclear power plant in east coastal areas	• Grid-access technology for novel renewable generation
		• Cogeneration of power, heat, and hydrogen
Transmission	• Long-distance large capacity energy transmission	• Advanced power transmission technology (Superconductors, FACTS, HVDC)
	• Resilience to disturbance, attack and natural disasters	• Wide-area monitoring and control technology
		• Information and communication integration technology
Distribution	• Integration of different generation and storage technologies (small wind turbine, roof solar, biogas, fuel/cell among others) and in different sizes	• Distributed generation and management technology
		• mini-grid technology
		• Step-down transformer technology
		• Advanced voltage control technology
Utilization	• Diversified power supply options for customers (based on both cost and product differentiation offered)	• Power consumption information feedback technology
	• More active involvement opportunity for customers as power operators and small generators	• Integrated design of green building and energy system
	• Customized services	• Energy efficiency and demand-side response management technologies
		• Electric vehicle charging load management technology
		• Bilateral energy management and smart billing system
		• Smart homes technology integration

5 Road Map for Developing SSG in China

5.1 Transition Pathways of Low-Carbon Power Sector in China

Geels and Schot [8] proposed a typology of transition pathways as, reproduction: ongoing processes of change within the socio-technical regime (i.e., not involving interaction with the landscape or a technological niche); transformation: processes of change that arise from the interaction of an evolving landscape with the socio-technical regime (but not with the technological niche level); substitution: replacement of one dominant technology within the socio-technical regime by another as a consequence of interaction between all three levels; de-alignment/realignment: interaction between the three levels resulting in competition between a dominant technology within the regime and a number of other competing options, which have different performance characteristics, eventually resolved through emergence of a new dominant option; and re-configuration: replacement of a set of interlocking technologies by an alternative array of inter-related technologies which fulfill the same or similar functions.

According to analysis on technology niches, we propose the transition pathways for low-carbon power system development in China (Table 5). In the table, we classify available and potential options into five streams, namely fossil fuel, nuclear, renewable, demand side and energy carriers and storage, and appraise the availability of options according to typical technology innovation stages: mature, being available at hand with affordable prices in large scale; early commercialization, though technically feasible but still being expensive and in need of a technological learning process to become technically effective; development and demonstration; and the research stage. The time line in the table is not strict, given the uncertainty in innovation and the inaccuracy of appraisals, and is calibrated based on technology launch stage.

5.2 Milestones and Key Tasks for Building SSG in China

Based on the overall transition pathways of China's power sector and the requirements on SSG, we propose the milestones for SG development developing SSG in China (Table 6). Again, even the timeline is not strict, the road map is divided into three periods: 2010–2015, with the emphasis on transmission system and large-scale integration of renewable generation; 2015–2020, with the emphasis on integration of middle-scale renewable generation and energy storage; and 2020–2030 and beyond, with the emphasis on integral of smart transmission, DG, and microgrid.

Table 5 Possible transition pathways of power system in China

Stage pathway	First stage: 2010–2020	Second stage: 2020–2030 (and beyond)
Reproduction: efficiency enhancement in coal-based large-scale power system	Regular efficiency enhancement; CHP; substitution of small-scale inefficient unit with UVC unit; power grid operation optimization	Same as first stage
Transformation: minor modifications to coal-based large-scale power system	Demand-side options as technology innovation in industrial process, popularization of energy-efficient appliances, house refurbishment such as insulation and double glazing; isolated distributed generation (DG) in remote areas	Low-carbon building
Substitution: technology substitution or power demand substitution within centralized power system	Gas power, waste (biogas) power, and/or gas + CHP; fission nuclear power; inland and onshore wind power; solar power in demonstration and commercialization; heat pump; passive solar	Gas + CHP power; GIF nuclear; offshore wind power; advanced solar power; fuel cells; Micro-CHP
Reconfiguration: major modification to coal-based large-scale power system	CCS research and experiment and demonstration; Fuel cells research and experiment and demonstration; Centralized power grid + solar/wind/hydro/biogas and other combined DG	Regional power grid + ultra voltage transmission connection + expansion of kinds of DG technologies including fuel cells
De-alignment/realignment: power system coevolves with hydrogen/fuel cell technologies	Research and development of hydrogen technology	Macrogird with large-scale UVC + CCS units and other traditional carbon-free units or large-scale fuel cell plants based on underground coal gasification + hydrogen production as carriers + DG and microgrid with fuel cell, wind, solar, and other renewable units

Table 6 Milestones and key tasks for developing SSG in China

Period	Milestones	Key tasks
2010–2015	• **Developing super transmission system** • Optimizing of generation mix • Integrating large-scale renewable generation (GW scale) • Cogenerating of power and heat and other low-cost energy storage form	• Optimizing generation mix with traditional technologies • Building smart transmission system • Integrating large-scale renewable generation • Piloting distributed generation • Piloting smart management of electric vehicle • Testing superconductor transmission system • R&D of next generation energy storage technology • Testing the feasibility of smart customer management and related technology R&D
2015–2020	• **Developing super transmission + distributed generation system** • Integrating medium-scale renewable generation (10 MW scale) and energy storage	• Piloting superconductor transmission system • Optimizing power generation with CCS units piloting and testing next generation energy storage technology • Building DG • Piloting smart customer management • Piloting microgrid operation and management technology
2020–2030 and beyond	• **Developing super transmission + distributed generation + microgrid system** • Integrating small-scale renewable generation (1 MW scale) • Preparing for integration of cogeneration of power, heat, and hydrogen	• Building smart superconductor transmission system • Optimizing DG operation and management • Building microgrid and all kinds of smart customer management technologies • Optimizing and reshaping the architecture of the power grid • Building infrastructure and architecture for the cogeneration of power, heat, and hydrogen • Preparing for the integration of smart grid and internet of things

6 Policy Implications

Transition to a smart and low-carbon power system represents multiple institutional challenges. For one thing, renewable energy in different forms is distributed and intermittent in nature. For another thing, without technically feasible energy storage options, power system must be balanced in real time. Therefore, historically, operators of the "dumb grid" were primarily concerned with stable and reliable operation of the system. The grid was operated using a combination of command-and-control actions and hierarchical decision making. Perhaps most importantly, the preferences of customers were largely left out of the planning and operations process in traditional power grids.

Integration of medium- and small-scale renewable generation will face huge challenge. As discussed above, SG based on renewable generation and energy storage technologies radically differs with traditional one. In China, renewable generation as wind and solar power is integrated into the system by building large-scale bases and then constructing transmission system to deliver it. In this way, wind farms are taken as bulk generation of coal-based units. However, the integration of medium- and small-scale renewables into the grid means that the renewables will be connected in the distribution system or directly built on customers' houses/buildings. So first of all, level playground for small and private investors is needed to attract millions of prospective individuals into the power sector. Then, regarding power grid as public-accessible public goods and ensuring fair grid access is another priority to overcome this barrier.

Pricing mechanism is the biggest challenge for smart management of customers and innovations in demand-side technologies. Though currently smart meter and the related information and management technologies are the focus of R&D and pilot projects, price reform is ultimately the final answer to it. For one thing, retail tariff must reflect the cost of energy service in consideration of the time served and quality requirement. For another thing, currently, there is only wholesale generation price, and final consumption tariff based on customer classification and grid companies' revenue is the difference between them multiplying by power supply to different customers. Under this situation, grid companies have no incentive to deliver demand-side management. Also, electricity price policy has historically assumed too much economic growth and social stability function in China, instead of economic function to signaling for the scarcity of resource. So to reflect the cost clearly, there must be price catalog of wholesale generation, distributed generation, and microgeneration and the separation of generation, transmission, distribution, and retail prices, which in turn calls for a new perspective on the regulation of T&D network. Understanding that there is deep-rooted origin of the current tariff policy, pricing reform can be regarded as a touchstone to see whether Chinese Government is determined to the *Scientific Development Perspective* and power sector deregulation.

The integration of medium- and small-scale renewable generation as well as pricing mechanism reform also calls for the gradual redefinition of the power grids and restructuring of the utility companies. It would pose the most difficult institutional puzzle for developing SG in China. Currently, SGCC, the biggest national grid company, is the first mover of SG. But the focus of SGCC is on ultra-high voltage and long-distant transmission system, which can serve as a physical infrastructure to maintain and strengthen its vertical monopoly power. However, the worldwide trend requires splitting the grid company into transmission operator, distribution operator, and retailers, among which T&D operator will provide public goods to generators, retailers, and customers. Without clear-cut separation of public goods and private goods, it is unlikely that SG can function as expected. The puzzle is that the grid company would have no incentive to develop SG if it was destined to be dismantled. Therefore, a subtle design to synchronize the SG road map with power sector deregulation is necessary.

Finally, there is another challenge to draft a clear national strategy and road map for developing SSG when trying to put all the elaborated components together and into operation. SG is perhaps the most complicated manmade system and involves at least electric equipment manufacturing industry, power industry, information and communication industry, and other novel industries which provide component solutions. Some components, such as smart substation, FACTS, HVDC, wide-area monitoring and control technology, are mature and available. Others, such as smart meters, DG, microgrid, next generation communication and control system, Internet of Things, are still in R&D or early pilot stage. Therefore, drafting a clear national strategy and road map based on comprehensive appraisals of numerous component technologies and their launch pathways poses another puzzle for developing SSG in China. An accompanied issue is how to integrate the international endeavors with domestic efforts.

7 Conclusion

This chapter adopts the framework of socio-technical systems transition to discuss the development of SG in China. As a starting point, we present the brief history and key features, landscape and regime of power sector in China for understanding the institution background. A low-carbon scenario for power sector into 2030 is compiled, and its salient features are discussed to understand the necessity and the requirements on SG. We then address the main functions of SSG in China and draft the transition pathways to a low-carbon power sector. A road map for developing SSG in China is proposed based on the overall transition pathways of the power sector. Institutional barriers on the road map are outlined and policy implications discussed. Based on the study in this chapter, we have three main findings.

Clear national strategy and road map is needed in China. To fill the institutional gaps, the government body (currently National Energy Administration) should be empowered to be responsible for national strategy formulation, a roadmapping think-tank should be formed to develop and update the road map, combining world-wide industrial, government, and academic expertise to assess technologies and identify obstacles and recommend appropriate policies, and an industry-government consortia should be formed to provide integrated translation research and strong front-end support.

Restructuring of the power sector should be synchronized with the development of SG. Dismantle the grid companies into regional transmission companies and provincial distribution companies to provide public goods to the private users. Set level playground in both technology and investor sense, encourage entry, and open the power sector to private investors. A national dispatch center is needed to take charge of national power system planning, regional interconnection, and the development of super smart transmission system. Power market could be established on both national and regional level.

Pricing mechanism reform should take the precedence of SG, in order to provide proper signal to power customers. A parallel institutional arrangement is to empower the costumers with rights to choose their own power retailers or energy management suppliers or to engage themselves as microgenerators.

Acknowledgment This chapter is an updated republication of the paper "Delivering power system transition in China" by Jiahai Yuan, Yan Xu, and Zhaoguang Hu, Energy Policy 50, 751–772. The copyright permission for reusing the paper has been granted by Elsevier.

References

1. Anderman M (2007) Status and prospects of battery technology for hybrid electric vehicles, including plug-in hybrid electric vehicles. Briefing to the U.S. senate committee on energy and natural resources, Jan 26. www.advancedautobat.com/press-articles/Anderman-Senate-Energy.pdf
2. APEC Energy Database (2011) http://www.ieej.or.jp/egeda/
3. Cai H (2010) An analysis of problems with the regulation of T&D costs in China (in Chinese). Mod Bus Trade Ind 4:171–172
4. Cherni JA, Kentish J (2007) Renewable energy policy and electricity market reforms in China. Energy Policy 35(7):3616–3629
5. China Electricity Council (CEC) (2010) China electricity annual development report. China Electric Power Press, Beijing
6. Foxon TJ, Hammond GP, Pearson PJG (2010) Developing transition pathways for a low carbon electricity system in the UK. Technol Forecast Soc Chang 77(8):1203–1213
7. Geels F (2002) Technological transitions as evolutionary reconfiguration processes: a multi-level perspective and a case-study. Res Policy 31(8–9):1257–1274
8. Geels FW, Schot J (2007) Typology of sociotechnical transition pathways. Res Policy 36(3):399–417
9. Hu Z et al (2010) Integrated resource strategic planning in China. Energy Policy 38(8):4635–4642
10. IEA (2006) China's power sector reforms: where to next?. OECD/IEA, Paris
11. IEA (2009) How the energy sector can deliver on a climate agreement in Copenhagen. www.iea.org/weo/docs/weo2009/climate_change_excerpt.pdf
12. Larsson M (2009) Global energy transformation. Palgrave Macmillan press, New York
13. Liu B, Chen D (2010) The scientific development of China's electricity system will rely on institutions. State electricity regulatory commission discussion Paper
14. National Bureau of Statistics of China (NBSC) (2011) China statistical yearbook 2010. China Statistics Press, Beijing
15. National Development and Reform Commission (NDRC) (2007) (b) Medium and long term plan for renewable energy development
16. Rip A, Kemp R (1998) Technological change. In: Rayner S, Malone EL (eds) Human choices and climate change, vol 2. Battelle Press, Columbus, Ohio

17. Shackley S, Green K (2007) A conceptual framework for exploring transitions to decarbonised energy systems in the United Kingdom. Energy 32(3):221–236
18. State Electricity Regulatory Commission (SERC) (2010) Electricity regulation annual report 2009
19. Weiss C, Bonvillian WB (2009) Structuring an energy technology revolution. The MIT Press, Cambridge, Massachusetts
20. Yu C, (2011) Recent growth of China's energy-intensive industries. Presentation on EC2 Seminar
21. Zhang X et al (2010) A study of the role played by renewable energies in China's sustainable energy supply. Energy 35(11):4392–4399
22. Zhou X et al (2011) An overview of power transmission systems in China. Energy 35(11):4302–4312

Exploring the Value of Distributed Energy for Australia

William Lilley, Luke Reedman and Anthony Szatow

Abstract This chapter examines the important role that distributed energy (DE) can play in smart grids in Australia to achieve a low carbon future. It discusses the findings of a recent major government-funded study which showed that uptake of DE could result in potential economic savings of around $130 billion by 2050. It also provides updates of more recent modelling that takes into account sensitivities in the price of natural gas and the price of solar photovoltaic (PV). This chapter further examines the key issues, opportunities and challenges of realising the value of DE in Australia. The analysis highlights the importance of integrating DE with the electricity grid and urban environment and discusses enablers and barriers for large-scale uptake of DE.

Keywords DE · Economic modelling · Carbon price · Electricity markets

1 Introduction

In a traditional electricity network such as the one shown in Fig. 1, electricity is produced by large centralised plant located remote to the user. These plants typically convert energy contained in a fuel, e.g. coal, gas or nuclear material into electricity, which is transformed to high voltage for efficient long-distance transport. When the electricity nears major load centres, e.g. a town, it enters the more widely spread distribution network for transport to numerous end-users. Here, it is brought to lower voltage levels, potentially a number of times, before reaching the final consumer. On average, in Australia, around 6.5 % of the electricity produced is lost as it is transferred from the point it is generated to the point it is used [16].

W. Lilley (✉) · L. Reedman · A. Szatow
Commonwealth Scientific and Industrial Research Organisation (CSIRO), Canberra, ACT, Australia
e-mail: william.lilley@aramco.com

D. Mah et al. (eds.), *Smart Grid Applications and Developments*, Green Energy and Technology, DOI: 10.1007/978-1-4471-6281-0_15,

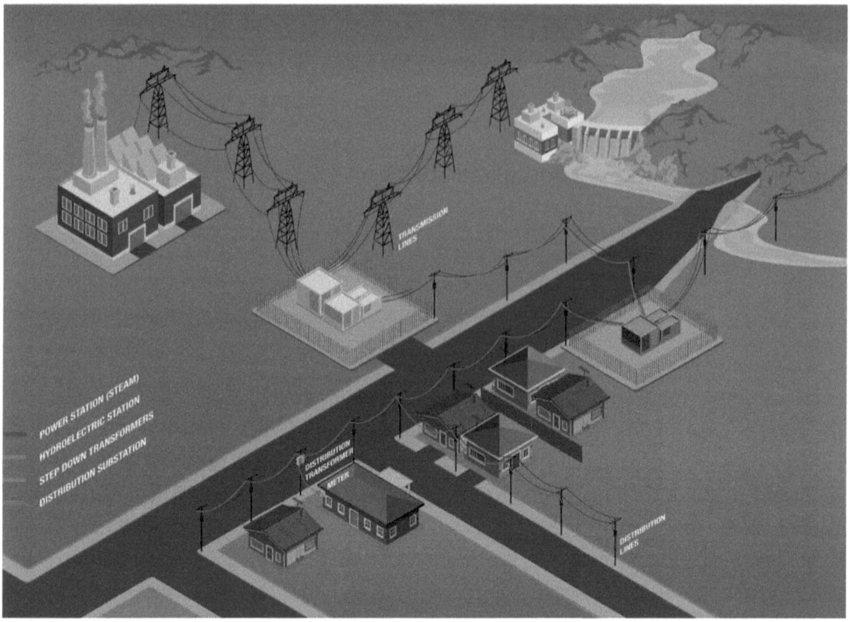

Fig. 1 A simplified view of electricity generation and transfer [10]

Depending on the type of plant, and type of fuel used, up to two-thirds of the chemical energy contained in the fuel can be lost as it is converted to electricity. In some cases, plant such as combined cycle gas turbines are installed to minimise this heat loss. In other instances, economic decisions result in the uptake of less efficient plant such as open-cycle gas turbines due to their lower capital and operation costs. By using distributed energy (DE) technologies, some of these inefficiencies can be reduced through minimising transmission and distribution losses and by utilising waste heat in secondary processes such as the delivery of hot water and steam.

In Australia, electricity is supplied to consumers primarily from the combustion of black coal, brown coal and natural gas (Fig. 2). Renewable sources are primarily from large hydroelectricity plant with a small but growing contribution from wind and rooftop solar photovoltaic (PV), driven through state-based feed-in-tariffs and government policies such as the large- and small-scale renewable energy target (RET) that ensures that 41,000 GWh of large-scale renewables enters the market in 2020. No yearly target is set for the small-scale generation [7, 8]. Over the period 2002–2003 to 2011–2012, electricity generation from solar energy grew at an average rate of 46 % a year, from 60 GWh in 2002–2003 to 1,400 GWh in 2011–2012 [6]. Electricity in the eastern states of Australia is supplied by the National Electricity Market (NEM). The NEM is the longest alternating current system in the world at over 5,000 km from far north Queensland to Tasmania, and west to Adelaide and Port Augusta and is long and linear in

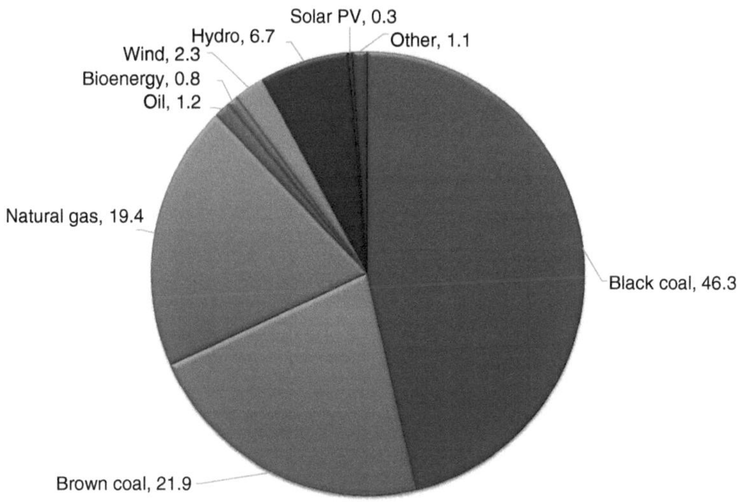

Fig. 2 Australia's electricity generation (%), by energy source [5]

comparison with Europe and North America [2]. In total, it contains around 785,000 km of overhead transmission and distribution lines and around 124,000 km of underground cables [6], which supply about 200 TWh of energy to businesses and households each year.

Rising electricity prices in many developed countries has been driven by expenditure on distribution networks to meet fast-growing demand, particularly in the domestic sector. Much of this demand growth has been driven by the installation of large consumer devices such as air conditioners (AC) as rising affluence leads to an increase in living standards. These power hungry devices can lead to high levels of demand at certain times of the year, in the case of AC on very hot days. The network must be designed to meet the very highest levels of demand. In countries such as Australia, this peak demand occurs on only a small number of days per year. Reducing or shifting peak demand can result in significant savings as smaller, less costly components can be used to deliver the overall needs in electricity. Installation of DE is one way to achieve this outcome. Figure 3 illustrates the components of residential electricity costs in the NEM states for 2012–2013 [6]. The figure clearly shows that network costs are the major component of the residential price followed by wholesale costs from generation. Retail margins and the impost of "green" and "carbon" costs make up the remainder.

While conceptually simple, the adoption of DE technologies can pose some technical issues particularly within distribution networks. Traditionally, electrical networks were designed to transfer electricity in essentially one direction, from the point of generation to the point of use. Protection equipment and operational regimes were built around this paradigm, and installing distributed generation (DG) equipment can lead to excess electricity being supplied into the network.

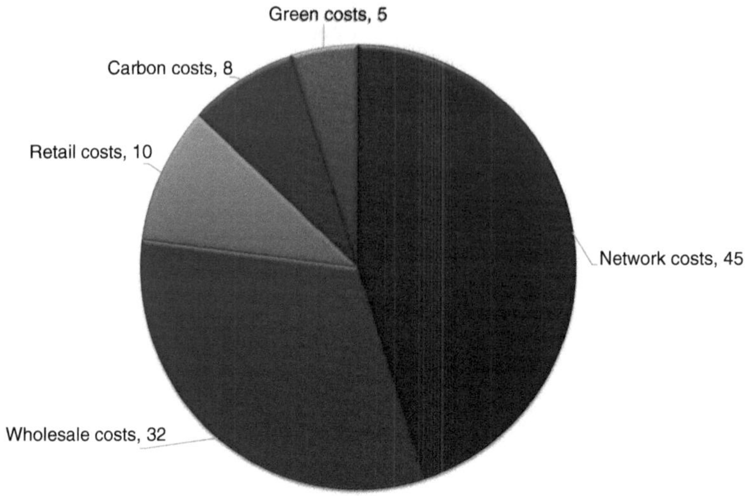

Fig. 3 Residential electricity cost components (%) in Australia's NEM 2012–2013 [6]

These new two-way flows need to be measurable and controllable to ensure that issues around safety and performance are not unduly affected by their introduction.

In many countries, there has been a growing trend towards the adoption of smart grid systems that can help among other things to minimise the impact of DG and reduce system losses through advanced monitoring and control. To date, these activities have been primarily government funded while the cost-effectiveness of various activities is assessed for potential wider-scale adoption. In Australia, the largest smart grid programme is a commercial-scale rollout in the city of Newcastle. The Australian Government provided a $100 million grant for the "Smart Grid Smart City" programme to assess the costs and benefits of adopting smart grid technologies and activities within the electrical network and within the property of consumers. The programme is trialling both grid-side (network) and customer-side activities that aim to deliver efficient energy services [4].

Other Australian network companies are also undertaking smart grid- and DE-related activities based on a combination of government incentives such as the demand management incentive scheme [3], federal and state government programmes such as Solar Cities [12], National Broadband Network [11], Victorian Advanced Metering Infrastructure programme [13], Victorian Electric Vehicle trials [14] as well as commercial investments by the companies themselves such as SPAusNet's [21] distribution feeder automation (DFA) programme to improve its service standards, or Energex's [15] tariff trial, cool change and energy conservation community activities that provides incentives for reducing peak load demands from pool pumps, AC and water heaters.

In this chapter, we examine the economic outcomes from modelling the potential impact of DE. Section 2 presents the methodology of the economic modelling. Section 3 presents results and discussion of the modelling that considers earlier analysis [10] and more recent modelling taking into account a change in key drivers including the price of natural gas and solar PV. Section 4 discusses barriers and enablers for wide-scale adoption and Sect. 5 presents conclusions on the potential role of DE in Australia.

2 Modelling Approach

This chapter considers the potential economic benefits that can be realised through the use of DE technologies in helping to reduce Australia's greenhouse gas emissions. This is addressed by considering the long-term uptake of technologies through partial equilibrium modelling of the stationary energy and transport sectors of the Australian economy. The model was used to examine the potential energy mix for Australia's electricity sector under the imposition of a price on carbon. This analytical framework was chosen because it is relatively less resource intensive than general equilibrium modelling of the economy and because it offers the best opportunity to study the detailed technological implications of alternative scenarios.

The partial equilibrium model employed is called the energy sector model (ESM). ESM is an Australian ESM co-developed by the CSIRO and the Australian Bureau of Agricultural and Resource Economics (ABARE) in 2006. Since that time, CSIRO has significantly modified and expanded the ESM. The ESM is a partial equilibrium (bottom–up) model of the electricity and transport sectors. It has a detailed representation of the electricity generation sector with substantial coverage of DG technologies. The transport module considers the cost of alternative fuels and vehicles as well as detailed fuel and vehicle technical performance characterisation such as fuel efficiencies and emission factors by transport mode, vehicle type, engine type and age. Competition for resources between the two sectors and relative costs of abatement are resolved simultaneously within the model.

ESM is solved as a linear programme where the objective function is to maximise welfare, defined as the discounted sum of consumer and producer surplus over time. See Graham and Williams [17] for an example of the equations required to construct a similar but non-linear partial equilibrium model.

The main components of ESM include:

- Coverage of all States and the Northern Territory (Australian Capital Territory is modelled as part of New South Wales)
- Trade in electricity between National Electricity Market States
- Sixteen (16) DG electricity plant types: internal combustion diesel; internal combustion gas; gas turbine; gas micro-turbine; gas-combined heat and power (CHP); gas micro-turbine CHP; gas micro-turbine with combined cooling, heat and power (CCHP); gas-reciprocating engine CCHP; gas-reciprocating engine CHP; solar PV; biomass CHP; biomass steam; biogas-reciprocating engine; wind; natural gas fuel cell and hydrogen fuel cell

- Centralised generation (CG) electricity plant types: black coal-pulverised fuel; black coal-integrated gasification combined cycle (IGCC); black coal with partial CO_2 capture and sequestration (CCS) (50 % capture rate); black coal with full CCS (90 % capture rate); brown coal-pulverised fuel; brown coal IGCC; brown coal with partial CCS (50 % capture rate); brown coal with full CCS (90 % capture rate); natural gas-combined cycle; natural gas-peaking plant; natural gas with full CCS (90 % capture rate); biomass; hydro; wind; solar thermal; hot-fractured rocks (geothermal) and nuclear
- Four electricity end use sectors: industrial; commercial and services; rural and residential
- Nine road transport modes: light, medium and heavy passenger cars; light, medium and heavy commercial vehicles; rigid trucks; articulated trucks and buses
- Thirteen road transport fuels: petrol; diesel; liquefied petroleum gas (LPG); natural gas [compressed (CNG) or liquefied (LNG)]; petrol with 10-per cent ethanol blend; diesel with 20-per cent biodiesel blend; ethanol and biodiesel at high concentrations; gas-to-liquids diesel; coal-to-liquids diesel with upstream CO_2 capture; hydrogen (from renewables) and electricity
- Four engine types: internal combustion; hybrid electric/internal combustion; hybrid plug-in electric/internal combustion and fully electric
- All vehicles and centralised electricity generation plants are assigned a vintage based on when they were first purchased or installed in annual increments
- Rail, air and shipping sectors are governed by much less detailed fuel substitution possibilities
- Time is represented in annual frequency (2010, 2011, …, 2050).

All technologies are assessed on the basis of their relative costs subject to constraints such as the turnover of capital stock, existing or new policies such as subsidies and taxes. The model aims to mirror real-world investment decisions by simultaneously taking into account:

- The requirement that the plant is profitable over the term of its investment
- That the actions of one investor or user affect the financial viability of all other investors or users simultaneously and dynamically
- That the consumers react to price signals
- That the consumption of energy resources by one user affects the price and availability of that resource for other users, and the overall cost of energy and transport services, and
- Energy and transport market policies and regulations.

In July 2012, the Australian government began pricing carbon emissions. Under regulation, at the time of writing, the price is fixed each year for the first three years and requires the country's largest polluters to pay $23 per tonne until 2015–2016. From then, the price will be set by the market [8]. The modelling performed here assumes that the price of carbon follows the trend displayed in Fig. 4. Two potential carbon price paths are shown. The "Low" case represents a

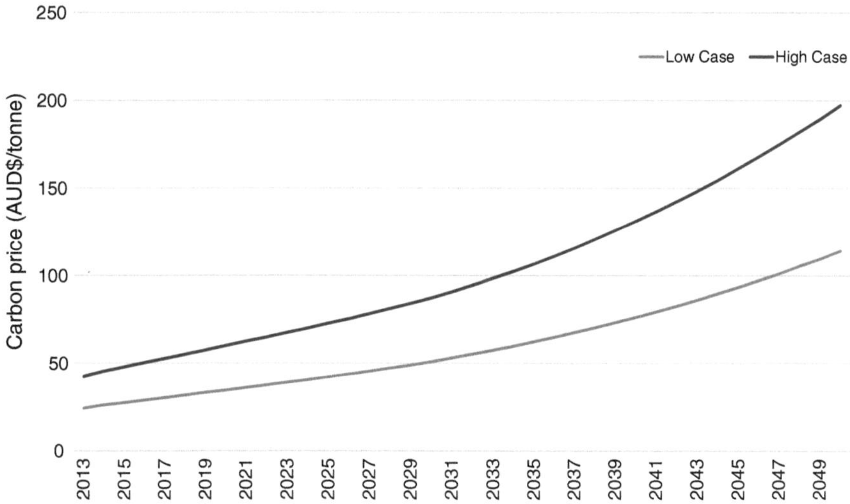

Fig. 4 Carbon price path assumed in the model

trajectory that sees Australia reducing its emissions from a base level in 2010 by 10 % in 2020 and 80 % in 2050, leading to an atmospheric concentration of 550 ppm CO_2e. The "High" case represents a trajectory that sees Australia reducing its emissions from a base level in 2010 by 25 % in 2020 and 90 % in 2050, leading to an atmospheric concentration of 450 ppm CO_2e. These nominal prices are provided as exogenous data to the model.

In the modelling shown here, three sets of projections are considered. The first is a base case that provides outputs from the earlier modelling by CSIRO [10] in the formulation of its "Intelligent Grid" (IG) report. The second case alters the assumptions of the base case for solar PV, and the third case reflects changes in the assumption of natural gas prices. The revised solar PV case reflects more recent cost and performance data collected after the earlier modelling. Figure 5 provides a graphical representation of the difference in price assumed between the base case and the revised prices used here. It is worth noting that while the price varies substantially before 2030, it remains essentially that same after that time as the technology reaches a high level of maturity and the price is governed primarily by its raw inputs rather than changes in performance through innovation.

The third case reflects differences in the assumed price of natural gas. Figure 6 shows the difference in price for the gas price assumed in the initial modelling and the price assumed in the revision presented here. The differences are illustrated by considering in this case the prices in the state of NSW. Prices are set for the model in each state or territory of Australia. Prices for Western Australia and the Northern Territory are assumed to track the export price for liquid natural gas (LNG). The projected east coast gas prices assume moderate LNG penetration in Queensland. Prices at the Gladstone port are predicted to reach export parity in 2025 with the southern state prices, converging with the Queensland price by around 2030.

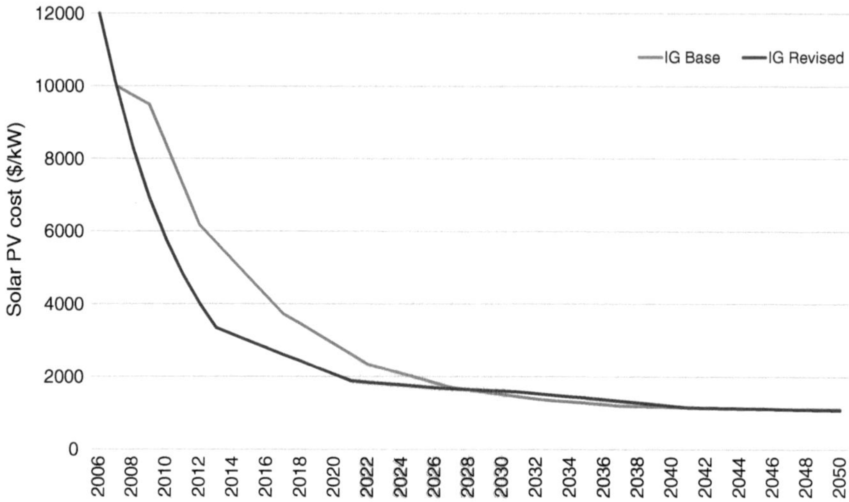

Fig. 5 Sensitivity of solar PV prices

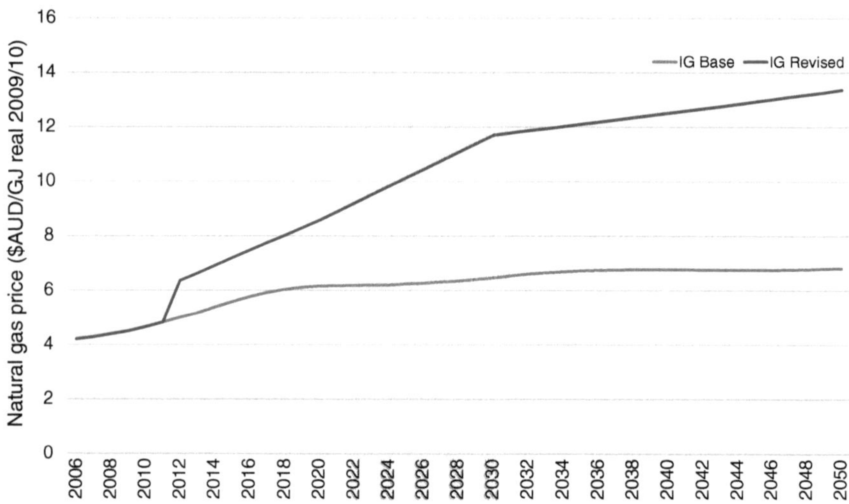

Fig. 6 Sensitivity of natural gas prices (NSW example)

The price of natural gas in other states is assumed to be largely driven by domestic demand and longer-term supply contracts, which do not completely track international market volatility. In the revised case, the price increase is due to the influence of high LNG netback prices in Australian dollar terms, resulting in additional LNG development in Gladstone occurring in the post-2020 period. Domestic electricity generators are required to compete with new LNG proposals, and as a result, domestic prices converge towards the higher netback level.

Greater detail and further discussion on the model assumptions are contained in CSIRO [10].

3 Modelling Results

Figures 7 and 8 provide the generation mix profile from ESM for all technologies and for DG technologies, respectively, under the "low" carbon price. These figures show the relative contribution of generation type by the amount of energy provided. They are based on the underlying assumptions used in earlier modelling by CSIRO [10]. Under these assumptions, the model predicts that DG can provide up to 81 TWh or equivalently around 18 % of total generation in 2050. In earlier years, the DG mix is dominated by gas-fired and biomass co-generation plants. Solar PV becomes more prevalent from 2017 with its initial impact in the residential sector followed in later years by non-urban use and then application in commercial buildings. Despite its high thermal efficiency, the decline in gas-fired CHP from around 2030 reflects the impost of higher carbon prices and more competitive costs for solar PV at this time.

In comparison, when using lower PV costs (Figs. 9 and 10) based on more recent cost and performance data, the model predicts a marginal increase in total DG output in 2050 of around 83 TWh. While the impact on DG remains similar in 2050, the charts clearly show that in this case, DG plays a stronger role in earlier years. The lower installation costs bring on solar around 5 years faster. The amount installed in later years remains relatively consistent between the two cases as the price from 2030 is essentially identical after the technology matures as discussed above. The increase in DG in earlier years results in a reduction of larger centralised renewables, in particular wind and solar thermal, as the model meets the constraint of the expanded renewable energy target (RET). This target ensures that 45,000 GWh of renewables enters the market in 2020, and in this case, solar PV provides a more cost-effective solution to meet that target. The change in PV price has marginal impact on the amount of traditional coal- and gas-fired centralised plant in early years due to their low capital costs and the relatively low price of carbon at this time.

Altering the assumed price of natural gas (Figs. 11 and 12) produces a significant impact on the generation profile. Overall demand in the sector is reduced by around 19 TWh in 2050 as consumers reduce energy use in response to the higher fuel prices for natural gas used in both centralised and decentralised plants. In the case of DG, demand for its supply drops around 11 % to 72 TWh in 2030 with large reductions in electricity created by gas-fired co-generation plant. Figure 13 provides the same chart but in this case, for the "high" carbon price. A similar trend is apparent for DG, but in this case, the stronger carbon price dampens DG in early years by around 4.5 TWh in 2020. By 2050, the difference is only 1 TWh. While the carbon price has a substantial impact, the greatest change shown by the model in 2050 results from the difference in assumed natural gas price rather than the impost of a carbon price.

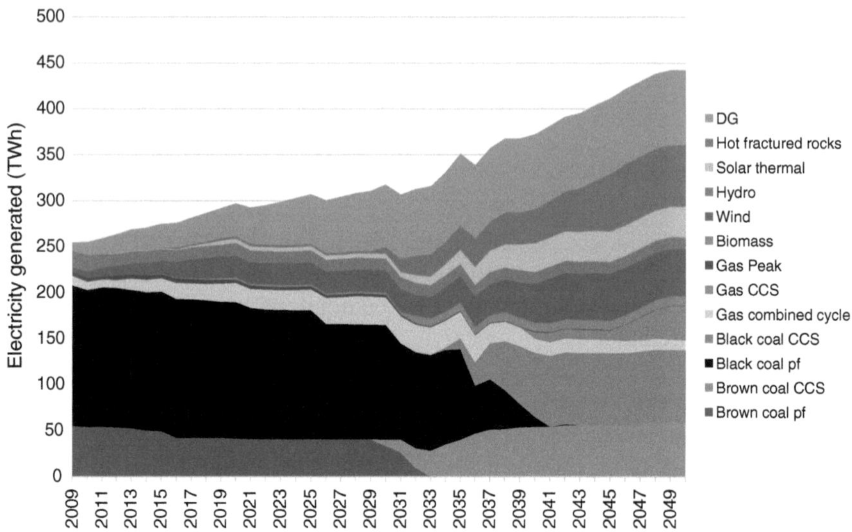

Fig. 7 Generation mix: base case—low carbon price

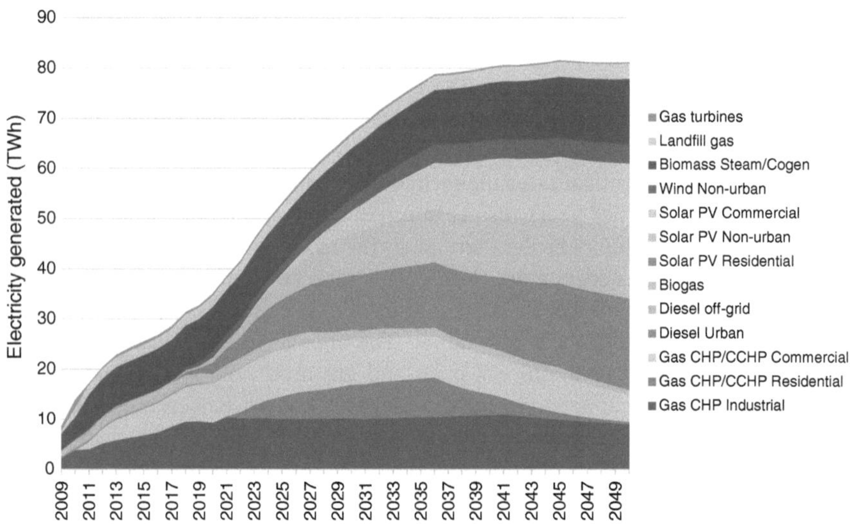

Fig. 8 Distributed generation mix: base case—low carbon price

4 Estimated Value of DE

Over the modelling period to 2050, the undiscounted value of energy efficiency, demand management, DG and structural change is around $800 billion. This saving is calculated by measuring the difference in weighted average prices multiplied by

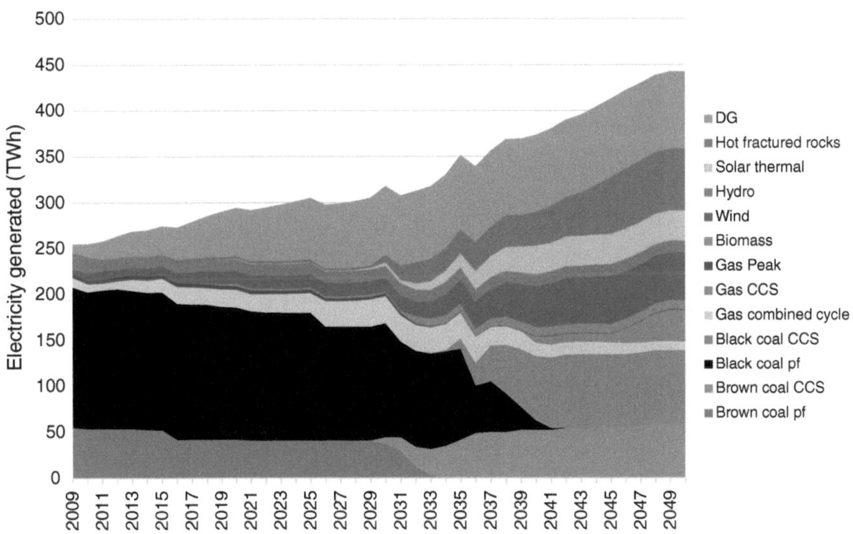

Fig. 9 Generation mix: lower PV cost—low carbon price

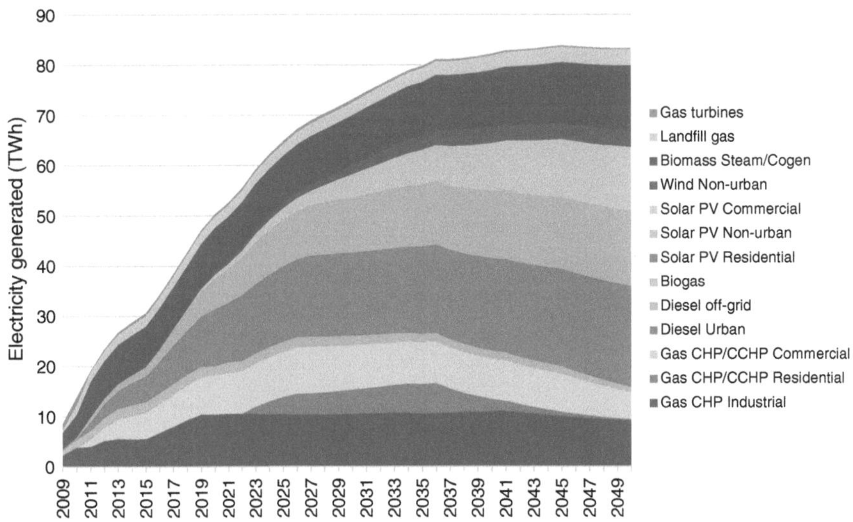

Fig. 10 Distributed generation mix: lower PV cost—low carbon price

energy consumed between scenarios modelled. Figure 14 below illustrates how this value accrues over time. The blue line represents total energy costs where DE is excluded as an option in the model, and the red line represents total energy costs where DE is included. The present value of the welfare gain is around $130 billion discounted at 7 % pa. Ultimately, these benefits are shared by all consumers of electricity. For the two sensitivity cases shown here, the values are $133 billion for

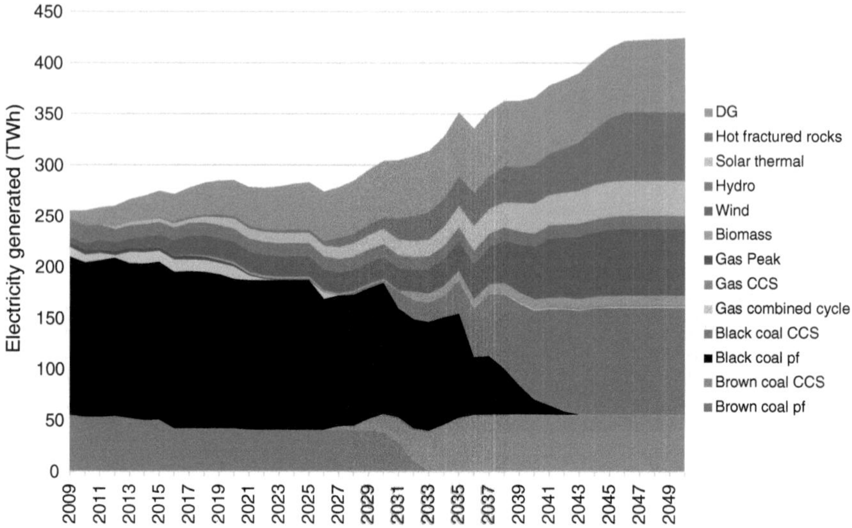

Fig. 11 Generation mix: higher natural gas price—low carbon price

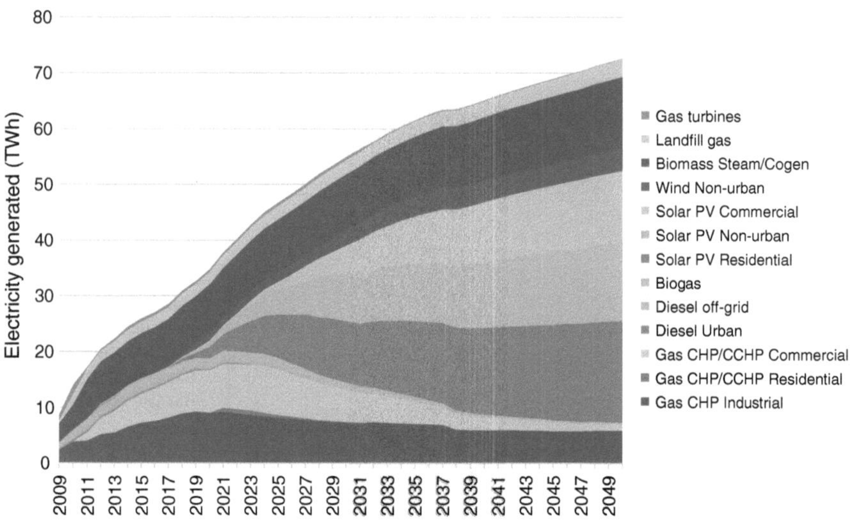

Fig. 12 Distributed generation mix: high natural gas price—low carbon price

the lower PV cost case and $154 billion for the increased gas price case. These results show that the value of DE is greatest for the increased gas price case mainly due to increased benefit of demand management.

It is important to note that the only major cost the ESM does not account for is the cost of structural change in the economy over time, such as the cost of unemployment or retraining that may occur. Costs associated with transforming

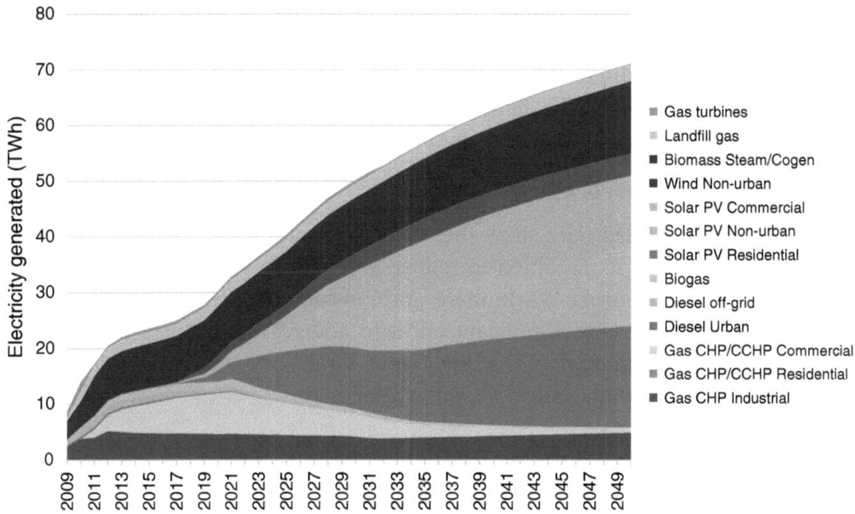

Fig. 13 Distributed generation mix: high natural gas price—high carbon price

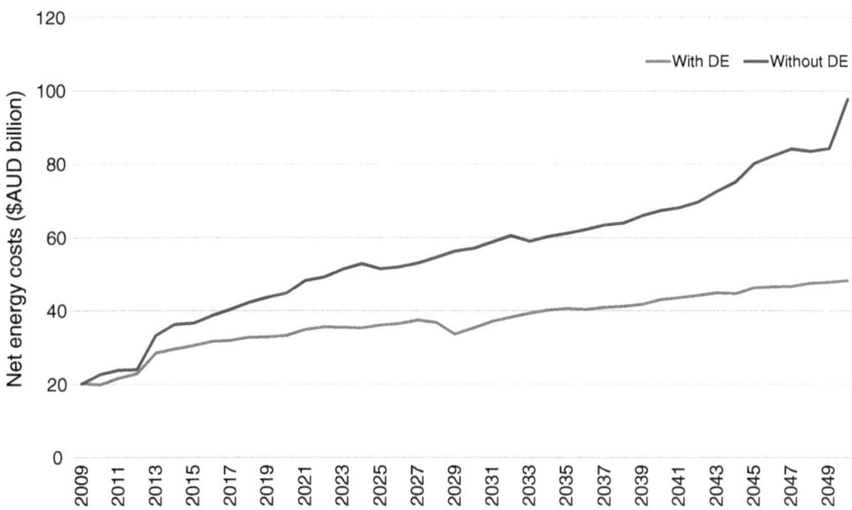

Fig. 14 Comparison of energy costs with and without DE

the energy supply chain are built into the model, and so can effectively be considered zero. The model optimised welfare by requiring a 7 % real rate of return on energy assets over their lifetime.

Fully valuing DE based on avoided or delayed spending on network and generation infrastructure is a complex exercise that is not fully captured by the ESM. The ESM captures avoided spending on peak generation infrastructure, and transmission networks to a degree, but imperfectly due to modelling limitations. Previous attempts have been made to quantify the market value of demand management specifically excluding network benefits, with estimates ranging from $363 M to $954 M over the period 2007–2025 [18]. Better analysis of these impacts is part of ongoing research in CSIRO's "Future Grid" programme [9].

1. ***Realising the value of DE***

The results presented here show that there is a significant potential for DE to help Australia realise its ambition to reduce greenhouse gas emissions in a cost-effective and safe manner. While there are considerable cost savings available, realising them will require significant effort to address the barriers and enablers that affect its full potential. The "Intelligent Grid" project [10] highlighted a number of these in detail, which we summarise here.

5 Integrating DE with the Electricity Grid and the Urban Environment

In Australia, network service providers (NSPs) operate, maintain and upgrade their infrastructure based on forecast growth over a five-year cycle. These forecasts are used to determine the amount of spend required to meet demand and are subject to regulatory checks. The Australian Energy Regulator (AER) determines how much distribution network service providers (DNSPs) can receive to deliver their services taking into account a fair profit. A vital component of their determination is a prediction of peak and base demand provided by the DNSP. Any changes to the network demand within a regulatory cycle that results from activities such as demand management or DG can affect this determination, and shortfalls in forecast throughput may operate as a real or perceived penalty for the business. Similarly, the introduction of a large-load mid-cycle could lead to significant concerns for the DNSP if their system does not have sufficient redundancy built in.

The process for connecting DG and allocating costs has been the subject of considerable review, and it is likely that any potential changes will take time to resolve due to the complex nature of the issue. As one example, a contentious point in the connection of DG has been the allocation of "shallow" or "deep" connection costs. This cost in essence is a charge from the DNSP for connection that takes into account the need for augmenting the network. The issue relates to the degree of augmentation, i.e., how far upstream of the connection does the effect occur and who bears the cost.

Dealing with these connection costs is an area of considerable and complex debate. In some cases, connection costs are seen as a barrier for introducing local generation. Connection costs are seen by some as contentious in part because most energy assets were built at a time of government-owned infrastructure, with costs shared across customers and taxpayers. Given these assets are now "sunk", any historical distortion of cost allocation is naturally impossible to undo. However, it

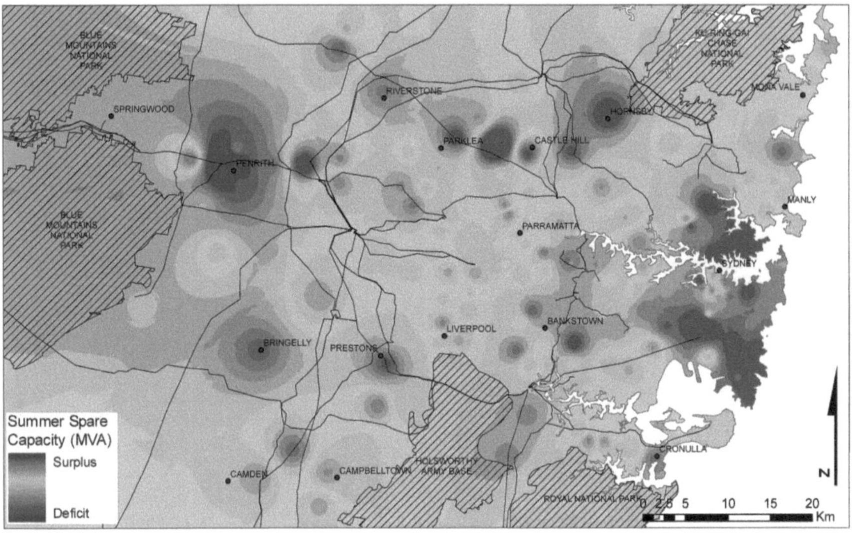

Fig. 15 Illustrative deficits and surpluses in substation capacity in the Greater Sydney region

is important to recognise the origins of the energy market structure, to realise that a centralised supply model dominates by virtue of historical circumstance. Perhaps what is most important today is that the process and methodology for calculating connection costs faced by all new generators connecting to distribution or transmission networks are consistent and cognisant of the potential for distortions to occur due to information or negotiating power asymmetry.

Furthermore, issues around reliability and safety are a significant concern for network operators who are responsible for the performance of their network and who are penalised for failing to meet reliability and service standards. Addition of generation (or demand reduction) within their network and outside their scope of control can lead to risk. Valuing the change (positive or negative) that local generation (or demand management) provides for the network is difficult to determine, is location specific and has no currently available standardised method for evaluation.

Figures 15, 16 and 17 graphically illustrate the potential complexity in optimally locating DE and a potential path forward by geographically mapping the potential opportunities. Figure 15 displays the locations of deficits and surpluses in the capacity of substations in the Sydney region. Areas of deficit provide a potential opportunity for DE to alleviate congestion on the substation. While the figure indicates there is considerable potential for non-network solutions particularly in the western areas of Sydney, it is interesting to compare it with the spread of commercial, industrial and residential areas in the region, illustrating the demographic implications on DE's ability to alleviate these network constraints.

Figure 16 provides an estimate of commercial DE in the Greater Sydney region taken from the uptake of DE for NSW predicted by the ESM. The commercial

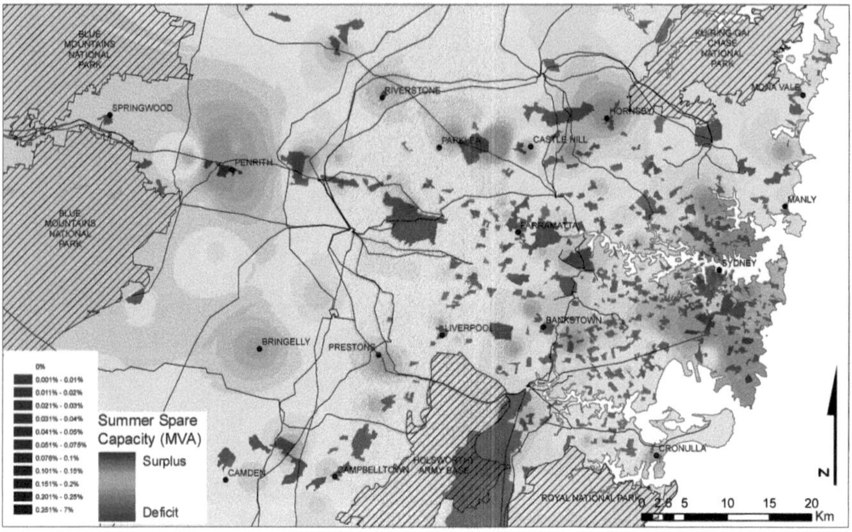

Fig. 16 Percentage of commercial DE installations from ESM for NSW in the Greater Sydney region overlaid with distribution network capacity

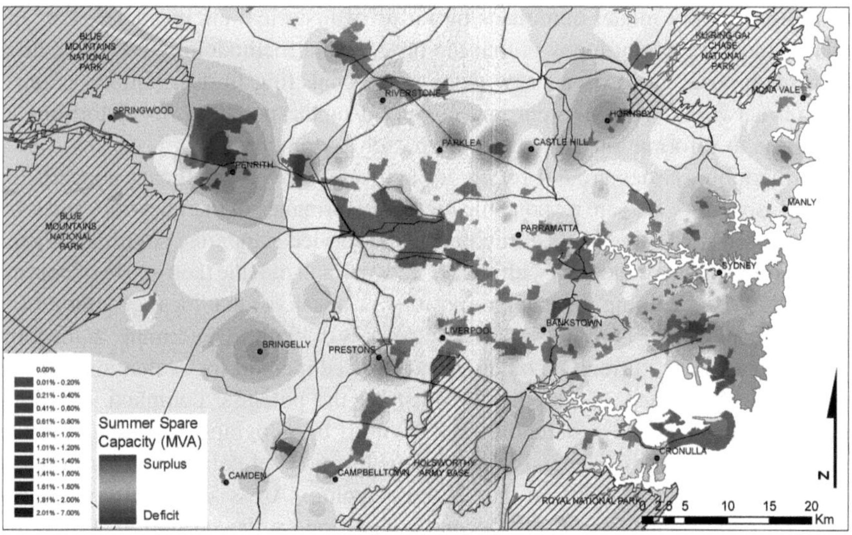

Fig. 17 Percentage of industrial DE installations for NSW from ESM in the Greater Sydney region overlaid with distribution network capacity

zones are marked in a mustard colour. The figure shows that DE in this sector is spread primarily throughout the eastern half of the city with some increased aggregation present around major centres such as Parramatta and the CBD.

This figure shows that in general, the network-constrained areas are not particularly well correlated to the commercial zones as such installations in this sector (in their present location) may not attract economic benefits beyond that predicted by ESM.

Figure 17 shows the placement of DE technologies across the Greater Sydney region in the industrial sector with an overlay with network capacity. These figures show a better correlation with network constraints. To an extent, this is not unexpected as industry tends to be located on the periphery of major urban centres, which in this case are the most under-serviced portion of the network. In this case, it might be expected that DE located within the industrial sector may be more economically viable than ESM predicts, particularly if it receives payment reflecting its value to the network. In many cases, industries have large process heat requirements and installation of co-generation may provide a mechanism to deliver the process heat while alleviating network constraints through the production of electricity for onsite use and for export to the grid.

While this relatively simple approach clearly identifies potential areas where network benefits can be gained, it does not account for more complex interactions such as the impact of potential DE installation on the environment for instance. In this case, the uptake of gas-fired generation in the west of Sydney might unduly affect local and regional air quality. Sydney's west and south-west occasionally experience elevated air pollution, and increasing emissions within this area may exacerbate the problem. Understanding this complex mix of economic, technical and environmental concerns highlights a major issue with DE technologies, which by their nature are located close to load and thereby close to population centres. A thorough analysis of these issues is required to assist regulatory authorities to make a value judgement on whether DE projects can proceed. For example, it is possible that the optimal placement of non-renewable DE technologies in this region might involve the requirement for additional pollution control equipment. While this provides an increased financial burden for the DE installation, it is possible this could be offset by financial gain if the unit is able to capture its true economic value to the network.

6 Enabling Large-Scale Uptake of DE

Realising the value of DE in an efficient way depends on many conditions being met, not just an effective integration of generation and the network. To inform an understanding of these issues, CSIRO [10] conducted a series of stakeholder interviews and undertook a meta-literature review of perceived barriers to DE. These were combined with issues determined in specific DE case studies.

The research suggests indicatively (based on interviews with 47 industry and government stakeholders) that a hierarchy of conditions need to be met before DE achieves wide-scale uptake:

- DE needs to be a commercially viable alternative to mains grid supply before it will have widespread uptake
- For DE to be commercially viable, policy and regulation needs to allow proponents to capture the value of DE where it reduces emissions or costs that are otherwise socialised—primarily seen as costs of peak demand infrastructure
- Policy and regulation must also have long-term certainty to give DE proponents and investors the confidence to implement DE
- Consumers, industry and governments all need to be educated on the value of DE and how it works to overcome cultural bias towards mains grid energy supply. This is also needed to inform appropriate policy and regulation development
- Technology and market development needs to be focussed on reducing cost and improving reliability.

A literature review of barrier studies broadly corroborated these findings and allowed a more fine grain understanding of issues to evolve. From these findings, a summary of key enablers (outcomes that should lead to an efficient deployment of DE technologies and systems) were developed. They are as follows:

- A long-term policy horizon with firm targets and commitments for uptake of DE that have widespread support across the political spectrum. Implicitly, that DE is a highly visible and important policy deliverable, and that the market has improved certainty about how DE is valued
- Data that allow more accurate valuation of different forms of DE-incorporating real-time market costs and a full suite of environmental and social externalities
- The use of a widely accepted, accurate, transparent, efficient and equitable DE valuation methodology across government agencies when developing DE-related policies, programmes and regulation including building standards, appliance standards, product rebates, feed-in-tariffs and so on
- Accurate, transparent-pricing methodologies, accounting for time- and location-specific environmental and social externalities, for energy exported by DG and/or when DG is used for demand management that allows value to be easily captured by a full range of market participants (small to large, with various technologies)
- Full and efficient access to markets for services provided by DG including the ability to easily aggregate small generating or load reduction units into wholesale markets
- A regulatory and policy framework and environment that effectively aligns the incentive of companies in the supply chain or encourages business model innovation, to provide efficient energy services to consumers, including conducting research, trials and continued innovation. These incentives must be compatible with market competition, have broad support, be we well understood and followed
- An efficient, transparent process for connecting DG standardised as far as possible and coupled with effective low-cost dispute resolution. Processes are needed for connecting multiple units and aggregating the costs based on aggregate impacts of connections

- A well-informed, trained/accredited, skilled workforce that understands the value of DE and can sell its benefits to consumers of all types using insights provided by decision-making science
- Improved information provision and framing of costs and benefits to consumers to allow easy and accurate valuations of DE options
- Tax, rebate and/or financing schemes that enable widespread access to cost-effective DE that would otherwise not be taken up due to high capital costs or lack of access to capital. That this be done by providing efficient, easily recoverable financial incentives and reframing decision-making biases (sticky budgets, incorrect weighting of probable outcomes, inefficiently high hurdle rates). This includes access to assistance for low-income and small-business market segments
- A comprehensive research and development programme that allows for overcoming technology lock-in at a scale in line with the need for efficient uptake of DE and complementary policies/programmes structured to move technologies efficiently through their development life cycle
- A system of state and local planning and environmental controls that allows for a full consideration of issues and ensures DE is not blocked without robust justification
- Education of relevant service sectors (designers, architects, engineers, builders, tradespeople, manufacturers) on the value of DE, and methods for better aligning their service incentives with long term, efficient supply of energy
- Continued and bolstered support for minimum performance standards (appliances, buildings), improved information provision (future energy prices, probable savings over time, etc.)
- Targeted, efficient incentives for landlords—structures for recovering cost savings from energy efficiency. Minimum efficiency standards can be stretched for the more expensive end of products/services
- Effective education around smart meters, tariff structures, how best to manage energy, processes that provide real-time feedback and rewards (internal and external) to customers for effective behaviour
- A policy and regulatory environment where experimentation can take place, but where best practice is quickly adopted consistently across the nation.

It is important to note that many of these enablers are either in place, or being worked towards by current and ongoing policy, regulatory, commercial and academic processes. These enablers should not be seen as a list of outcomes that either have not, or will not be achieved. Rather they can serve as a checklist for policy makers, regulators, industry and researchers to guide their actions as the market for DE evolves.

Also, due to the disaggregated centralised energy supply chain in Australia, no one business in this supply chain can capture the full value of DE. This acts to dilute the incentive to invest and has the potential to result in significant investments that do not achieve socially efficient energy supply. How this can be best overcome is not simple, but ultimately, enabling DE is likely to require the bringing together of many complementary policies. Split incentives, access to finance, renewable energy policies, energy prices, skill and industry capacity, from architects, through to builders and trades people, can all impact on the uptake of DE and must be addressed simultaneously.

7 DE Business Models

There is a natural interplay between business models for the delivery of energy and the ability to align consumer decision-making with efficient energy service outcomes. A business that can capture the full value of social and environmental externalities associated with energy consumption and production has naturally aligned incentives to deliver private and socially efficient levels of DE.

A relatively successful example of directly aligning the incentives across a range of functions that affect energy outcomes often referred to is the Woking Borough Council (WBC) energy service company model in the United Kingdom. In this model, WBC sought to accelerate emission reductions in its jurisdiction through setting up a special purpose vehicle (SPV) called Thameswey Ltd in 1999. The company's purpose was to form public/private partnerships to deliver projects targeting the Council's broader climate change strategy, including providing clean energy, tackling fuel poverty, water security, waste minimisation and clean transport. Generated revenues are channelled back to the council to reinvest in specific projects, such as improvements to housing, retrofitting solar PV and heating systems for low-income families [22]. By 2006, WBC had achieved a 33 % energy efficiency improvement and 21 % emissions reduction on residential property, against a 1991 benchmark [20].

The success of the model is that it aligns environment, social and economic objectives, delivered through a single entity that can capture the full benefit of its action either directly through revenue recovery, or indirectly through socialised value it provides to the community. Significantly, the vehicle is also empowered to deploy a full suite of DE options and is resourced with the necessary technical capability. The entity also aligns incentives that are often weakened due to disaggregation between generators, network companies and retailers, and the regulatory and market structures within which they operate.

In this way, WBC was able to overcome many overlapping and related barriers that typically impede a wide range of DE options. The business model also allows a number of key enablers identified in this chapter to be realised because it does not have to directly recover the value of social and environmental externalities that it reduces, as the value is spread across the community.

Ultimately, it is likely to be a policy choice whether such a business model can be deliberately constructed by government jurisdictions in Australia, for example by local governments, or whether through policy, market settings and regulation, such models are encouraged for others to develop and implement.

Another model for maximising the uptake of DE where it is efficient includes community collaborations. The Maines Power project [19] was example where a programme of work was initiated by a local community and environment group who worked in collaboration with four major businesses within their small town. With the help of CSIRO, MASG gathered external expertise and facilitated external funding through government-funding agencies. The businesses and other project participants discussed an ambitious goal of 30 % greenhouse gas reductions by 2010. The value selected was based on what the businesses believed they could achieve and to match the same target set by the local council. The project was developed as a non-legal partnership model to enable four businesses to work together with government agencies, peak industry bodies, energy retailers, distribution network owner operators and environmental organisations.

This partnership model was advantageous in two ways. First, it helped facilitate and guide business decisions, particularly for those businesses not directly employing an energy and operation specialist. While energy costs may be significant to most businesses, in general, the cost is considerably smaller than other processes such as labour and material costs. Second, the partnership model allows government agencies to continue their skill development, to retain information learnt, and to collate and pass on relevant knowledge to businesses and the community in the future.

The partnership model used in this project provides an excellent structure to improve project outcomes for future community-driven projects. While technologies may play a significant role in innovation, there are potentially more gains to be had in developing new business and engagement models that allow risk and gains to be spread across all participants. Ultimately, this is needed to overcome reluctance to deploy technology.

The project highlighted that the local network businesses are vital stakeholders when trying to reduce consumption in the stationary energy sector. It is their asset base that allows the flow of energy, and it is their asset in which local generation or demand reduction activities may be located. Network businesses are highly regulated and subject to severe penalties when their service falls below specified standards. As such, these businesses have a propensity to adopt well-understood practices, as opposed to continually innovate, to ensure their business operates within regulated guidelines. However, new regulatory incentives are helping to change this.

8 Integrating DE with Energy Markets

When considering business models for energy delivery, how they can effectively align incentives within the energy supply chain, and across property and appliance ownership structures, it is useful to consider commercial and institutional structures that influence Australia's energy markets.

The Australian NEM has relatively formal governance, commercial, security and technical decision-making regimes. In summary, these organisations have the following roles:

- The Council of Australian Governments (COAG) brings together federal and state governments in a forum to coordinate policy development and set policy principles at a high level
- The Ministerial Council on Energy (MCE) coordinates federal and state policy and has oversight for rule and regulation development
- Uniform industry-specific legislation, the National Electricity Law (NEL) defines decision-making constraints for the electricity industry including commercial, technical, security and regulatory arrangements. The specific details of these arrangements are set out in the National Electricity Rules
- The AEMC manages the National Electricity Rules and the rule change process by which they can be further developed
- The AER enforces regulatory requirements and manages particular regulatory processes such as the review and approval of network investment plans. It also monitors compliance with the National Electricity Rules by market participants as well assessing the overall effectiveness of these rules
- The Australian Energy Market Operator (AEMO) is both the market and system operator and thus has responsibility for implementing and managing both the security regime and the short-term aspects of the commercial regime.

The institutional and legislative framework of the NEM has been developed over many years and has been reinforced by the dominant centralised supply model. Small-scale energy and demand management have fulfilled relatively niche roles and, in some instances, have been used to reinforce the dominance of the centralised supply model, for example, the use of off-peak electric resistance hot-water systems.

It is important to note there are processes underway that will enable better integration of DE into the NEM, including potential rule changes to allow aggregated generation and load participation in wholesale market operations.

Design of regulation to meet environmental or social objectives and integration of this regulation with energy market operation is relatively new, and integrating them with the market is not always easy. Environmental and social objectives are not always immediately compatible with business models that operate in the existing energy supply chain. For example, very generally, various sources of cost and value for solar hot water and PV may impact on different businesses in the supply chain as follows:

- Baseload generators could be negatively affected by solar hot water at significant levels of deployment
- Peaking generators could be negatively affected by PV (at significant levels of deployment) to the extent that it correlates with times of network-wide peak demand

- Network companies could be negatively affected by reduced energy volumes if they result in demand growth slower than forecast as part of their network investment plans. Or could benefit from PV to the extent it provides network support, reduced losses and power quality benefits that outweigh any unrecovered connection costs
- Retailers whose net impact may vary depending on any generation assets owned, exposure to peak wholesale prices, or even levels of integration into solar PV and hot-water markets.

Supply-side participants in the energy market are generally large, well resourced, focussed almost exclusively on the electricity industry and have considerable shared interests. End-users are far more diverse, typically less well resourced and may have interests beyond electricity itself. In this environment, effective representation of end-user interest in NEM governance and broader policy decision-making is a difficult process. Formal governance processes must be able to manage these asymmetries between supply and demand-side stakeholders in order to represent environmental and social interests in NEM design and operation.

Ultimately, successful introduction of any new technology into the NEM requires the effective support of these institutional decision makers as well as supporting organisational infrastructure. In the case of DE, this includes not only the organisations and people directly involved with the technology such as designers, retailers and installers, but also those who have to manage the impacts of that technology on the rest of society. This support infrastructure does not automatically emerge in response to market signals and so highlights a role for government, not only in education and training but in developing the necessary institutional decision-making structures. This is an important consideration for policy makers seeking to harness the value of DE.

Integrating and valuing new technology, specifically DE into the NEM, also requires explicit, transparent methodologies for signalling, motivating and optimising end-user participation to facilitate effective decision-making. Active participation by the majority of end-users (residential and commercial) will require that the uncertain time- and location-varying value of energy is better reflected in the prices these end-users see, or that end-users and/or third parties will be able to capture the value of DE where it supplies energy services below time- and location-specific costs. The use of interval meters coupled with price deregulation goes some way to achieving this. However, it must be recognised that price is a limited tool if used in isolation. Without access to information, financial support and potentially specialist skills to facilitate behaviour change as well as new physical infrastructure, customers may have a limited response to price signals. This decision-making complexity highlights the potentially important role of energy service companies that can optimise delivery of energy services to customers.

Investment decision-making is also a critical component of the NEM. Forward-looking prices and planning documents can help signal where future investment is needed. For example, investment in CG is largely driven by the Electricity

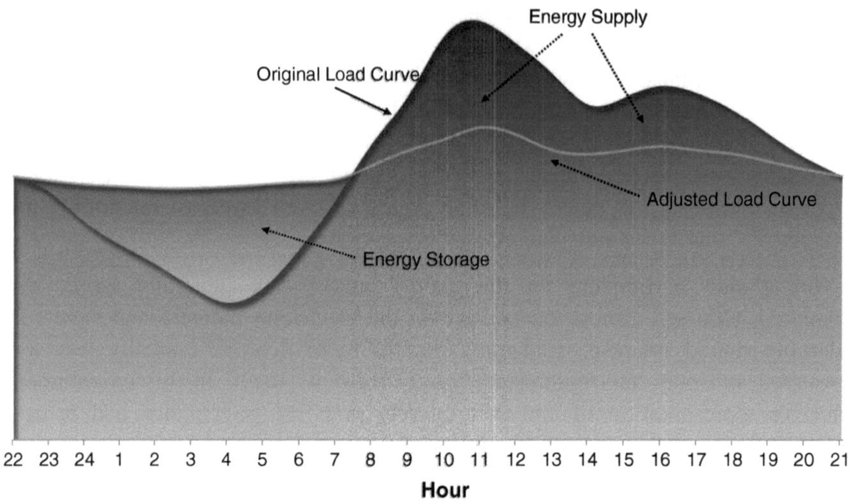

Fig. 18 A hypothetical daily demand profile including storage

Statement of Opportunities (ESOO) report. Released by the market operator (AEMO), it details historical demand, demand projections and projections in energy shortfalls. Efforts are being made to replicate a comparable process at a distribution network level, with distribution companies required to release network-planning details and signal opportunities for demand-side proponents to offer alternatives to network building. However, signalling for investment in DE is naturally a more complex process due to the fine grain nature of location and price signals it can be driven by, as well as the sometimes competing interests of the energy supply chain and DE proponents. Furthermore, to capture a full suite of DE opportunities, these decision-making signals must incorporate the intersection between natural gas and electricity, a potentially significant issue given the relative immaturity of natural gas markets and the predicted impacts of natural gas-based technologies as detailed in the economic modelling above.

Demand management (DM), one important element of DE, relies heavily on energy market structures to realise value. This is because it primarily operates to resolve time-specific network or generation constraints. DM refers to a suite of technologies and techniques used to alter demand profiles over time. Active control measures can smooth or shift demand peaks, or substitute local generation for CG. Passive control measures such as energy efficiency can reduce total demand over time. Figure 18 provides an illustrative demand profile, indicative of a network area dominated by commercial energy demand and hence a midday peak in summer. In this example, the blue curve shows demand ramping up from 6 am when people begin to wake and get ready for work. Demand grows during the day peaking before lunch as air-conditioning demand grows in commercial buildings. The demand then drops off slowly briefly peaking again in the late afternoon as people return home.

DM offers potential value to network companies, energy retailers, energy customers and the market operator. Most simply, it can defer spending on assets for network companies, reduce wholesale prices particularly at peak times and reduce retailer exposure to wholesale market price volatility. However, it can also offer more complex services, for instance, controlling loads in emergency situations (such as power shortages or outages), providing frequency control and ancillary services, or managing customer exposure to generation or network charges according to stated preferences. By way of example, Read et al. (1998) in Ackerman et al. [1] found that ancillary service costs were reduced by 75 % in the first year that interruptible load was allowed to participate in the New Zealand energy market.

A number of DM programmes have been run or are in operation in Australia. Most of these are operated by distribution companies and often in response to a policy action such as the AER's demand management incentive scheme (DMIS). While these trials have provided significant insight into the potential short- and long-term impacts of DM and DE more widely, it is important to note that DM led by distribution network businesses is only one model. DM could be performed by retailers or third parties, to the extent they can capture the value of doing so and thereby make a commercial return. Providing sufficient but not exaggerated incentive for participants to explore efficient alternatives remains a significant challenge to policy makers.

9 Conclusion

Economic modelling clearly shows that there is considerable potential for Australia to adopt DE in its suite of options to reduce its greenhouse gas emissions and met its global obligations. Research by CSIRO [10] shows that incorporating larger amounts of DE is technically feasible but that determining its optimal location taking into account the economic, technical, social and environmental issues is highly complex. Some of these issues are naturally being addressed by advances in smart grid technology that marries communication and electrical infrastructure to better measure and control a range of traditional and non-traditional network devices. Critically, however, the research shows that despite many years of debate and considerable progress in some areas, there is still much work required to fully address the barriers and enablers to allow DE to compete freely and allowing it to unlock its full potential.

Acknowledgments This chapter is an updated republication of the paper "Intelligent Grid—A value proposition for DE in Australia" by Commonwealth Scientific and Industrial Research Organisation ("CSIRO"). The copyright permission for reusing the paper has been granted by CSIRO.

References

1. Ackerman TG, Anderson G, Soder L (2000) Electricity market regulations and their impact on distributed generation: electric utility deregulation and restructuring and power technologies. In: Proceedings of international conference on electric utility deregulation and restructuring and power technologies. http://ieeexplore.ieee.org/xpls/abs_all.jsp?arnumber=855735
2. AEMO (2013) National Electricity Market (NEM). http://www.aemo.com.au/About-the-Industry/Energy-Markets/National-Electricity-Market
3. AER (2013) Demand management incentive scheme (DMIS) assessment 2011–12 and 2012. http://www.aer.gov.au/node/19952
4. AusGrid (2013) Smart grid smart city. http://www.smartgridsmartcity.com.au/
5. BREE (2012) 2012 Australian energy update. http://www.bree.gov.au/documents/publications/aes/bre0133energyupdate2012.pdf
6. BREE (2013) Energy in Australia. http://www.bree.gov.au/documents/publications/energy-in-aust/bree-energyinaustralia-2013.pdf
7. Clean Energy Regulator (2013) Renewable energy target. http://ret.cleanenergyregulator.gov.au/about-the-schemes
8. Clean Energy Regulator (2013) Carbon pricing mechanism. http://www.cleanenergyregulator.gov.au/Carbon-Pricing-Mechanism/Pages/default.aspx
9. CSIRO (2013) Australia's future grid: evaluating whole-of-system options for Australia's future electricity system. http://www.csiro.au/future-grid-forum
10. CSIRO (2009) Intelligent grid—a value proposition for DE in Australia. CSIRO Report ET/IR 1152. http://www.csiro.au/resources/IG-report.html
11. DBCDE (2013) Department of Broadband, Communication and Digital Economy—National Broadband Network. http://www.nbn.gov.au/
12. DIICCSRT (2013) Department of Industry, Innovation, Climate Change, Science, Research and Tertiary Education—Solar Cities Program. http://www.climatechange.gov.au/energy-efficiency/solar-cities
13. DPI (2013) Victorian Department of Primary Industry—Smart Meters. http://www.dpi.vic.gov.au/smart-meters
14. DPTL (2013) Victorian Department of Transport, Planning and Local Infrastructure—Electric Vehicle Trial. http://www.transport.vic.gov.au/projects/ev-trial
15. Energex (2013) Sustainability. http://www.energex.com.au/sustainability
16. IEA (2013) Electricity/heat in Australia in 2009. http://www.iea.org/stats/electricitydata.asp?COUNTRY_CODE=AU
17. Graham P, Williams D (2003) Optimal technological choices in meeting Australian energy policy goals. Energy Econ 25(6):691–712
18. Hoch L, Chattopadhyay D, Fazio J, Fulford B, McCall K (2006) Assessing the value of demand response in the NEM. Prepared for the Australian IEA Task XIII Team
19. MASG (2009) Maine's power. http://masg.org.au/projects/csiro/
20. Resource Smart (2008) Smart energy zones case study: Woking, UK, Melbourne: Sustainability Victoria. www.resourcesmart.vic.gov.au/documents/SEZ_Case_Study_Woking.pdf
21. SP AusNet (2013) Distribution System Planning Report 2013–2017. http://www.sp-ausnet.com.au/CA2575630006F222/Lookup/Procedures/$file/DSPReport.pdf
22. WBC (2009) Sustainable Energy in Development, Woking Borough Council. http://www.woking.gov.uk/planning/service/energy

Part V
Postscript

Postscript

Daphne Mah, Peter Hills, Victor O. K. Li and Richard Balme

Abstract The smart grid landscape is extremely dynamic as the contributions in this book have demonstrated. We have focused on some of the latest technological developments and the importance of various stakeholder perspectives. We have also provided a variety of international case studies that have allowed us to explore the latest policy developments from Europe to the US and the Asia-Pacific region. This postscript explores the future trajectories of SGs, highlighting three major trends that point the way towards future advances and progress in the years ahead. Future research agenda is also discussed.

1 Future Trajectories for Smart Grids

The smart grid landscape is extremely dynamic as the contributions in this book have demonstrated. We have focused on some of the latest technological developments and the importance of various stakeholder perspectives. We have also

D. Mah (✉)
Department of Geography, Hong Kong Baptist University, Hong Kong
People's Republic of China
e-mail: daphnemah@hkbu.edu.hk

P. Hills
The Kadoorie Institute, The University of Hong Kong, Hong Kong
People's Republic of China

V. O. K. Li
Department of Electrical and Electronic Engineering, The University of Hong Kong, Hong Kong, People's Republic of China
e-mail: vli@eee.hku.hk

R. Balme
Paris School of International Affairs and Centre d'Etudes Europeennes, Sciences Po, Paris, France

D. Mah et al. (eds.), *Smart Grid Applications and Developments*,
Green Energy and Technology, DOI: 10.1007/978-1-4471-6281-0_16,

provided a variety of international case studies that have allowed us to explore the latest policy developments from Europe to the USA and the Asia–Pacific region. The 15 chapters of this book are intended to provide a basis for a better understanding of emerging SG technologies, related issues, and policy developments. A number of chapters have pointed to the critical issues that need to be resolved to achieve expanded deployment of SG. As well as further technological advances, other critical issues include regulatory reforms, the challenge of securing public support, and the urgent need for the development of appropriate business and financial models.

Preparation of this book commenced in mid-2012. This book provides insights into, and an account of, the development of smart grids as of the end of 2013. Since we started to work on the book, these technologies have continued to evolve and develop rapidly across many important fronts. Three major trends point the way towards future advances and progress in the years ahead.

The first relates to the move from SG pilots to full deployment which will occur in the near future. A number of countries, including South Korea and the UK, have committed to rolling out nation-wide metering infrastructure in the next few years. New policies are emerging to improve the institutional, financial, and operational environments for the growth of these technologies. One example of these initiatives is the smart grid plan in the UK. The British government has recently published a modified smart grid plan in preparation for the 2015 launch of its smart meter roll-out [1, 4]. In the USA, the New York Independent System Operator—which plans, coordinates, controls, and monitors the operation of the electricity systems of the city—has recently started to implement its 2014–2018 Strategic Plan in which smart grid technologies are key components to improve system integration and enhance demand-side markets [12]. New York City, like California, is focusing on the potential for enhanced energy storage as part of a broader SG strategy. Thermal energy storage systems for cooling systems designed to shift load away from peak demand periods to off-peak are seen as a mechanism to enhance network reliability during peak hours when system reserve capacity is at its lowest [9].

As smart grid roll-outs continue in many parts of the world, the global smart grid market is projected to be scaled up by an order of magnitude in the near future. Recent reports have projected that the cumulative value of the smart grid market is expected to reach US$400 billion by 2020 [2, 5]. Transnational financial institutions such as the World Bank have recently made commitments to finance smart grid initiatives, indicating the global interest in these growing markets [15].

One major development in 2013 was the emergence of China as the largest investor in SG technologies. While SG spending in the USA declined by nearly one-third to $US3.6 billion, investment in China reached $US4.3 billion representing more than 25 % of global SG spending. At the global level, spending reached an estimated $US14.9 billion (up from $US14.2 billion in 2012) with approximately half of this figure spent on metering projects and the remaining on distribution automation and grid-related projects [8].

The second major trend relates to the emergence of numerous business models. These emerging models seek to identify who the customers are, what customer values can be created, how a business can make money from it, and what revenue models can work well to deliver value to customers at reasonable cost while ensuring profitability [10, 16]. In Japan, following the completion of the four high-profile demonstration projects in Kyoto, Yokohama, Kitakyushu, and Toyota City in 2014, the next phase of seven new smart grid pilots is scheduled to be launched also in 2014 with a new focus on piloting business models [10]. Hitachi City, which hosts one of these seven pilots, for example, is piloting how to create and validate markets for electric vehicles [11]. The success or otherwise of these emerging business models will be an important factor determining the pace of systemic change in the power sector.

The third important trend relates to consumer responses to smart grids, metering, and dynamic pricing. Customer acceptance is central to the full and effective deployment of SGs. Emerging pilots have been launched in various countries to explore what combinations of pricing programmes, in-home technology, and engagement methods can facilitate customer acceptance [7]. However, a smart meter backlash has continued to slow smart grid deployments in some pilots while some dynamic pricing programmes recorded a relatively high drop-out rate [3, 14]. Utilities and regulators need to address opposition from customers as they implement new pricing and service options.

Taken together, these emerging trends suggest that SGs are moving to the next level of development and deployment. The next stage will see the deployment of commercial-scale SG roll-outs, in which consumer engagement and viable business models will be critical to allow for the necessary systemic changes. Despite the momentum created by these emerging developments, new challenges are appearing as a result of fundamental changes in the relationships between utilities, regulators, and customers. These challenges, such as financing SG R&D and infrastructure, controlling costs, managing consumer backlash, and promoting market acceptance through pricing strategies and consumer education, will require new forms of and much closer collaboration between utilities, new market players, government, and communities. Major demonstration trials such as the Smart Grid, Smart City Program in Australia and the four major demonstration projects in Japan are some of the growing number of examples of such collaboration in which traditional utilities are developing new forms of collaboration with other stakeholders in innovative ways [6].

2 Refining the Research Agenda

Given the significance and dynamism of the SG landscape, we feel that it will offer numerous opportunities to extend the boundaries of interdisciplinary research in the broad area of energy systems and the environment. Here, we highlight just a few possible areas which appear to offer the potential for new and significant research engagement.

We have, for example, briefly examined the plans and status of renewable energy resource development and energy policy in Europe. The future European power system will be characterized by the high penetration of renewable energy resources which will present significant challenges to system operation and control. Smart grid provides an efficient and effective tool to address this challenge, and future research will be needed to examine how technologies and government policies can best be developed to address the challenges arising from the high penetration of renewable energy.

Another area concerns dynamic and time-of-use pricing which can yield significant load reductions. Future research work is needed on the effect of customer awareness and education, and the types of customers which might be involved in dynamic pricing. Microgrids offer another possible research focus. As microgrids may be operated in grid-connected mode, they may affect the stability of the existing grid and future research is needed on the stability issues related to this operation mode. We have also highlighted communication requirements, and the security and privacy requirements of smart grids. This will continue to represent a critical issue for SG acceptance, and future research will need to focus on developing smart grid communication architectures and protocols which satisfy the identified requirements. Experience in the USA has indicated that a significant barrier to SG deployment has been the absence of standards allowing devices connected to the grid (e.g. meters and energy distribution systems) to actually communicate with each other. Coding systems (i.e. communication protocols), though existing, are not yet harmonized, and this can deter utility companies from joining the SG [13]. The development of appropriate standards and their effective diffusion is therefore an area requiring urgent attention from both research and application perspectives.

The case studies presented in this book demonstrate that different countries have tackled issues of SG development and deployment in often very different ways. Aspirations, approaches, and the focus of SG efforts have varied. Existing national experiences suggest that SG promotion and deployment involve a variety of complex issues many of which relate not so much to technological concerns but rather more to economic, social, and institutional factors. There is clearly an ongoing need for studies of SG experiences at all levels from individual utilities, community test beds, and city initiatives to national efforts. Such case studies will help to illuminate both opportunities for and constraints on SG deployment. In particular, there is a need for more detailed and extensive studies that probe both the financing and regulatory environments within which SG developments are and will be set. In the USA, for example, the Department of Energy is actually winding down its SG investment programme, a large part of which was funded by a $US3.4 billion award from the American Recovery and Reinvestment Act [13]. Thus far, in the USA and indeed in a number of other countries, a high proportion of SG funding has gone into smart metering but the provision of such meters represents only a part of the SG revolution. More research is needed on funding and financing requirements, the cost-effectiveness of different funding mechanisms and how the scaling-up of SGs in the future will require new and perhaps innovative approaches to pricing and the regulation of electricity markets in the broader context of the development of viable business models for SG.

Another potentially interesting and fruitful research focus will be the relationship between SGs and the state itself. Traditional grids, together with other utility and transportation networks, have been a critical instrument for the territorial integration of the nation state during the industrial age. Public services provision and regulation still remain fundamental competencies of public authorities. However, the decentralized and hybrid public–private nature of smart grids seriously challenges the vertical and top-down aspects of regulation as we usually think of it. We have only limited knowledge so far of the relations between smart grids and smart city parallel developments for instance. More generally, SG developments potentially suggest the emergence of entirely new relations between local communities and national regulation. They could also transform interactions among local communities themselves, possibly across borders, in electricity production and distribution. What will be the appropriate scales, instruments, and regulatory activities in a world of greater fluidity and less standardization? These issues will require research attention.

Consumer responses to the SGs will also merit continued research attention. While consumer resistance in countries like the USA appears to be relatively low, elsewhere, including the UK, it has at times caused some controversies as concerns about data privacy continue to grow. Providing consumers with real-time information regarding their energy consumption patterns and the actual costs involved is a necessary prerequisite to support energy-saving behaviour. However, there has been growing concern and dissatisfaction regarding the tariffs of major energy suppliers in countries like the UK, and this may reinforce doubts about the motives for SG deployment unless consumers feel that they will benefit through lower electricity bills. Thus, the development of business models that can reconcile business risks with consumer needs and concerns will remain a key priority. In this context, the future role of energy regulators and associated regulatory frameworks will also be a key research focus.

Electricity grids that emerged and developed from the early 20th century onwards are on the verge of a remarkable transformation. These grids have in many cases served their customers well over many decades. But they are now outdated, inefficient and often vulnerable to a variety of factors which render them potentially unstable. Using energy in all its forms more efficiently has become an issue of fundamental concern as the global community confronts the challenges of energy resource availability and costs on the one hand and climate change on the other. Smart grids will not of themselves provide a solution to these global challenges, but they can be a key building block that help to ensure the electricity supply systems that we all depend on will be more efficient and reliable in coming decades.

Acknowledgments We gratefully acknowledge the funding of our research by the Hong Kong Baptist University (through the Faculty Research Grants from the Faculty of Social Sciences, FRG2/12-13/057 and FRG1/13-14/051) and the University of Hong Kong (through the Initiative on Clean Energy and Environment).

References

1. ANAO (2014) Administration of the smart grid, smart city program. Australian National Audit Office, Canberra. Retrieved 25 Feb 2014, from http://www.smartgridnews.com/artman/uploads/2/Australian_Audit_of_Smart_Grid_Smart_City_program.pdf
2. BusinessGreen (2013) Global smart grid market to more than double by 2020. GreenBiz Group. Retrieved 25 Feb 2014, from http://www.greenbiz.com/news/2013/06/18/global-smart-grid-market-more-double-2020, 18 June 2013
3. DECC (2012) Smart metering implementation programme: government response to the consultation on the consumer engagement strategy. Department of Energy and Climate Change, London. Retrieved 25 Feb 2014, from https://www.gov.uk/government/uploads/system/uploads/attachment_data/file/43042/7224-gov-resp-sm-consumer-engagement.pdf
4. DECC (2013) Smart metering implementation programme: second annual report on the roll-out of smart meters. Department of Energy and Climate Change, London. Retrieved 25 Feb 2014, from http://www.smartgridnews.com/artman/uploads/2/UK_second_annual_report_smart_meters.pdf
5. Groarke D, Pollock Z, Kellison B (2013) Global smart grid technologies and growth markets 2013–2020. GTM Research. Retrieved 25 Feb 2014, from http://www.greentechmedia.com/research/report/global-smart-grid-technologies-and-growth-markets-2013-2020
6. IEEE (2013) Australia. IEEE Smart Grid. Retrieved 25 Feb 2014, from http://smartgrid.ieee.org/resources/public-policy/australia
7. John JS (2013) Using dynamic pricing to balance solar power with the smart grid. GTM Research. Retrieved 25 Feb 2014, from http://www.greentechmedia.com/articles/read/dynamic-pricing-to-balance-solar-power-with-the-smart-grid
8. John JS (2014) Report: China outspent US on smart grid in 2013. The Energy Collective. Retrieved 25 Feb 2014, from http://theenergycollective.com/jeffstjohn/343491/report-china-outspent-us-smart-grid-2013
9. MacCracken M (2013) New York joins California as leader in energy storage (and what it means for your state). SmartGridNews. Retrieved 25 Feb 2014, from http://www.smartgridnews.com/artman/publish/Technologies_Storage/New-York-joins-California-as-leader-in-energy-storage-and-what-it-means-for-your-state-6371.html/
10. Mah D, Wu Y-Y, Ip J, Hills P (2013) The role of the state in sustainable energy transitions: a case study of large smart grid demonstration projects in Japan. Energ Policy 63:726–737
11. METI (2013) The summary of the 14th meeting of societal system study committee (in Japanese). Retrieved 25 Feb 2014, from http://www.meti.go.jp/committee/summary/0004633/014_haifu.html
12. NYISO (2012) 2014–2018 strategic plan. New York independent system operator, New York. Retrieved 25 Feb 2014, from http://www.nyiso.com/public/webdocs/company/strategic_plan/2014-18_NYISO_Strategic%20Plan_final.pdf
13. Ravindranath M (2014) As smart-grid funding winds down, energy leaders assess progress. Washington Post, Washington D.C. Retrieved 25 Feb 2014, from http://www.washingtonpost.com/business/on-it/as-smart-grid-funding-winds-down-energy-leaders-assess-progress/2014/02/23/05743a72-9992-11e3-80ac-63a8ba7f7942_story.html

14. Sullivan M (2013) National grid meters opposed. Telegram.com. Retrieved 25 Feb 2014, from http://www.telegram.com/article/20130423/NEWS/104239969/1237
15. World Bank (2013) Toward a sustainable energy future for all: directions for the World Bank group's energy sector. World Bank, Washington D.C. Retrieved 25 Feb 2014, from http://www.smartgrid.com.br/eventos/smartgrid2013/062.pdf
16. Zott C, Amit R, Massa L (2011) The business model: recent developments and future research. J Manag 37:1019–1042

MIX
Papier aus verantwortungsvollen Quellen
Paper from responsible sources
FSC® C105338

If you have any concerns about our products,
you can contact us on
ProductSafety@springernature.com

In case Publisher is established outside the EU,
the EU authorized representative is:
Springer Nature Customer Service Center GmbH
Europaplatz 3, 69115 Heidelberg, Germany

Printed by Libri Plureos GmbH
in Hamburg, Germany